Communications
in Computer and Information Science 2060

Rationale
The CCIS series is devoted to the publication of proceedings of computer science conferences. Its aim is to efficiently disseminate original research results in informatics in printed and electronic form. While the focus is on publication of peer-reviewed full papers presenting mature work, inclusion of reviewed short papers reporting on work in progress is welcome, too. Besides globally relevant meetings with internationally representative program committees guaranteeing a strict peer-reviewing and paper selection process, conferences run by societies or of high regional or national relevance are also considered for publication.

Topics
The topical scope of CCIS spans the entire spectrum of informatics ranging from foundational topics in the theory of computing to information and communications science and technology and a broad variety of interdisciplinary application fields.

Information for Volume Editors and Authors
Publication in CCIS is free of charge. No royalties are paid, however, we offer registered conference participants temporary free access to the online version of the conference proceedings on SpringerLink (http://link.springer.com) by means of an http referrer from the conference website and/or a number of complimentary printed copies, as specified in the official acceptance email of the event.

CCIS proceedings can be published in time for distribution at conferences or as post-proceedings, and delivered in the form of printed books and/or electronically as USBs and/or e-content licenses for accessing proceedings at SpringerLink. Furthermore, CCIS proceedings are included in the CCIS electronic book series hosted in the SpringerLink digital library at http://link.springer.com/bookseries/7899. Conferences publishing in CCIS are allowed to use Online Conference Service (OCS) for managing the whole proceedings lifecycle (from submission and reviewing to preparing for publication) free of charge.

Publication process
The language of publication is exclusively English. Authors publishing in CCIS have to sign the Springer CCIS copyright transfer form, however, they are free to use their material published in CCIS for substantially changed, more elaborate subsequent publications elsewhere. For the preparation of the camera-ready papers/files, authors have to strictly adhere to the Springer CCIS Authors' Instructions and are strongly encouraged to use the CCIS LaTeX style files or templates.

Abstracting/Indexing
CCIS is abstracted/indexed in DBLP, Google Scholar, EI-Compendex, Mathematical Reviews, SCImago, Scopus. CCIS volumes are also submitted for the inclusion in ISI Proceedings.

How to start
To start the evaluation of your proposal for inclusion in the CCIS series, please send an e-mail to ccis@springer.com.

Hai Jin · Yi Pan · Jianfeng Lu
Editors

Computer Networks and IoT

First International Artificial Intelligence Conference, IAIC 2023
Nanjing, China, November 25–27, 2023
Revised Selected Papers, Part III

 Springer

Editors
Hai Jin
Huazhong University of Science
and Technology
Wuhan, Hubei, China

Yi Pan
Chinese Academy of Science
Shenzhen, China

Jianfeng Lu ⓘ
Nanjing University of Science
and Technology
Nanjing, China

ISSN 1865-0929 ISSN 1865-0937 (electronic)
Communications in Computer and Information Science
ISBN 978-981-97-1331-8 ISBN 978-981-97-1332-5 (eBook)
https://doi.org/10.1007/978-981-97-1332-5

This Springer imprint is published by the registered company Springer Nature Singapore Pte Ltd.
The registered company address is: 152 Beach Road, #21-01/04 Gateway East, Singapore 189721, Singapore

Paper in this product is recyclable.

Preface

These conference proceedings are a collection of the papers accepted by IAIC 2023 — the 2023 International Artificial Intelligence Conference, held on November 25–27, 2023 in Nanjing, China.

The conference was organized by Nanjing University of Science & Technology, and Tech Science Press. IAIC 2023 aimed to provide a platform for the exchange of ideas and the discussion of recent developments in artificial intelligence. The conference showcased a diverse range of topics, including machine learning, natural language processing, computer vision, robotics, and ethical considerations in AI, among others.

The reviewing process for IAIC 2023 was meticulous and thorough. We received an impressive number of qualified submissions, reflecting the growing interest and engagement in the field of artificial intelligence. The number of the final accepted papers for publication is 86. The high standard set for acceptance resulted in a competitive selection, with a commendable acceptance rate that attests to the caliber of the contributions presented at the conference.

We extend our gratitude to the authors for their outstanding contributions and dedication, as well as to the reviewers for ensuring the selection of high-quality papers, which made these conference proceedings possible.

We also would like to thank the organizers and sponsors whose generous support made IAIC 2023 possible. Their commitment to advancing the field of artificial intelligence is commendable, and we acknowledge their contributions with sincere appreciation. The logos of our esteemed sponsors can be found on the following pages.

We hope this volume serves as a valuable resource for researchers, academics, and practitioners, contributing to the ongoing dialogue that propels the field forward.

December 2023 IAIC 2023 Organizing Committee

Organization

General Chairs

Hai Jin — Huazhong University of Science and Technology, China

Yi Pan — Shenzhen Institute of Advanced Technology, Chinese Academy of Sciences, China

Jianfeng Lu — Nanjing University of Science and Technology, China

Technical Program Chairs

Yingtao Jiang — University of Nevada Las Vegas, USA

Q. M. Jonathan Wu — University of Windsor, Canada

Technical Program Committee Members

Yudong Zhang — University of Leicester, UK

Shuwen Chen — Jiangsu Second Normal University, China

Xiaoyan Zhao — Nanjing Institute of Technology, China

Wentao Li — Southwest University, China

Chao Zhang — Shanxi University, China

Huiyan Zhang — Chongqing Technology and Business University, China

Tao Zhan — Southwest University, China

Muhammad Attique Khan — HITEC University, Pakistan

Tallha Akram — COMSATS University Islamabad, Pakistan

Zhewei Liang — Mayo Clinic, USA

Yi Ding — University of Electronic Science and Technology of China, China

Xianhua Niu — Xihua University, China

Yingjie Zhou — Sichuan University, China

Dajiang Chen — University of Electronic Science and Technology of China, China

Fang Liu — Hunan University, China

Zhiping Cai	National University of Defense Technology, China
Zongshuai Zhang	Chinese Academy of Sciences, China
Daniel Xiapu Luo	Hong Kong Polytechnic University, China
Jieren Cheng	Hainan University, China
Xinwang Liu	National University of Defense Technology, China
Qiang Liu	National University of Defense Technology, China
Xiangyang (Alex X.) Liu	Michigan State University, USA
Wei Fang	Nanjing University of Information Science and Technology, China
Victor S. Sheng	Texas Tech University, USA
Jinwei Wang	Nanjing University of Information Science and Technology, China
Leiming Yan	Nanjing University of Information Science and Technology, China
Jian Su	Nanjing University of Information Science and Technology, China
Zheng-guo Sheng	University of Sussex, UK
Si-guang Chen	Nanjing University of Posts and Telecommunications, China
Yanchao Zhao	Nanjing University of Aeronautics and Astronautics, China
Hao Han	Nanjing University of Aeronautics and Astronautics, China
Hao Wang	Ratidar Technologies LLC, China

Publication Chair

Zhihua Xia	Jinan University, China

Publicity Chairs

Lei Chen	Shandong University, China
Yuan Tian	Nanjing Institute of Technology, China

Organization Committee Members

Laith Abualigah	Al Al-Bayt University, Jordan
Muhammad Azeem Akbar	LUT University, Finland
Farman Ali	Sejong University, South Korea
Shuwen Chen	Jiangsu Second Normal University, China
Chien-Ming Chen	Nanjing University of Information Science and Technology, China
Dajiang Chen	University of Electronic Science and Technology of China, China
Ting Chen	University of Electronic Science and Technology of China, China
Ke Feng	National University of Singapore, Singapore
Honghao Gao	Shanghai University, China
Xiaozhi Gao	University of Eastern Finland, Finland
Ke Gu	Changsha University of Science and Technology, China
Mohammad Kamrul Hasan	Universiti Kebangsaan Malaysia, Malaysia
Celestine Iwendi	University of Bolton, UK
Heming Jia	Sanming University, China
Deming Lei	Wuhan University of Technology, China
Peng Li	Nanjing University of Aeronautics and Astronautics, China
Huchang Liao	Sichuan University, China
Mingwei Lin	Fujian Normal University, China
Anfeng Liu	Central South University, China
Xiaodong Liu	Edinburgh Napier University, UK
Niancheng Long	Shanghai Jiao Tong University, China
Jeng-Shyang Pan	Shandong University of Science and Technology, China
Danilo Pelusi	University of Teramo, Italy
Kewei Sha	University of Houston, USA
Shigen Shen	Huzhou University, China
Xiangbo Shu	Nanjing University of Science and Technology, China
Adam Slowik	Koszalin University of Technology, Poland
Jin Wang	Changsha University of Science and Technology, China
Kun Wang	Fudan University, China
Changyan Yi	Nanjing University of Aeronautics and Astronautics, China
Yudong Zhang	University of Leicester, UK
Chengwen Zhong	Northwestern Polytechnic University, China

Junlong Zhou   Nanjing University of Science and Technology,
                            China
Xiaobo Zhou   Tianjin University, China
Fa Zhu   Nanjing Forestry University, China

Contents – Part III

A Critical Server Security Protection Strategy Based on Traffic Log Analysis

Haiyong Zhu[✉], Chengyu Wang[✉], Bingnan Hou, Yonghao Tang, and Zhiping Cai

National University of Defense Technology, Changsha 410073, China
{zhy,chengyu}@nudt.edu.cn

Abstract. Traditional perimeter-based security systems provide a level of defense against external attacks. However, with the increasing prevalence of advanced network attacks and the continual emergence of novel attack methodologies, these conventional security mechanisms are witnessing diminishing effectiveness. Attackers frequently shift their focus to the core assets within an organization's internal network, such as database servers, file servers, and email servers. By breaching the external perimeter, they execute lateral movement within the internal network, searching for high-value assets to achieve the goal of data theft. The potential consequences stemming from an assault on core assets can be monumental, underscoring the paramount importance of safeguarding them. Nevertheless, existing measures for the protection of critical core assets exhibit several deficiencies. In response, we propose a security protection strategy for critical servers based on the analysis of traffic logs. We establish an integrated micro-boundary on the critical servers, comprising four constituent modules. A micro-boundary intrusion detection system (IDS) module, a micro-boundary traffic collection module, a micro-boundary dynamic access control module, and an agent module. This security protection strategy encompasses three core security functionalities. Network intrusion detection, network access relationship analysis, and dynamic management of access control policy. It facilitates timely and effective detection of internal threats, significantly bolstering the security of critical servers. We have implemented this security protection strategy in two real-world scenarios, assessed the feasibility of its implementation, and uncovered potential security vulnerabilities and network threats.

Keywords: Critical Server Protection · Network Intrusion Detection · Network Traffic Analysis · Dynamic Access Control

1 Introduction

In the digital era of ubiquitous connectivity, the field of network security faces numerous new challenges. Vulnerabilities are widespread, and the use of cyber-attacks to compromise critical information systems, exfiltrate military and technological intelligence, and disrupt essential infrastructures such as the Internet,

H. Jin et al. (Eds.): IAIC 2023, CCIS 2060, pp. 1–18, 2024.
https://doi.org/10.1007/978-981-97-1332-5_1

transportation, energy, and finance has become a primary choice in modern conflict across various domains. Virtually all valuable data in the modern world is stored in various forms on servers, making the overall stability of the network heavily contingent upon the security of these servers. Consequently, servers are often the primary targets of attacks.

Currently, the majority of enterprise network security architectures primarily focus on perimeter defense. Various types of protective measures, such as firewalls, intrusion detection systems, and malware detection tools, are continuously added at the network edge with the aim of isolating threats outside the network. This approach often results in network administrators neglecting the monitoring of the internal network. The Lockheed Martin Cyber Kill Chain model [1] provides a straightforward framework for understanding the process by which cybercriminals compromise the network edge, as shown in Fig. 1.

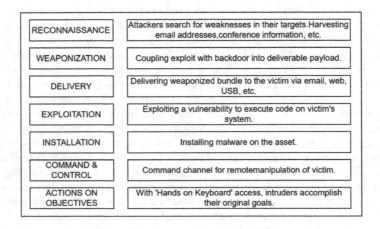

Fig. 1. Lockheed Martin's Cyber Kill Chain Model

Despite substantial efforts and resources invested in safeguarding the network edge (corresponding to the first three steps in Fig. 1), complete prevention of intrusions remains unattainable. Once attackers breach the network's interior, they can exploit vulnerabilities, deploy malicious software and proceed with lateral movement throughout the entire intranet system. The 'EternalBlue' attack [2], which emerged in 2017, resulted in widespread disruptions and impacts on the intranet through flooding via port 445. Power stations in Iran and Venezuela were subjected to attacks, leading to extensive network failures. According to VMware's quantitative analysis in "Lateral Movement in the Real World (2022)", 45% of intrusions involve lateral movement events [3].

Moreover, many enterprises today grapple with the challenge of network boundary ambiguity. A single change in permissions, a configuration alteration, a personnel shift, or a version update can all signify changes in an organization's network assets. Each of these alterations has the potential to redefine the network boundary, introducing new risks. Traditionally, access control at the network boundary has been achieved using firewalls, which are passive defensive measures that inspect and filter data traffic passing through them. However, traditional, static firewalls struggle to cope with the dynamic nature of network boundary.

To address these issues, we propose an integrated security protection strategy designed specifically for critical servers. This strategy employs an agent-server architecture, creating a micro-boundary on critical servers. It offers multi-dimensional security protection for critical servers. Subsequently, we deploy this strategy in real-world scenarios, and the validation results confirm the feasibility and effectiveness of threat detection in this strategy.

The main contributions of this paper are as follows:

1. Proposed an novel integrated security protection strategy tailored for critical servers. This strategy demonstrates effectiveness in detecting malicious activities within the intranet.
2. A pioneering method for analyzing network traffic logs across different time periods is proposed. This method serves as the basis for dynamic adjustments to micro-boundary access control policies, thereby enhancing overall network security.
3. Through real-world deployments and verifications, the paper successfully identifies and addresses security risks. These findings underscore the practical efficacy of the proposed security protection strategy.

2 Related Work

Network intrusion detection systems (NIDS) are employed to monitor network traffic to detect potentially malicious activities. These systems are generally categorized into two main types based on the detection techniques they utilize.

Signature-based intrusion detection [4–7]. This approach relies on predefined attack patterns or signatures. While effective in detecting known attacks, it may struggle to adapt to new attack forms. Behavior-based intrusion detection [8–13]. This method concentrates on monitoring the normal behavioral patterns of network traffic. It utilizes techniques such as machine learning and statistical analysis to identify activities that deviate from established behavior patterns.

Depending on the auditing target, intrusion detection can be categorized into two types. Host-based intrusion detection [14–16]. This primarily involves gathering data from sources such as system logs and application logs for analysis to determine if there are any intrusion attempts. Network-based intrusion detection [17–21]. This is used to monitor data packets transmitted across the entire network, analyzing and identifying suspicious activities. Immediate alerts are generated upon detection of any suspicious behavior or intrusion attempts.

Wazuh is a robust open-source host-based security monitoring platform based on OSSEC [22]. It extends the core functionality of OSSEC and supports integration with other security tools. Stefan et al. provided a detailed summary of Wazuh's log-based attack detection capabilities in [23].

Suricata is a multi-threaded intrusion detection system supporting both IDS (Intrusion Detection System) and IPS (Intrusion Prevention System). Compared to traditional Snort [24], Suricata's multi-threaded and modular design surpasses the original Snort in efficiency and performance. It parallelizes CPU-intensive deep packet inspection tasks across multiple concurrent threads, fully leveraging the advantages of multi-core hardware to enhance the throughput of the intrusion detection system [25]. The working principle of Suricata is shown in Fig. 2.

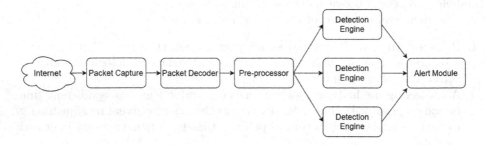

Fig. 2. Working principle of Suricata

Zeek is a globally leading passive network security monitoring platform designed and developed by Vern Paxson in 1999. It operates by passively monitoring network traffic to detect network intruders in real-time. Key features of Zeek include high-speed monitoring, real-time alerts, separation of mechanism and policy, and robust scalability. Zeek is structured into an 'event engine' and a 'policy script interpreter.' The event engine simplifies network flows filtered through the kernel into a series of high-level events, while the policy script interpreter consists of event handlers written in a specialized language [26]. The working principle of Zeek is shown in Fig. 3.

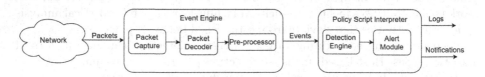

Fig. 3. Working principle of zeek

Filebeat serves as a lightweight shipper designed to efficiently forward and centralize log data. Installed as an agent on servers, Filebeat actively monitors specified log files or locations, collects log events, and seamlessly forwards them to indexing systems. Elasticsearch stands as a highly scalable full-text search and analytics engine that offers near-real-time data search and analysis capabilities. Within Elasticsearch, an index functions as a collection of related documents. These documents are strategically distributed across different containers known as shards. The distribution of documents into multiple shards, spread across various nodes, ensures redundancy. Kibana is an robust data analysis and visualization platform. With Kibana, users can effortlessly search, view, and interact with data stored in Elasticsearch indices. It provides diverse display options, including charts, tables, and more, facilitating comprehensive data exploration and understanding.

3 Model Design

3.1 Security Model Description

In this section, we present a comprehensive explanation of our proposed integrated security protection strategy. We begin by introducing a security protection model centered on traffic log analysis. Within this model, we adopt an agent-server security architecture, allowing the agent to establish either a one-to-one or one-to-many relationship with the server. We employ a dedicated security center as the server component, which receives the security data transmitted by the agents. The received data undergoes analysis within the security center, traversing through a data forwarding tool and a search engine before being presented in a dashboard, providing a clear and concise overview of the results.

We implement a security micro-boundary on critical servers, comprising four essential components. A micro-boundary intrusion detection module, a micro-boundary traffic collection module, a micro-boundary dynamic access control module, and a agent module. The micro-boundary intrusion detection module ensures continuous monitoring of incoming and outgoing traffic on critical servers, actively detecting potential network threats. Simultaneously, the network traffic collection module gathers all traffic entering and leaving the servers, forming a foundational dataset for subsequent threat analysis. The dynamic access control module plays a crucial role by receiving access control policies from the security center and dynamically adjusting micro-boundary access control policies in real-time. This adaptability enhances the system's responsiveness to evolving security needs. The agent module is responsible for collecting and transmitting security data from the critical servers to the security center. This integrated security micro-boundary, composed of these four modules, proves highly effective in threat detection. Furthermore, the scalability of our approach is evident - if another critical server within the internal network necessitates protection, a similar micro-boundary can be established, and its security data seamlessly uploaded to the same security center. The overall framework of our proposed security protection model for critical servers is depicted in Fig. 4.

In Fig. 4, external internet traffic first enters the enterprise intranet and undergoes detection by security devices like the enterprise border firewall. Within the intranet, we identify "Critical Server 1" and "Critical Server 2" as objects requiring substantial protection. The "Compromised Server" represents a server

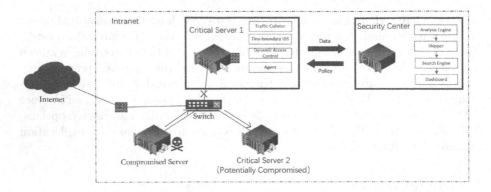

Fig. 4. Integrated security protection model for critical servers.

successfully compromised by attackers through vulnerabilities, social engineering, phishing attacks, etc. Additionally, there is a security center server provided by us. Once a server is compromised, attackers seek to gather intranet information, including details about network topology, hostnames, and IP addresses, to strategically select their next targets. Subsequently, they employ the compromised server as a pivot point, attempting to log into other servers. To fortify "Critical Server 1" against such threats, we've implemented a micro-boundary (depicted by the red firewall in Fig. 4) and utilized an agent to upload security data to the security center for analysis. This proactive strategy enables the prompt and effective detection of malicious activities within the intranet. In contrast, due to a lack of necessary protective measures, "Critical Server 2" remains a potential target for attackers.

After careful consideration, we have selected tools such as Wazuh, Suricata, Zeek, IPtables, Filebeat, Elasticsearch, and Kibana to implement the integrated security protection solution for critical servers as described in the model. One reason for choosing these tools is their widespread use, mainstream status, and excellent scalability. The feasibility of this solution is discussed in the following sections.

We installed Wazuh agents on the critical servers to act as agents in the model, responsible for collecting and uploading IDS alerts and traffic logs. Additionally, we fully leveraged the additional security features provided by Wazuh agents, such as file integrity checking, vulnerability scanning, security configuration assessment, and other functions. In practice, we have also configured these security features.

In comparison to Snort, Suricata supports multi-threading and can optimally utilize multi-core processors, resulting in faster intrusion detection processing. Therefore, we employed Suricata to fulfill the micro-boundary IDS role in our model. We configured Suricata's alert rules using the open-source rule set from Emerging Threats [27], which comprises over thirty thousand rules.

We employed Zeek as the traffic collection tool in the model, utilizing its extensibility through custom scripts for creating tailored analysis tools and plugins to meet specific requirements. For instance, a customized Zeek script was developed to correlate network traffic with system process information, enhancing the richness of security insights. Additionally, Zeek's capability to parse data packets into events, rather than storing raw pcap data, contributes to a reduction in data storage volume.

A dynamic micro-boundary access relationship model is crafted through cross-temporal analysis on the micro-boundary traffic data generated by Zeek. The resultant policies are deployed to critical servers, optimizing micro-boundary access control policies. The system's built-in iptables tool is employed to implement the functions of the micro-boundary access control module. Furthermore, we augment the context of alert information by associating Suricata's alert data with Zeek's event information. Leveraging Elastic Stack components, including Filebeat, Elasticsearch, and Kibana, we establish them as data forwarding tools, full-text search engines, and dashboards. In the following sections, we will dive into our methods for alert context correlation and the cross-temporal construction of the micro-boundary dynamic access relationship model.

3.2 Alert Context Correlation Method

we adopted the "Community ID" flow hash implemented in [28], which standardizes the generation of string identifiers representing a given network flow, simplifying the correlation into straightforward string comparisons. Additionally, we utilized Zeek's unique IDs (UIDs) for further correlation. We illustrate the alert correlation method in a graphical format, as shown in Fig. 5.

In Fig. 5, "alert.log" denotes an alert generated by Suricata. The alert is named "ET HUNTING SUSPICIOUS SMTP Attachment Inbound PPT attachment with Embedded OLE Object M6". This Suricata alert signals an incoming SMTP email with a PPT attachment that includes an embedded OLE object. Given the potential for OLE objects to be utilized for malicious software distribution, further investigation is advisable to ensure network security. We correlate Suricata's alert with Zeek's logs using the community ID and additionally associate multiple logs using Zeek's UID. By correlating "alert.log" with "conn.log", "file.log", and "smtp.log", we can extract more specific information, including source address, source port, destination address, destination port, sender's email name, recipient's email name, CC email addresses, and the file hash of the email attachment. This supplementary information proves invaluable for assessing the severity of the alert.

3.3 Cross-Temporal Micro-boundary Dynamic Access Relationship Model Construction Method

We established a cross-temporal micro-boundary dynamic access relationship model based on Zeek logs. Through statistical analysis on Zeek's conn.log, we identified both inbound and outbound access relationships for hosts, ultimately

Fig. 5. Alert context correlation method schematic diagram.

shaping the micro-boundary access control policies. The detailed methodology is outlined below.

Zeek possesses the capability to parse captured traffic and organize it into session flows. Fundamental session connection relationships are recorded in the conn.log, with its data items illustrated in Fig. 6. In Zeek's conn.log, the fields are defined as follows: ts: The start time of the connection. uid: The unique identifier for the connection. id: The 4-tuple of endpoint address/port for the connection. proto: The transport layer protocol used for the connection. service: The application layer protocol used over the connection. duration: The duration of the connection. orig_bytes: The payload bytes sent by the originator. resp_bytes: The payload bytes sent by the responder. conn_state: This field records the connection state. missed_bytes: Represents the number of bytes missing in content gaps, indicating packet loss situations. history: Records the history of the connection state, represented by a series of letters.

During the analysis, we identified noise in the original logs. Zeek employs a default TCP idle timeout of 5 min to optimize system resource usage. If a TCP session experiences no data transmission within this interval, Zeek terminates the session, considering it concluded, and logs the session information. This mechanism can pose a challenge when a specific TCP session exhibits a time gap between two packets exceeding Zeek's TCP idle timeout. Consequently, Zeek may interpret this as two distinct sessions, generating two log entries. Moreover, the first packet sent after the timeout is deemed the initial packet of the second session, potentially causing a misidentification of the initiator for the second session.

```
"ts": 1698062522.217638,
"uid": "COGI2q3e1PffpDvRC7",
"id.orig_h": "███ ██▄▄ ▄▄█",
"id.orig_p": 58934,
"id.resp_h": "███ ████ ██▄▄█",
"id.resp_p": 1900,
"proto": "udp",
"duration": 3.0279009342193604,
"orig_bytes": 700,
"resp_bytes": 0,
"conn_state": "S0",
"local_orig": false,
"local_resp": false,
"missed_bytes": 0,
"history": "D",
"orig_pkts": 4,
"orig_ip_bytes": 812,
"resp_pkts": 0,
"resp_ip_bytes": 0,
"process": "37405/sshd:"
```

Fig. 6. Fields in Zeek's conn.log

To mitigate this issue, session merging becomes imperative. This process relies on the insight that each session commences with the first SYN packet during the handshake phase. Initially, sessions with the history field containing "S" (indicating the presence of a SYN packet) are isolated, and the corresponding five-tuple (source IP address, source port, transport layer protocol, destination address, destination port) for these sessions is preserved. Then, sessions lacking "S" in the history field are examined. Two five-tuple hashes, namely hash1 = [source IP address, source port, transport layer protocol, destination address, destination port], and hash2 = [destination address, destination port, transport layer protocol, source address, source port], are constructed and matched against the stored five-tuple. Upon discovering a match, the two sessions are merged. The pseudo-code for the network connections processing method is presented in Algorithm 1.

By employing this approach, we can reconstruct the entire session information for sessions where the time interval between packets exceeds Zeek's TCP idle timeout, irrespective of the timeout setting. Subsequently, we categorize the connections into inbound and outbound, distinguishing between the internal and external networks.

For inbound connections, we determine the establishment of a connection based on whether the local port is open, signifying a successful TCP three-way handshake. Ports with connection attempts but no successful handshake are considered closed ports, indicative of network scanning behavior. The presence of "ShA" in the history field determines a successfully established connection. This method effectively identifies ports that are briefly open and then closed, a pattern commonly associated with high-risk behavior. Utilizing statistical methods, we calculate daily averages for access frequency, duration, request quantity, and response quantity for each local port. This allows us to generate a list of open ports, aiding in identifying frequently accessed services, rarely accessed ones, and even shadow ports-ports that are open but receive minimal traffic, possibly

---

Algorithm 1: Process Network Connections

 Data: all_conn.log - List of network connection records
 Result: syn_data - Set of SYN connections, back_data - List of non-SYN
    connections

1 **for** *each conn in all_conn.log* **do**
2 **if** *conn.proto is "tcp"* **then**
3   **if** *"S" is in conn.history* **then**
4   hash ← conn.orig_h + "#" + conn.orig_port + "#" + conn.proto
     + "#" + conn.resp_h + "#" + conn.resp_port;
5   syn_data.add(hash);
6   **end**
7   **else**
8   back_data.add(conn);
9   continue;
10   **end**
11 **end**
12 StatisticalOperation(*conn*);
13 **end**
14 **for** *each conn in back_data* **do**
15 hash_r ← conn.resp_h + "#" + conn.resp_port + "#" + conn.proto +
   "#" + conn.orig_h + "#" + conn.orig_port;
16 **if** *hash_r is in syn_data* **then**
17   Transfer flow direction;
18 **end**
19 StatisticalOperation(*conn*);
20 **end**

---

warranting closure. For outbound connections, we define the field [external IP address, transport layer protocol, destination port] as an outbound connection channel. In an enterprise's critical server, outbound behavior tends to remain consistent. Through continuous monitoring across different time scales, we compile a whitelist of legitimate outbound connection channels. Any newly appearing outbound connection channels are then flagged as significant security events for further verification.

4 Case Study

In this section, we implement the proposed strategy in real-world scenarios and validate its feasibility through two case studies. In both instances, we successfully identified potential security threats and confirmed malicious activities as actual attack events.

4.1 Security Monitoring for a Large-Scale Email Server

We implemented the security protection strategy on a real-world, large-scale email server and gathered data over a seven-day period, from September 16, 2023, to September 23, 2023. The email server had a system configuration with 128 GB of memory, a minimum of 2 TB of disk capacity, a 64-core CPU, an x86_64 architecture, and ran CentOS Linux 7 as the operating system.

An assessment of the system resources consumed by the micro-boundary was conducted, focusing on four main aspects: CPU usage of processes, memory usage of processes, virtual memory size of processes, and resident set size of processes. Five random time points were selected, and the averages were calculated. The evaluation results are presented in Table 1.

Table 1. System resource utilization after deploying security protection strategy.

Process	%CPU	%MEM	VSZ (KB)	RSS (KB)
Suricata	11.8	0.6	1115924	876116
Zeek	43.9	0.3	3131788	459096
Wazuh-execd	0.0	0.0	36168	1380
Wazuh-agentd	0.1	0.0	257768	4872
Wazuh-syscheckd	0.2	0.0	417376	6500
Wazuh-logcollector	0.0	0.0	478656	2776
Wazuh-modulesd	0.1	0.0	5384396	40544

The column "%CPU" indicates the percentage of CPU usage by each process, "%MEM" represents the percentage of memory usage, "VSZ" denotes the virtual memory size of each process in kilobytes, and "RSS" reflects the resident set size in kilobytes, indicating the actual physical memory usage. Examining Table 1, it is noticeable that the Zeek process displays a relatively high CPU usage of 43.9%. In contrast, the Wazuh processes exhibit lower CPU resource consumption, approximately 0.4%.

For the security center, which utilizes an Ubuntu 20.04 system with 32 GB of memory, an 8-core CPU, and 1 TB of disk space, an evaluation of system overhead was conducted. The results are outlined in Table 2.

Analyzing Table 2, among the ten Wazuh processes, Wazuh-modulesd incurs a relatively higher system overhead, accounting for 18.1%. Filebeat and Kibana, on the other hand, consume fewer system resources. Elasticsearch utilizes approximately 49.5% (around 15.84 GB) of memory space, which falls within an acceptable range. Furthermore, it is possible to limit the memory consumption of Elasticsearch by adjusting its configuration files.

We have compiled the trend chart for the IDS alerts in the micro-boundary over seven days, from September 16, 2023, to September 22, 2023, as shown in Fig. 7.

Table 2. System resource utilization of the security center

Process	%CPU	%MEM	VSZ (KB)	RSS (KB)
Wazuh-authd	0.2	0.0	181040	1320
Wazuh-db	0.1	0.0	762016	16860
Wazuh-execd	0.0	0.0	25328	1368
Wazuh-maild	0.0	0.0	25360	1704
Wazuh-analysisd	0.8	0.3	4276020	105264
Wazuh-syscheckd	0.0	0.0	191488	6068
Wazuh-remoted	0.5	0.0	5384396	6484
Wazuh-logcollector	0.0	0.0	467804	2404
Wazuh-monitord	0.0	0.0	25404	2712
Wazuh-modulesd	18.1	4.6	2987836	1518788
Filebeat	0.0	0.1	1777780	40176
Elasticsearch	11.1	49.5	27129148	16265068
Kibana	1.5	1.0	1226136	350428

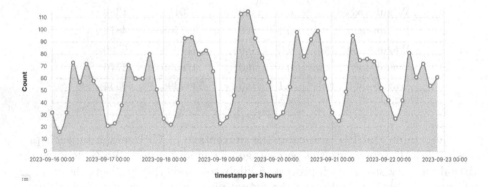

Fig. 7. Trend in the number of alerts for the micro-boundary intrusion detection system from September 16, 2023, to September 22, 2023.

Analyzing the trend curves in Fig. 7, we observed a clear pattern in the variation of alerts counts. There was an increase in alerts during the daytime, while the number decreased during the nighttime. In total, 3306 alerts were generated by the micro-boundary IDS. The top ten alerts by quantity are listed in Table 3.

Table 3. Top ten alerts by quantity on the mail server from September 16, 2023, to September 22, 2023.

Alert Description	Count
ET DNS Query for .cc TLD	547
ET POLICY Vulnerable Java Version 1.8.x Detected	531
ET DROP Dshield Block Listed Source group 1	496
ET INFO Observed DNS Query to .world TLD	403
ET INFO Observed DNS Query to .life TLD	266
ET DNS Query to a *.top domain - Likely Hostile	123
ET DROP Spamhaus DROP Listed Traffic Inbound group	98
ET 3CORESec Poor Reputation IP group 6	91
ET DNS Query for .to TLD	86
ET 3CORESec Poor Reputation IP group	54

In Table 3, the top ten alerts collectively contribute to 2,685 out of a total of 3,306 alerts over the seven-day period, representing approximately 81.27% of the total alert count. The most frequently occurring alert is "ET DNS Query for .cc TLD", indicating the detection and logging of DNS requests containing the ".cc" domain in queries. ".cc" stands for the country code top-level domain (ccTLD) of the Cocos (Keeling) Islands. Typically, DNS queries translate domain names into their corresponding IP addresses. However, improper ".cc" TLD queries, as detected by this alert, could signify potential malicious activities, including online fraud, malware propagation, or communication with Command and Control (C&C) servers. As an illustration, let's examine a specific instance of the "ET DNS Query for .cc TLD" alert. Through our alert correlation method, we identified that the associated DNS query address is "count..cc," and the query returned an IP address of 43..*.107. This website is operated by a technology company in Shenzhen, specializing in email marketing services.

Our analysis of micro-boundary traffic data from September 16, 2023, to September 22, 2023, utilizing the previously outlined cross-temporal micro-boundary dynamic access relationship model construction method, yielded a comprehensive list of ports on the email server, as presented in Table 4.

Table 4 reveals that the following ports were open: tcp/110, tcp/143, tcp/25, tcp/443, tcp/465, tcp/80, tcp/8025, tcp/993, tcp/995, and tcp/22. Using this information, we collaborated with network administrators to verify and catalog assets. Additionally, we pinpointed accesses to closed ports such as tcp/585, tcp/933, tcp/994, udp/443, tcp/587, udp/53, and udp/137, which could indicate potential malicious scanning activities. We conducted a thorough investigation into these ports, identifying corresponding IP addresses and performing additional verification. For instance, a detailed analysis of the closed port tcp/994 revealed external IP addresses accessing it. From a security standpoint, even if a port is closed, it should not be exposed to the public internet. We identified this

Table 4. Port list for the email server from September 16, 2023, to September 22, 2023.

Port	State	Total Connection	Average Daily Request Size (Unit: MB)	Average Daily Respond Size (Unit: MB)
tcp/110	open	251461	85.78	3075.35
tcp/143	open	1374260	422.16	4123.42
tcp/25	open	34230574	2914.02	8426.71
tcp/443	open	3420482	22547.46	30191.76
tcp/465	open	543293	253.47	366.39
tcp/80	open	169396	1340.86	2156.7
tcp/8025	open	91	0.01	0.01
tcp/993	open	4601186	2069.56	13921.95
tcp/995	open	150372	101.4	3035.17
tcp/22	open	12	0.06	0.13
tcp/585	close	72	0	0
tcp/933	close	1120	0.01	0.01
tcp/994	close	4565	0.03	0.02
udp/443	close	452	0.4	0
tcp/587	close	3263	0.03	0.02
udp/53	close	452	0.04	0
udp/137	close	1	0	0

as a port forwarding policy issue. It seemed that, after discontinuing the service associated with port tcp/994, the network administrator promptly closed the host's tcp/994 port but overlooked updating the port forwarding policies, introducing potential security risks.

4.2   Security Monitoring of an Enterprise Intranet Server

In the second case, security monitoring was implemented on an internal OA server within an enterprise. The micro-boundary was deployed on this OA server, and security data was transmitted to the security center, which operated in a bypass mode. We compiled statistics on the alerts captured by the micro-boundary IDS from June 18, 2023, to June 24, 2023. Over these seven days, a total of 6,429 alerts were generated, with the top ten alerts by quantity listed in Table 5.

In Table 5, it is evident that the alert "ET POLICY Successful Non-Anonymous LDAPv3 Bind Request Outbound" has been reported 5650 times, constituting approximately 87% of the total alerts over the seven days. This alert signifies a successful, non-anonymous LDAPv3 bind request originating from the internal network to the external network. The occurrence of a successful LDAPv3

Table 5. Top ten alerts by quantity on the OA server from June 18, 2023, to June 24, 2023.

Alert Description	Count
ET POLICY Successful Non-Anonymous LDAPv3 Bind Request Outbound	5650
ET DROP Dshield Block Listed Source group 1	316
ET MALWARE Possible Upatre Downloader SSL certificate (fake loc)	57
ET JA3 Hash - [Abuse.ch] Possible Adwind	24
ET INFO User-Agent (python-requests) Inbound to Webserver	24
ET INFO Request to Hidden Environment File - Inbound	23
ET 3CORESec Poor Reputation IP group 11	22
ET HUNTING Suspicious IFS String Observed in HTTP URI	21
ET CINS Active Threat Intelligence Poor Reputation IP group 42	21
ET SCAN Laravel Debug Mode Information Disclosure Probe Inbound	16

bind request raises concerns about potential security vulnerabilities, as it implies successful authentication by an entity. If an attacker gains valid credentials, there is a risk of unauthorized access to restricted resources. Consequently, a thorough examination of the corresponding IP hosts is necessary to mitigate potential network attacks.

Employing the cross-temporal micro-boundary dynamic access relationship model construction method outlined in the preceding section, we scrutinized the micro-boundary traffic of the OA server. Utilizing Zeek logs from June 18, 2023, to June 24, 2023, we identified ten open ports on the OA server, as detailed in Table 6.

Table 6. Port list for the OA server from June 18, 2023, to June 24, 2023.

Port	Total Connection	Average Daily Request Size (Unit: MB)	Average Daily Respond Size (Unit: MB)
tcp/10050	507513	21.49	20.3
tcp/443	419226	717.66	2366.84
tcp/5225	249	0.13	0.3
tcp/7443	1046	0.1	0.42
tcp/80	26379	0.95	1.43
tcp/81	350	0.16	7.64
tcp/8999	268	0.05	1.51
tcp/9000	14818	6.64	18.37
tcp/9081	56	0.23	0.5
tcp/9922	12	0.73	26.7

Upon careful inspection, no external network access to the closed ports has been detected, affirming that these closed ports are not externally exposed and comply with our security policy requirements. However, internal IP addresses accessing 16 closed ports have been observed, indicative of abnormal port scanning behavior. It is imperative to conduct further verification of the corresponding IP addresses' hosts to mitigate the potential risk of intranet attacks.

Furthermore, an analysis of outbound connections to foreign countries over the seven days revealed a total of 12 outbound connection channels to foreign countries. Among these, 10 initiated requests to the udp/53 port, and 2 to the tcp/80 port. Upon closer examination, all 10 addresses that sent requests to the udp/53 port were identified as root domain name server addresses. The 2 addresses initiating requests to the tcp/80 port were determined to be domain addresses of the official software repository mirror list for the CentOS Linux operating system. The details of outbound connection channels to foreign countries are outlined in Table 7.

Table 7. Port list for the OA server from June 18, 2023, to June 24, 2023.

Dest Address	Dest Port	Region	Total Connection	Average Daily Request Size (Unit: MB)	Average Daily Respond Size (Unit: MB)
192.*.*.201	udp/53	America	3	0.0001	0.0012
198.*.*.53	udp/53	America	2	0.0001	0.0011
199.*.*.42	udp/53	America	2	0.0001	0.0011
202.*.*.33	udp/53	Japan	2	0.0001	0.001
192.*.*.4	udp/53	America	1	0.0001	0.0011
192.*.*.12	udp/53	America	1	0.0001	0.0011
192.*.*.17	udp/53	Sweden	2	0.0001	0.0009
198.*.*.4	udp/53	America	1	0.0001	0.0011
199.*.*.13	udp/53	America	1	0.0001	0.0011
199.*.*.201	udp/53	America	2	0.0001	0.0011
18.*.*.18	tcp/80	America	14	0.0002	0
35.*.*.213	tcp/80	France	15	0.0002	0

Continuous monitoring outbound connection channels across various time scales facilitates the effective detection of abnormal proactive outbound behavior.

5 Conclusion

In summary, we proposed an innovative, integrated security protection strategy tailored to critical servers. This strategy involves the direct implementation of a micro-boundary on critical servers, aiming to bolster security across multiple dimensions. To assess the efficacy of this strategy, we applied it in real-world scenarios, examining additional factors such as system overhead, intrusion detection capabilities, traffic analysis capabilities, and dynamic management of micro-boundary access control policies. The practical results demonstrate the high feasibility of the proposed strategy and its success in detecting security risks.

Acknowledgements. This work is supported by the National Natural Science Foundation of China (No. 62102425) and the Science and Technology Innovation Program of Hunan Province (Nos. 2022RC3061, 2021RC2071).

Conflicts of Interest. The authors declare that they have no conflicts of interest to report regarding the present study.

References

1. Yadav, T., Rao, A.M.: Technical aspects of cyber kill chain. In: Abawajy, J.H., Mukherjea, S., Thampi, S.M., Ruiz-Martínez, A. (eds.) SSCC 2015. CCIS, vol. 536, pp. 438–452. Springer, Cham (2015). https://doi.org/10.1007/978-3-319-22915-7_40
2. Fairfield-Sonn, J.: WannaCry, EternalBlue, SMB Ports, and the Future (2017)
3. https://blogs.vmware.com/security/2022/06/lateral-movement-in-the-real-world-a-quantitative-analysis.html
4. Kumar, V., Sangwan, O.P.: Signature based intrusion detection system using SNORT. Int. J. Comput. Appl. Inf. Technol. 1(3), 35–41 (2012)
5. Masdari, M., Khezri, H.: A survey and taxonomy of the fuzzy signature-based intrusion detection systems. Appl. Soft Comput. 92, 106301 (2020)
6. Li, W., Tug, S., Meng, W., Wang, Y.: Designing collaborative blockchained signature-based intrusion detection in IoT environments. Futur. Gener. Comput. Syst. 96, 481–489 (2019)
7. Le Jeune, L., Goedeme, T., Mentens, N.: Machine learning for misuse-based network intrusion detection: overview, unified evaluation and feature choice comparison framework. IEEE Access 9, 63995–64015 (2021)
8. Nitin, T., Singh, S.R., Singh, P.G.: Intrusion detection and prevention system (IDPS) technology-network behavior analysis system (NBAS). ISCA J. Engineering Sci. 1(1), 51–56 (2012)
9. Moon, D., Im, H., Kim, I., Park, J.H.: DTB-IDS: an intrusion detection system based on decision tree using behavior analysis for preventing APT attacks. J. Supercomput. 73, 2881–2895 (2017)
10. Abusafat, F., Pereira, T., Santos, H.: Proposing a behavior-based IDS model for IoT environment. In: Wrycza, S., Maślankowski, J. (eds.) SIGSAND/PLAIS 2018. LNBIP, vol. 333, pp. 114–134. Springer, Cham (2018). https://doi.org/10.1007/978-3-030-00060-8_9
11. Soltani, M., Ousat, B., Siavoshani, M.J., Jahangir, A.H.: An adaptable deep learning-based intrusion detection system to zero-day attacks. J. Inf. Secur. Appl. 76, 103516 (2023)

12. Saba, T., Rehman, A., Sadad, T., Kolivand, H., Bahaj, S.A.: Anomaly-based intrusion detection system for IoT networks through deep learning model. Comput. Electr. Eng. **99**, 107810 (2022)
13. Martins, I., Resende, J.S., Sousa, P.R., Silva, S., Antunes, L., Gama, J.: Host-based IDS: a review and open issues of an anomaly detection system in IoT. Futur. Gener. Comput. Syst. **133**, 95–113 (2022)
14. Deshpande, P., Sharma, S.C., Peddoju, S.K., Junaid, S.: HIDS: a host based intrusion detection system for cloud computing environment. Int. J. Syst. Assur. Eng. Manag. **9**, 567–576 (2018)
15. Liu, M., Xue, Z., Xu, X., Zhong, C., Chen, J.: Host-based intrusion detection system with system calls: review and future trends. ACM Comput. Surv. (CSUR) **51**(5), 1–36 (2018)
16. Besharati, E., Naderan, M., Namjoo, E.: LR-HIDS: logistic regression host-based intrusion detection system for cloud environments. J. Ambient. Intell. Humaniz. Comput. **10**, 3669–3692 (2019)
17. Zheng, K., Cai, Z., Zhang, X., Wang, Z., Yang, B.: Algorithms to speedup pattern matching for network intrusion detection systems. Comput. Commun. **62**, 47–58 (2015)
18. Yu, Y., Long, J., Cai, Z.: Session-based network intrusion detection using a deep learning architecture. In: Torra, V., Narukawa, Y., Honda, A., Inoue, S. (eds.) MDAI 2017. LNCS (LNAI), vol. 10571, pp. 144–155. Springer, Cham (2017). https://doi.org/10.1007/978-3-319-67422-3_13
19. Min, E., Long, J., Liu, Q., Cui, J., Cai, Z., Ma, J.: SU-IDS: a semi-supervised and unsupervised framework for network intrusion detection. In: Sun, X., Pan, Z., Bertino, E. (eds.) ICCCS 2018. LNCS, vol. 11065, pp. 322–334. Springer, Cham (2018). https://doi.org/10.1007/978-3-030-00012-7_30
20. Yu, Y., Long, J., Cai, Z.: Network intrusion detection through stacking dilated convolutional autoencoders. Secur. Commun. Netw. **2017**, 1–10 (2017)
21. Zheng, K., Zhang, X., Cai, Z., Wang, Z., Yang, B.: Scalable NIDS via negative pattern matching and exclusive pattern matching. In: 2010 Proceedings IEEE INFOCOM, pp. 1–9. IEEE, March 2010
22. Maraş, A.O.R.: Host-based intrusion detection systems OSSEC open source HIDS. Mil. Secur. Stud. **2015**, 43 (2015)
23. Stanković, S., Gajin, S., Petrović, R.: A review of Wazuh tool capabilities for detecting attacks based on log analysis (2022)
24. Awal, H., Hadi, A.F., Zain, R.H.: Network security with snort using IDS and IPS. J. Dyn. (Int. J. Dyn. Eng. Sci.) **8**(1), 32–36 (2023)
25. Boukebous, A.A.E., Fettache, M.I., Bendiab, G., Shiaeles, S.: A comparative analysis of Snort 3 and Suricata. In: 2023 IEEE IAS Global Conference on Emerging Technologies (GlobConET), pp. 1–6. IEEE, May 2023
26. Paxson, V.: Bro: a system for detecting network intruders in real-time. Comput. Netw. **31**(23–24), 2435–2463 (1999)
27. https://rules.emergingthreats.net/OPEN_download_instructions.html
28. https://github.com/corelight/zeek-community-id

Stable Monocular Visual Odometry Based on Optical Flow Matching

Yujia Liu[1], Chaoxia Shi[1(✉)], and Yanqing Wang[2(✉)]

[1] Nanjing University of Science and Technology, Nanjing 210094, China
scx@njust.edu.cn
[2] Nanjing Xiaozhuang University, Nanjing 211171, China
wyq0325@126.com

Abstract. A monocular visual odometry method combining deep learning and geometry is proposed to address the sparsity and position discontinuity problems. An unsupervised optical flow estimation network is constructed to obtain high-quality dense optical flow. Next, optical flow masks and depth masks are proposed for filtering key points. Finally, position estimation and scale recovery are performed based on multi-view geometry. Experiments are extensively validated on the KITTI dataset, and the visual odometry method achieves 3.48 (%) translation error and 0.67 rotation error ($°/100\,\mathrm{m}$) outperforming the baseline DF-VO.

Keywords: Monocular Visual Odometry · Unsupervised Optical Flow Estimation · Deep Learning

1 Introduction

It is important for a robot to accurately estimate its position and perceive its environment. Vision methods have the advantages of low cost, low power consumption, and can provide rich texture information to support subsequent Simultaneous Localization and Mapping (SLAM). In recent years, monocular vision has made significant progress in estimating the six-degree-of-freedom motion of a moving camera. Algorithms for visual odometry are divided into two main categories: the feature-based method and the direct method. Feature-based method [13] is based on feature point extraction and matching, which estimate camera motion by detecting and tracking significant feature points. This method works well for scenes with significant textures and features. For low texture areas or scenes with insignificant features, it may be difficult to extract or match, resulting in performance degradation. The direct method [2] is a pixel photometric based method that utilizes the brightness changes of all pixels in an image to estimate the camera motion. It does not need to extract feature points, it processes the entire image and is more suitable for low texture scenes or image sequences that lack distinct features. The traditional method to compute the dense optical flow takes a lot of inference time and cannot be used for real-time computation.

H. Jin et al. (Eds.): IAIC 2023, CCIS 2060, pp. 19–29, 2024.
https://doi.org/10.1007/978-981-97-1332-5_2

With hardware advancements and the continuous development of neural network computations, dense optical flow estimation has generally improved in terms of inference speed and accuracy. Learning-based methods can be categorized into supervised and unsupervised approaches. Supervised methods train networks by learning from labeled ground truth flow. However, obtaining accurate ground truth labels is challenging. Also, models may exhibit high accuracy on the training dataset but lack generalization capabilities. Moreover, the optical flow generated by the supervised method has a small area that can be filtered for features, as shown in Fig. 1. This makes the selection of matching points few and concentrated in some areas, which greatly affects the robustness of the odometry system. Unsupervised methods do not require real optical flow labels to assist in training, and train optical flows based on photometric consistency assumptions. It can be adapted to different scenes and environments. Thus, it can provide stable quality correspondence, which makes it more robust in dealing with complex scenes or challenging conditions.

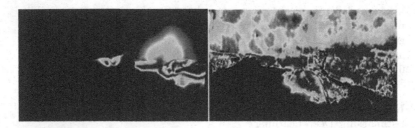

Fig. 1. Comparison of optical flow consistency (RAFT on the left and the proposed method on the right. The red part indicates to differ greatly). (Color figure online)

Currently, unsupervised methods suffer from the problem of poor accuracy due to upsampling ambiguity. An unsupervised optical flow estimation network based on deformable convolution and self-attention mechanism is proposed. Texture information is enhanced in the extracted features by fusing multi-scale input images. Deformable convolution is utilized to learn the spatial structure in the flow residuals, which is up-sampled according to the shape of the objects in the image. In addition, a self-attention mechanism is used to learn the long-term dependencies in the flow.

Forward-and-backward consistency [18] is widely used in direct odometry for feature selection. This method eliminates erroneous flow in occluded regions and ranks the reliability of flow. However, it does not effectively remove poorly matched points. Unsupervised optical flow is trained using photometric loss. By calculating the similarity between the pixels in the previous image and the pixels wrapped by the optical flow to the target image, the photometric loss can effectively screen the accuracy of the optical flow. In addition, the previous methods suffer from discontinuity in position estimation in the 01 sequence of the KITTI dataset [3]. This is because the depth of the selected feature points is far away, and the optical flow at these positions remains unchanged. When most of the points in the image belong to this type, the camera motion is considered to be small, which leads to discontinuous bit-position estimation. To solve this problem, this paper proposes a depth mask to remove points whose estimated depths are too far away.

2   Method

Overview. Figure 2 presents the proposed method, consisting of an unsupervised optical flow estimation network and a monocular depth estimation network monodepth2 [4]. Two consecutive images at time steps $t-1$ and t are taken as an image pair to estimate the forward and backward optical flows ($t->t-1$, $t-1->t$) and the depth at time step t. The input images at $t-1$ and t, along with the forward optical flow, are used to compute the predicted photometric loss. The depth at time step t is used to create a depth mask. Based on forward-and-backward consistency, photometric loss, and the depth mask, dense optical flow is filtered. Sparse optical flow is obtained and used for matching image correspondences to estimate camera pose. Multi-view geometry is employed to calculate the transformation between two images. The predicted depth is used to recover the scale. The key features of this method include using the unsupervised network to estimate flow and depth mask to improve visual odometry estimation. The proposed unsupervised optical flow estimation network will be described first, followed by an explanation of feature selection and pose estimation.

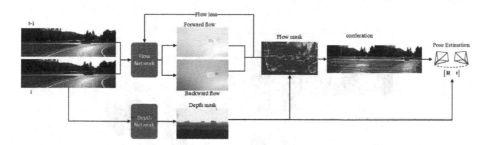

Fig. 2. Overall structure of the proposed methodology.

Unsupervised Optical Flow Estimation. A new unsupervised optical flow estimation network based on feature pyramid network structure [7] is proposed as shown in Fig. 3. A module for multi-input feature fusion and a deformable up-sampling module are proposed to address the problem of blurring and artifacts during up-sampling that exists in the current unsupervised approaches. Training

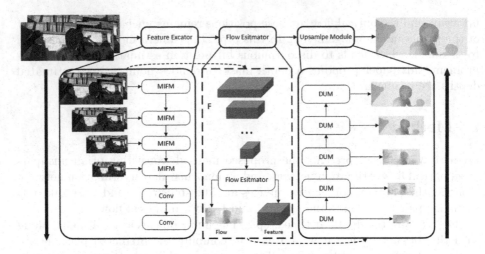

Fig. 3. The overall architecture of the unsupervised optical flow estimation network. We add a multi-input feature fusion module (MIFM) in the feature extraction part and introduce a deformable upsampling module (DUM).

is performed using photometric coherence, smoothing loss, and self-supervised loss.

Multi-scale image feature extraction is an effective approach for handling objects of different sizes. We incorporate features extracted from the original image into the feature extraction process and merge them with the pyramid features.

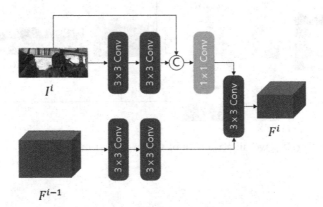

Fig. 4. Illustration of the multi-input feature fusion module of our network.

For the $i-th$ layer of the optical flow pyramid, $i = 0, 1, 2, 3, 4$. $I^i \in R^{\frac{H}{2^i} * \frac{W}{2^i} * 3}$ is the image obtained by scaling the input image to i-th layer, from where the feature is extracted using two $3 * 3$ convolution layers. Then, after joining the features from the previous step and the down-sampled image I^i, use a $1 * 1$ convolution layer fusing to generate the down-sampling feature. The feature $F^{i-1} \in R^{\frac{H}{2^{i-1}} * \frac{W}{2^{i-1}} * 3}$ obtained from the front layer are used in two $3 * 3$ convolutional layers to derive size that is consistent with the down-sampling features $F^{i-1\downarrow}$. F^i is mixed through a $3 * 3$ convolution which contains the down-sampling feature and $F^{i-1\downarrow}$, as shown in Fig. 4.

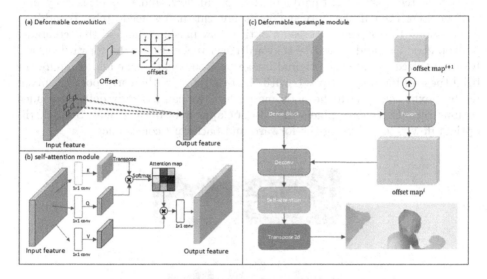

Fig. 5. Architecture of the proposed DUM.

Our approach combines deformable convolution [1] and self-attention mechanisms [14] to form a deformable upsampling module, as shown in Fig. 5. Deformable convolution can change the shape of the convolution kernel by learning an offset map for each location and then applying the resulting offset map to the feature map to resample the corresponding features for weighting. The self-attention mechanism [14] can capture long-distance interdependent features in an image. During the calculation process, it directly connects two pixels in the image through a single step, which is conducive to the effective use of features.

By using dilated convolution, the optical flow residual feature Res^i at the $i - th$ layer of the pyramid is set to a fixed dimension which can reduce the heavy computational burden. Then, use Res^i to calculate the offset map and update through fusion it with the sampled one from the upper layer, which is briefly described in Fig. 5(c). Perform deformable convolution on Res^i according to the offset map for a further refined optical flow feature $Refine^i$. We use the self-attention mechanism to compute the relationship in $Refine^i$ and transposed convolution to amplify the features and retain the flow relationship figured by the self-attention mechanism to estimate optical flow at this scale.

Correspondence Selection. Through the above unsupervised optical flow network, it computes the forward optical flow, backward optical flow, and the optical flow mask. Forward-backward consistency detection means that a pixel in the original image, first arrives at a position in another image along the forward optical flow and then returns along the reverse optical flow, and eventually the pixel can return to its initial position in the original image. However, points beyond the range of the image cannot be consistently detected. In addition, relying only on the consistency of the forward and backward optical flow does not accurately assess the quality of the optical flow, and the consistency may still be high even if the optical flow is not calculated accurately. To address these issues, this paper proposes an optical flow masking method that combines photometric loss and forward-backward consistency. First, the forward optical flow is utilized to map the original image onto the next image and compare the brightness of the mapped pixels with the brightness of the corresponding pixels in the next image. With the photometric loss, we can not only detect occluded regions but also remove pixels that do not appear in both images. Second, the optical flow is sorted using the forward and backward consistency.

Fig. 6. Deep mask example (Yellow region denotes the relatively distant area where feature points are not selected. The dark region represents the area where feature selection is performed). (Color figure online)

In outdoor environments, there are some extreme far points. A small number of these points can stabilize the system, but too many can cause the odometer to think that the camera is not moving, thus affecting the continuity of the odometer. To solve this problem, this method uses a monocular depth estimation network to predict the relative depth of each pixel in the image. Then, the points with predicted depth values larger than 40 are considered extremely far points and are removed. As shown in the figure above (Fig. 6), the yellow part is the masked area, and when selecting the key points, the points as close as possible to the camera are chosen for the calculation. Next, in the optical flow field obtained after the dense optical flow calculation, we filter and select the sparse optical flow points. Then, these optical flow points are sorted according to the forward-and-backward consistency criterion, and arranged in the order of smallest to largest. To avoid over-concentration of optical flow points in certain regions, we divide the image into 12 image blocks and select up to 100 optical flow points in each image block for position estimation.

Pose Estimation. The aforementioned steps involve filtering the obtained dense optical flow into the sparse optical flow. By matching the sparse optical flow, a two-dimensional to two-dimensional correspondence is obtained to predict the pose between two consecutive frames. First, the method defines the geometric relationship between two consecutive images using the essential matrix E provided by the epipolar constraint:

$$P_j^T K^{-T} E K^{-1} P_i = 0 \tag{1}$$

where (P_i, P_j) are the corresponding pixel coordinates of the two frames, and K is the known camera intrinsic matrix. The camera motion $[R, t]$ is solved by decomposing the essential matrix E = $[t] \times R$, where $[t] \times$ denotes the matrix representation of the cross product with t.

The essential matrix alone cannot recover the scale of the pose. The method employs the depth map estimated from the depth network to recover the scale. The 3D positions of the points were calculated using triangulation, and the scale was then recovered based on the ratio to the estimated depth.

3   Experimental Results and Analysis

Experimental Setup. The experiment utilized the KITTI 2015 dataset [12] as the standard setup for training and evaluating the unsupervised optical flow network. The learning rate was set to 1e−5, and the training process involved 100,000 iterations with one image per iteration. The experiment was conducted using a single GPU (3090Ti), and the training time took approximately 48 h. For depth estimation, the monocular depth estimation network was employed. In the odometry experiments, the KITTI dataset [3] was used for testing, which consists of 11 driving sequences. Each sequence is composed of images with a size of 640×192 pixels. The experiments were conducted on sequences 00 to 10.

Results for Unsupervised Optical Flow Estimation Networks. Comparisons were made between the proposed method and advanced unsupervised methods on the KITTI dataset, as shown in Table 1. The results obtained from the KITTI dataset demonstrate that our method can accurately estimate optical flow. Endpoint error (EPE) and the percentage of erroneous pixels (ER) were used as evaluation metrics for comparison. On the KITTI 2015 dataset, the ER achieved by our method was 9.04%, outperforming the UpFlow method which had an ER of 0.34%. Additionally, visual comparisons were conducted between our method and the classical method UFlow (Fig. 7).

Table 1. Comparison with unsupervised optical flow estimation methods.

method	train	test (F1-all)
DDFlow [9]	5.72	14.29%
SelFlow [10]	4.84	14.19%
UFlow [5]	2.71	11.13%
Upflow [11]	2.45	9.38%
CoT-AMFlow [15]	-	10.34%
MDFlow-fast [6]	3.02	11.43%
ours	2.54	9.04%

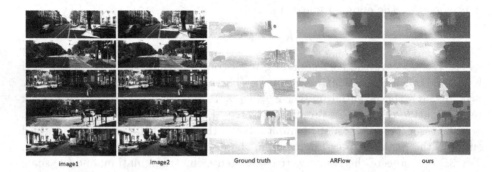

image1   image2   Ground truth   ARFlow   ours

Fig. 7. Visualization comparison with ARFlow on KITTI 2015 training dataset.

Comparison on KITTI Odometry Dataset. The experiments compare the trajectory accuracy of the proposed method with other methods for 11 sequences on the KITTI dataset. This paper is compared with the traditional method orbslam2, the deep learning methods depth-vo-feat [16], sc-sfmlearner [8], and DF-VO [17] which is a mixture of deep learning and geometric methods.

Table 2. Result on KITTI Odometry Seq.00-10. The best result is in bold and the second best is underlined.

method	0		1		2		3		4		5		6		7		8		9		10	
	terr	rerr	terr	rerr	terr	rerr	terr	rerr	terr	rerr	terr	rerr	terr	rerr	terr	rerr	terr	rerr	terr	rerr	terr	rerr
depth-vo-feat [16]	6.23	2.44	23.78	1.75	6.59	2.26	15.8	10.6	3.14	2.02	4.94	2.34	5.8	2.06	3.56	2.39	5.45	2.39	11.9	3.70	12.82	3.41
sc-sfmlearner [8]	11.01	3.39	27.09	1.31	6.74	1.96	9.22	4.93	4.22	2.01	6.70	2.38	5.36	1.65	8.59	4.53	8.11	2.61	7.64	2.19	10.74	4.58
orbslam2	11.43	**0.58**	107.6	**0.89**	10.3	**0.26**	**0.97**	**0.19**	1.30	**0.27**	9.04	**0.26**	14.6	**0.26**	9.77	**0.36**	11.46	**0.28**	9.31	**0.26**	2.66	**0.39**
dfvo	**1.74**	0.79	63.5	19.38	**2.48**	0.78	3.34	1.02	<u>0.90</u>	0.84	**1.69**	0.81	**1.57**	<u>0.54</u>	**2.14**	0.99	**2.05**	0.76	3.36	0.63	2.70	0.89
ours (wo mask)	<u>1.95</u>	0.65	<u>18.41</u>	<u>1.07</u>	2.78	0.77	3.94	1.16	1.40	0.70	2.35	0.99	2.74	1.05	2.7	<u>0.65</u>	2.48	0.73	<u>2.60</u>	0.56	<u>2.40</u>	0.61
ours (w mask)	1.98	<u>0.60</u>	**16.37**	1.28	<u>2.49</u>	<u>0.68</u>	<u>3.27</u>	<u>0.80</u>	**0.89**	<u>0.48</u>	<u>1.76</u>	<u>0.60</u>	<u>2.10</u>	0.67	<u>2.60</u>	0.69	<u>2.11</u>	<u>0.58</u>	**2.39**	<u>0.47</u>	**2.32**	<u>0.53</u>

The experiments use translational error (terr) and rotational error (rerr) as evaluation metrics. Translational error measures the difference between the estimated camera translation vectors and the true translation vectors. The smaller the translational error, the closer the estimated camera translation vector is to the true translation vector. The rotation error measures the difference between the estimated camera rotation matrix or quaternion and the true rotation matrix or quaternion. A smaller rotation error indicates that the estimated camera rotation is closer to the true rotation, the estimated camera pose is more accurate. The relative trajectory error (REP) is used to evaluate the accuracy of the relative bit pose estimation between neighboring frames. It calculates the difference between the estimated relative bit pose and the true relative bit pose. It can provide a localized error picture about the pose estimation and can detect cumulative error or drift.

Table 3. Comparison between the proposed method, ORB-SLAM2, and DF-VO in Relative pose error (without seq01).

	REP (m)	REP (°)
DF-VO	0.033	0.05
orbslam2	0.131	0.07
ours	0.029	0.06

Table 2 shows the comparison between the proposed method and the classical method in terms of translation error and rotation error using KITTI odometry dataset. The translation error and rotation error of the proposed method are improved compared to DF-VO. On the 01 sequence, thanks to the use of a depth mask, the proposed method shows great advantages. The average translation error decreases from 7.77 to 3.48, and the average rotation error decreases from 2.49 to 0.67. In other sequences, the present method is also competitive.

The method conducted an ablation study on the depth mask for sequences 00 to 10. The results are shown in Table 2. The depth mask proves to be effective in scenarios with a high density of extremely distant points, as it helps improve the estimation accuracy. Interestingly, even in scenarios with fewer extremely distant points, the depth mask still contributes to enhanced estimation precision. Figure 8 shows the visual comparison of trajectory estimation between our method and some classical methods.

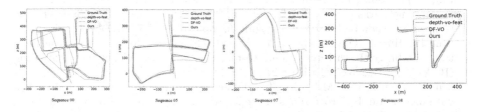

Fig. 8. VO results on KITTI benchmark dataset.

Excluding the interference from sequence 01, Table 3 presents the comparison in terms of Relative Pose Error (REP) on translational and rotational, using the KITTI odometry dataset. The proposed method achieves improvement in translational error compared to orbslam2 and DF-VO. Our method still demonstrates strong competitiveness, with REP errors consistently within this range.

4   Conclusion

In this paper, we propose a new monocular visual odometry method based on optical flow matching obtained by an unsupervised method, which can well estimate the camera position by depth mask and flow mask. The visual odometry method achieves on average 3.48 (%) translation errors and 0.67 rotation errors ($°/100\,$m). However, the method also suffers from cumulative errors, which have been difficult to solve for odometry and may need to be explored in-depth in subsequent pairs.

References

1. Dai, J., et al.: Deformable convolutional networks. In: Proceedings of the IEEE International Conference on Computer Vision, pp. 764–773 (2017)
2. Forster, C., Pizzoli, M., Scaramuzza, D.: SVO: fast semi-direct monocular visual odometry. In: 2014 IEEE International Conference on Robotics and Automation (ICRA), pp. 15–22. IEEE (2014)
3. Geiger, A., Lenz, P., Urtasun, R.: Are we ready for autonomous driving? The KITTI vision benchmark suite. In: Conference on Computer Vision and Pattern Recognition (CVPR) (2012)
4. Godard, C., Mac Aodha, O., Firman, M., Brostow, G.J.: Digging into self-supervised monocular depth estimation. In: Proceedings of the IEEE/CVF International Conference on Computer Vision, pp. 3828–3838 (2019)
5. Jonschkowski, R., Stone, A., Barron, J.T., Gordon, A., Konolige, K., Angelova, A.: What matters in unsupervised optical flow. In: Vedaldi, A., Bischof, H., Brox, T., Frahm, J.-M. (eds.) ECCV 2020. LNCS, vol. 12347, pp. 557–572. Springer, Cham (2020). https://doi.org/10.1007/978-3-030-58536-5_33
6. Kong, L., Yang, J.: MDFlow: unsupervised optical flow learning by reliable mutual knowledge distillation. IEEE Trans. Circuits Syst. Video Technol. **33**(2), 677–688 (2023). https://doi.org/10.1109/TCSVT.2022.3205375
7. Lin, T.Y., Dollár, P., Girshick, R., He, K., Hariharan, B., Belongie, S.: Feature pyramid networks for object detection. In: Proceedings of the IEEE Conference on Computer Vision and Pattern Recognition, pp. 2117–2125 (2017)
8. Liu, H., Huang, D.D., Geng, Z.Y.: Visual odometry algorithm based on deep learning. In: 2021 6th International Conference on Image, Vision and Computing (ICIVC), pp. 322–327. IEEE (2021)
9. Liu, P., King, I., Lyu, M.R., Xu, J.: DDFlow: learning optical flow with unlabeled data distillation. In: Proceedings of the AAAI Conference on Artificial Intelligence, vol. 33, pp. 8770–8777 (2019)
10. Liu, P., Lyu, M., King, I., Xu, J.: SelFlow: self-supervised learning of optical flow. In: Proceedings of the IEEE/CVF Conference on Computer Vision and Pattern Recognition, pp. 4571–4580 (2019)

11. Luo, K., Wang, C., Liu, S., Fan, H., Wang, J., Sun, J.: UPFlow: upsampling pyramid for unsupervised optical flow learning. In: Proceedings of the IEEE/CVF Conference on Computer Vision and Pattern Recognition, pp. 1045–1054 (2021)
12. Menze, M., Heipke, C., Geiger, A.: Joint 3D estimation of vehicles and scene flow. In: ISPRS Workshop on Image Sequence Analysis (ISA) (2015)
13. Mur-Artal, R., Tardós, J.D.: ORB-SLAM2: an open-source SLAM system for monocular, stereo, and RGB-D cameras. IEEE Trans. Rob. **33**(5), 1255–1262 (2017)
14. Vaswani, A., et al.: Attention is all you need. In: Advances in Neural Information Processing Systems 30 (2017)
15. Wang, H., Fan, R., Liu, M.: CoT-AMFlow: adaptive modulation network with co-teaching strategy for unsupervised optical flow estimation. In: Conference on Robot Learning, pp. 143–155. PMLR (2021)
16. Zhan, H., Garg, R., Weerasekera, C.S., Li, K., Agarwal, H., Reid, I.: Unsupervised learning of monocular depth estimation and visual odometry with deep feature reconstruction. In: Proceedings of the IEEE Conference on Computer Vision and Pattern Recognition, pp. 340–349 (2018)
17. Zhan, H., Weerasekera, C.S., Bian, J.W., Garg, R., Reid, I.: DF-VO: what should be learnt for visual odometry? arXiv preprint arXiv:2103.00933 (2021)
18. Zou, Y., Luo, Z., Huang, J.-B.: DF-Net: unsupervised joint learning of depth and flow using cross-task consistency. In: Ferrari, V., Hebert, M., Sminchisescu, C., Weiss, Y. (eds.) ECCV 2018. LNCS, vol. 11209, pp. 38–55. Springer, Cham (2018). https://doi.org/10.1007/978-3-030-01228-1_3

End-to-End Motion Planning Based on Visual Conditional Imitation Learning and Trajectory-Guide

Yiming Gao[1], Chaoxia Shi[1(✉)], and Yanqing Wang[2(✉)]

[1] Nanjing University of Science and Technology, Nanjing 210094, China
scx@njust.edu.cn
[2] Nanjing Xiaozhuang University, Nanjing 211171, China
wyq0325@126.com

Abstract. End-to-end motion planning based on visual conditional imitation learning includes control prediction and trajectory planning. The former lacks global environment perception, the latter does not consider vehicle kinematic modeling, and both lack interpretability. This paper proposes a method that combines the two. Interpretability is provided for end-to-end networks by generating a semantic bird's eye view. The high-level features of the bird's eye view are fed into the trajectory planning branch and the control prediction branch, while the control prediction branch obtains trajectory guidance, making the two tightly integrated and mutually beneficial. In comparison experiments with vision-advanced methods LBC, TCP, and DLSSCIL on the NoCrash benchmark of the Carla simulation, the proposed method obtains an average gain of 13% in complex scenarios.

Keywords: Autonomous driving · Motion planning · Semantic bird's eye view · Conditional imitation learning · End-to-end learning

1 Introduction

End-to-end motion planning based on visual conditional imitation learning provides promising solutions for autonomous driving, which can map images directly to the trajectory or the low-level commands of the vehicle, thus avoiding cascading errors caused by complex module design and human-designed autonomous driving rules. However, mapping images directly to the expected output lacks perception of the global environment and interpretability [12,16], which will inevitably lead to failures of the task during testing. In addition, the output form of models has an important impact. Mainstream methods are categorized into two main groups according to the output form: trajectory planning, and control prediction.

Trajectory planning methods predict the waypoints that the vehicle expects to reach in the future. LBC [4] uses a two-stage training strategy, where the first

H. Jin et al. (Eds.): IAIC 2023, CCIS 2060, pp. 30–43, 2024.
https://doi.org/10.1007/978-981-97-1332-5_3

stage trains the privileged network with map information to generate trajectories. The second stage uses images as inputs, and the outputs of the first stage as training labels for predicting trajectories. Semantic Bird's Eye View (SBEV) provides scale information and interpretable references for end-to-end motion planning. The optimal trajectory is selected in the paper [3,17] by calculating the consumption of different paths on the SBEV. This approach requires additional controllers (e.g. PID controllers) to transform the planned trajectories into commands, which have the advantage of fully utilizing the global environment information, generating multiple waypoints, and reducing the risk of unexpected accidents. However, tracking the trajectory through the controller is not always correct. Simple controllers like PID do not perform well at intersections with large turns or red light activation scenarios, as shown in Fig. 1.

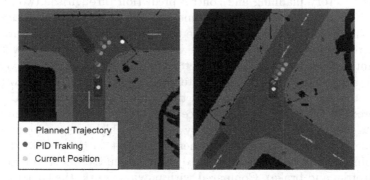

Fig. 1. Example of a turning scenario where the PID controller does not accurately track the predicted trajectory

Control prediction methods directly map images to low-level commands through optimized models. Among them, Imitation Learning (IL) approaches [2,10] can accomplish road tracking and obstacle avoidance, but it can only learn the expert's driving actions, not the expert's driving intentions, so it cannot accomplish the driving task in scenarios such as intersections. To solve this problem, Conditional Imitation Learning (CIL) [6–8] is proposed, which inputs conditional branch commands (left turn, right turn, straight, etc.), and the model is capable of generating correct driving actions according to different expert intentions. The CIL method suffers from difficulty in starting after stopping, CIL extension with a ResNet [9] and speed prediction network (CILRS) [7] predicts the speed through independent branches. Based on the difference between the actual speed and the predicted speed calculate the angle of the throttle, which solves the problem. The advantage of this approach is that it can use an optimized model to directly predict commands, but its success relies on the assumption that the command at the current moment satisfies the assumption of independent and identically distributed with the command at other moments, which doesn't always hold in testing.

A combined trajectory planning and control prediction approach is explored in paper [15]. Intermediate features of the trajectory branch are provided to the control prediction branch by means of multitask learning [1], and the control prediction branch obtains guidance from it by learning the attention graph. The model predicts both the trajectory and the commands at multiple timestamps. However, the approach lacks interpretability. To address this problem, the DLSSCIL approach [14] uses the LS [11] model to predict pixel depths, generates SBEV, and combines the SBEV with conditional imitation learning to directly predict the commands. However, this approach performs poorly in challenging unknown environments in the absence of trajectory planning and with partially known environments as a prerequisite.

In this paper, an End-to-End Motion Planning Based on Conditional Imitation Learning and Trajectory-guide (TLSSCIL) method is proposed. TLSSCIL performs trajectory planning and control prediction through the two branches, respectively. Surrounding view images, and measurement vectors (velocity, target point, conditional branch command) are used as inputs. The images are passed through a prediction network to generate visual features, bird's eye view features, and semantic bird's eye view. The trajectory branch encodes the state vectors to generate state features, which are concatenated with the bird's eye view features to predict the waypoints of the next K timestamps through the GRU [5]. The control prediction branch seeks guidance from the trajectory planning branch. It converts the waypoints into trajectory features, which are concatenated with visual and bird's eye view features. Then it is input to the conditional imitation learning motion planning module to generate multistep control commands (steer, throttle, and brake). Compared with previous work, the contributions of this paper are:

1) A trajectory planning method based on semantic bird's eye view is proposed, which, unlike image-based trajectory planning, provides interpretable references for end-to-end networks.
2) An end-to-end motion planning method fusing image features, semantic bird's eye view features, trajectory features, and conditional imitation learning is proposed.
3) A control prediction framework based on trajectory-guide is designed, and the control prediction branch predicts commands through the guidance of trajectory, which fully integrates the advantages of the trajectory planning method and the direct control prediction method.

2 Method

2.1 Problem Setting

The end-to-end motion planning based on conditional imitation learning interacts with the environment at discrete moments. At each time stamp t, the model inputs the observation data o_t acquired by the sensors and outputs the commands

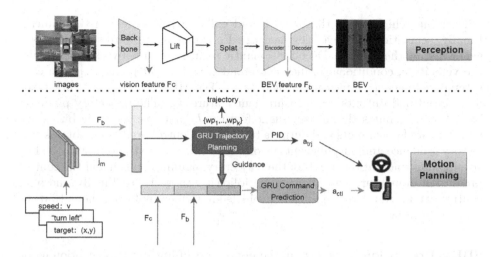

Fig. 2. Overview of the TLSSCIL model architecture. The surrounding view image is passed through the perception module to generate visual features and bird's eye view features (up). The perceptual features are combined with the measurement features to generate trajectories through the trajectory planning branch, and the control prediction branch seeks guidance from the trajectory planning branch to generate commands. The outputs of the two branches are fused to generate the final commands.

a_t. The observation-commands form the training dataset $D = \{\langle o_i, a_i \rangle\}_{i=1}^{N}$, which is obtained by recording the expert driving behavior. The model obtains a driving effect close to expert driving by imitating the expert driving behavior. This is a supervised learning problem, and the optimal parameter θ of the mapping $F(o_i, \theta)$ from observations to commands can be described as:

$$\theta^* = \arg\min_{\theta} \sum_i l(F(o_i, \theta), a_i) \tag{1}$$

The TLSSCIL method input surround view images I and camera calibration information T, measurement data m consisting of the speed v, conditional branch commands c, and target point (x, y) generated by the global path planner, the model outputs a final control command consisting of longitudinal control commands throttle $t \in [0, 1]$, brake $b \in [0, 1]$, and lateral control command steer angle $s \in [-1, 1]$.

2.2   Architecture Design

As shown in Fig. 2, the network framework includes a surround view perception module and a motion planning module consisting of two sub-branches. Surrounding view images through the Backbone network of EfficientNet-B0 [13] to extract the visual image features F_c. The frustum point cloud is then generated from the pixel coordinates via the Lift depth estimation network. After Splat downsampling, the pixel features are projected into a local map centered on the self-driving vehicle using the internal and external parameters of the camera. Finally, semantic bird's eye view features F_b and semantic bird's eye view are generated by the Encoder-Decoder semantic segmentation network. Meanwhile, the velocity v, conditional branch command c, and target point (x, y) are concatenated to generate measurement data m. The MLP encoding network takes m as input and outputs the measurement feature J_m. The trajectory planning branch concatenates the measurement feature J_m and the semantic bird's eye view feature F_b as inputs and outputs trajectory. Control prediction obtains trajectory guidance and outputs control commands for the next K timestamps. The specific implementation details of the trajectory planning branch and the control prediction branch are described in the following subsections. Finally, the final control commands of the self-driving vehicle are formed by fusing the outputs of the two branches.

SBEV Preception. As shown in the upper part of Fig. 2, the perception module takes RGB images $\{I_k \in R^{3 \times w \times h}\}_n$, the internal matrix $\{i_k \in R^{3 \times 4}\}_n$ and the external matrix $\{e_k \in R^{3 \times 3}\}_n$ as inputs, and generates SBEV features and SBEV through four stages: image feature extraction, generation of frustum point cloud, feature projection, and semantic segmentation, respectively.

First, the feature extraction stage extracts the visual feature $F_c \in R^{c \times U \times V}$ through the Backbone network of EfficientNet-B0 [13], where U, V are the width and height of the visual feature map, and c is the feature dimension of each pixel. The original image pixel coordinates (h, w) corresponding to each coordinate (u, v) of the visual feature F_c are correlated with the depth d to generate a point cloud $\{(h, w, d)_i \in R^3 | d \in D_{n \times |D| \times U \times V}\}$, where D is the set of all possible depths of the pixel. For feature b of the visual feature map F_c located in coordinates (u, v), the depth distribution $\alpha \in R^{|D|-1}$ is predicted, and the pixel feature b is extrapolated with the depth distribution of the pixel to generate b_d:

$$b_d = \alpha_d \cdot b \tag{2}$$

Thus, the point cloud with depth information is associated with each pixel feature in the visual feature map, at each point (h, w, d) in the point cloud there is a pixel feature b_d associated with it to form the frustum point cloud. In the feature projection stage, the pixel coordinates of the frustum cloud with depth information are transformed into world coordinates centered around the vehicle using the camera's internal and external parameters. The corresponding pixel features are projected onto a grid with width W and height H to generate the grid feature map. Finally, in the semantic segmentation stage, the grid feature map is input into a Resnet18-based Encoder-Decoder network for semantic segmentation, where the encoder downsamples the input data to generate SBEV features B. The decoder performs an upsampling operation and predicts the distribution of pixel categories through a softmax network, resulting in a SBEV map with width W and height H.

The output of the SBEV is ultimately the category of each pixel, so the loss is calculated using the cross-entropy loss function for the perception module:

$$loss = - \sum_{0}^{N-1} \sum_{i=1}^{C-1} y_i \log \hat{y}_i \tag{3}$$

where N is the total number of pixels in the SBEV, C is the number of pixel categories, y_i is the true category of the i-th pixel, and $\hat{y}_i$ is the predicted category of the i-th pixel.

Trajectory Planning Branch. Unlike direct prediction commands, the trajectory planning branch first outputs the waypoints. Reasoning about the environment changes is important for predicting the waypoints, in order to simulate the interaction between the vehicle and the environment at each moment, TLSSCIL concatenates semantic bird's eye view features F_b and measurement features J_m to generate the input features j^{trj} for the trajectory planning branch:

$$j^{trj} = \text{MLP} \left(\text{Concat} \left(F_b, J_m \right) \right) \tag{4}$$

The j^{trj} is then inputted into the GRU [26] to generate the waypoints w $= \{wp_1, \dots, wp_k\}$ of future K timestamps. Inspired by [1], the longitudinal controller calculates the expected velocity v^* of the ego vehicle for the waypoints:

$$v^* = \frac{\|wp_k - wp_{k-1}\|_2}{\delta t} \tag{5}$$

where δ_t is the time step between adjacent waypoints. The difference between the expected velocity v^* and the current velocity v is used to calculate the angle a_t of the throttle. Since the angle of the throttle cannot be negative, a_t is set to zero when the expected velocity v^* is less than a certain threshold ϵ. The difference between the expected velocity v^* and the current velocity v is only used to calculate the angle a_b of the brake. The lateral controller obtains the angle a_s of the steer by calculating the angle of the waypoint with the vertical direction.

Control Prediction Branch. As described in Sect. 1, the control prediction method maps directly from the input of the sensor to commands, relying on the commands at different moments to satisfy the assumption of independent identical distributions, but this assumption does not always hold during the testing process, so our control prediction branch outputs the multi-step command $a = \{a_1, \dots, a_k\}$ based on the current perception, as shown in Fig. 3. However, it is difficult to predict multi-step commands only from the data obtained at the current moment. To solve this problem, the control prediction branch obtains the trajectory guidance at each time step. It first transforms the waypoints generated by the trajectory planning branch into a heat map, generates the trajectory features F_w through the MPL, then concatenates the semantic bird's

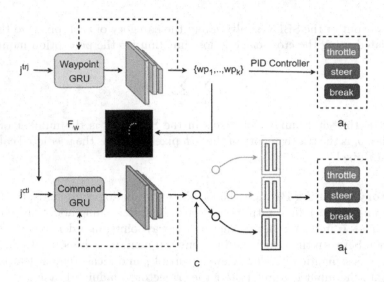

Fig. 3. The trajectory guide control prediction process. The trajectory branch predicts waypoints, which are transformed into commands by the PID controller. The control prediction branch obtains trajectory guidance from the trajectory planning branch and predicts multi-step commands.

eye view features F_b, the image features F_c, and the trajectory features F_w as the input features j^{ctl} of the control prediction branch:

$$j^{ctl} = \mathrm{MLP}\left(\mathrm{Concat}\left(F_b, F_c, F_w\right)\right) \tag{6}$$

The j^{ctl} is fed into the GRU, inspired by [8], a different MLP network is used to predict the command for different conditional branch instructions c (straight ahead, left turn, etc.).

To better estimate the current state, the control prediction branch predicts the speed using an independent fully connected network. The speed prediction network inputs j^{ctl} and outputs the speed of the vehicle s:

$$s = \mathrm{MLP}\left(j^{ctl}\right) \tag{7}$$

Loss Design. The loss function of the TLSSCIL method includes the trajectory loss $loss_{trj}$, control loss $loss_{ctl}$, and speed loss $loss_{aux}$. For the trajectory branch, the loss is represented as:

$$loss_{trj} = \sum_{i=0}^{k-1} ||wp_i, \hat{wp}_i||_1 \tag{8}$$

where wp_i, $\hat{wp}_i$ represent the real and predicted waypoints at $k = i$, respectively. For the control prediction branch, Codevilla et al. [9] demonstrated that L1 loss is more suitable for autonomous driving tasks than L2 loss. Therefore, the LSTCP method uses L1 loss function to calculate the control loss as follows:

$$loss_{\text{ctl}} = \sum_{i=0}^{k-1} ||a_i, \hat{a}_i||_1 \tag{9}$$

where a_i and $\hat{a}_i$ represent the real control and predicted control at $k = i$, respectively. The loss function for speed prediction is shown as follows:

$$loss_{\text{aux}} = \sum_{i=0}^{k-1} ||s, \hat{s}|| \tag{10}$$

where s and $\hat{s}$ represent the actual and predicted vehicle speeds, respectively. α_{trj}, α_{ctl}, and α_{aux} represent the weights of different losses, and the final loss is shown as follows:

$$loss = \alpha_{\text{trj}} \cdot loss_{\text{trj}} + \alpha_{\text{ctl}} \cdot loss_{\text{ctl}} + \alpha_{\text{aux}} \cdot loss_{\text{aux}} \tag{11}$$

2.3   Output Fusion

Algorithm 1   Output Fusion

Input: images I; calibration T; speed v; conditional command c; target (x, y);
Output: command a
1: $\{wp_i\}_{i=0}^k, a_{\text{ctl}} \leftarrow \text{LSTCP}\,(\text{I}, \text{T}, v, c, (x, y))$
2: $a_{\text{trj}} \leftarrow \text{PID}\left(\{wp_i\}_{i=0}^k\right)$
3: get conditional branch command c
4: **if** $c = $ turning **then**
5:     $a = \delta \times a_{\text{trj}} + (1 - \delta) \times a_{\text{ctl}}$
6: **else**
7:     $a = \delta \times a_{\text{ctl}} + (1 - \delta) \times a_{\text{trj}}$
8: **end if**

The two branches of the TLSSCIL have two kinds of outputs: the trajectories and the commands. The trajectory planning branch outputs the trajectory and uses a PID controller to transform it into the command a_{trj}, and the control prediction branch directly predicts the command a_{ctl}. In the test, TLSSCIL uses Algorithm 1 to fuse the outputs of the two branches. For the outputs of the two branches, a weight factor $\sigma \in [0, 0.5]$ is used to balance the output from the two branches, and different weights are selected according to different conditional branch commands. Since the PID controller is prone to tracking failure in the case of turning, the TLSSCIL method is dominated by command output from the control prediction branch in the case of turning, and the command output from the trajectory planning branch in other navigation scenarios.

3   Experimental Results and Analysis

3.1   Dataset

The data collection and experimental validation of the TLSSCIL are performed on the open-source autonomous driving simulation platform Carla [10]. The dataset mainly records the surround view images during the expert's driving process, which is acquired by six cameras, namely, front, left front, right front, front rear, left rear, and right rear, with the size of 480×320 and the FOV of $90\,°C$. Semantic bird's eye view annotated data with the size of 200×200, which represents the 20-m actual area around the vehicle. The data being collected at the same time are the vehicle's speed, expert commands (throttle, steering, brake), target points, and conditional branch commands. The 117K sets of data recorded from 25 routes driven by the expert in Town1 are used as training data for semantic bird's eye view and motion planning. The data are collected at a frequency of 2 fps, and online tests are done on 25 routes each in Town1 and Town2.

3.2   SBEV Visualization

Unlike the single-category SBEV generation in [11], the TLSSCIL perception module predicts nine categories for each pixel point: vehicle, lane, road, sidewalk, pedestrian, red light, green light, yellow light, and other, as shown in Fig. 4. Considering the uneven distribution of the nine categories of samples in the training data, different weight coefficients are used for the loss of different categories during training, and the weight coefficients of each category are inversely proportional to the total number of pixels of each category in the whole dataset. As shown in Table 1, SBEV created by the TLSSCIL method has an average accuracy of 85% and an average IOU of 47%, which provide effective environment information for the motion planning module.

Fig. 4. Visualization results of SBEV.

3.3   Motion Planning Comparison Experiment

The motion planning module sets the initial learning rate lr = 0.001 and uses the learning rate decay strategy during training. To address the problem of the

Table 1. Results of SBEV creation.

Class	vehicle	street	road	walk side	pedestrian	red	green	yellow	other	avg
ACC (%)	90	97	83	75	57	98	93	92	82	85
IoU (%)	41	20	87	62	15	56	35	27	77	47

uneven sample distribution of throttle, brake, and steer, similar to the semantic bird's eye view training, different weights are given to the loss of the three commands. The motion planning module is tested on the NoCrash Benchmark [9] of the Carla in version 0.9.10. Among the test routes are 25 in Town1 and Town2 respectively, which include various traffic scenarios such as going straight and turning. Six kinds of weather are included in the test route, four of which are known weather "ClearNoon", "WetNoon", "HardRainNoon" and "ClearSunset", and two unknown weather conditions not included in the training data, "WetSunset" and "SoftRainSunset". Due to the difference in the number of pedestrians and vehicles, there are three tasks for each route: empty, regular, and dense. The number of vehicles is 0, 20, and 100, and the number of pedestrians is 0, 50, and 250, respectively.

The NoCrash benchmark considers a vehicle successful when it reaches the target point from the start point of the given route within the specified time, and no serious collision (collision with pedestrians, collision with a vehicle, rollover, etc.) occurs in the course of the test. The specified time is obtained by dividing the total length of the route by the driving speed of 5 km/h. The final statistic for the experiment was the route completion rate (%), i.e. the number of routes completed divided by the total number of routes.

Offline training of LBC, TCP, and other methods is performed on the dataset introduced in Sect. 3.1. To compare the generalization performance of the different models, in addition to online testing in unknown Town2, online testing is also performed in the Town1 involved in the training. The comparison of the TLSSCIL with the visual state-of-the-art CIL [6], CILRS [4,7], TCP [15], and DLSSCIL [14] methods is realized, as shown in Tables 2 and 3.

Table 2. Carla NoCrash benchmark Town1 (Train)

Task	weather	CIL	CILRS	LBC	TCP	DLSSCIL	TLSSCIL
Empty	Train	69	82	97	84	96	**97**
Regular		60	76	93	85	**98**	95
Dense		18	39	71	88	92	**94**
Empty	Test	78	79	87	84	100	**100**
Regular		49	70	87	92	99	**99**
Dense		11	58	63	86	92	**93**

Table 2 shows that in Town1 participating in offline training, there is no significant degradation in the performance of the various methods with changing weather conditions. The TLSSCIL and the DLSSCIL performed slightly better than the other methods overall.

Table 3. Carla NoCrash benchmark Town2 (Test)

Task	weather	CIL	CILRS	LBC	TCP	DLSSCIL	TLSSCIL
Empty	Train	45	49	**100**	76	92	96
Regular		21	43	89	79	88	**90**
Dense		9	35	51	70	67	**73**
Empty	Test	23	73	70	72	76	**78**
Regular		11	58	62	74	69	**78**
Dense		1	42	39	70	62	**70**

Table 3 shows that the TCP achieves well by using only one monocular camera. The LBC achieves decent results by trajectory prediction with known weather but the performance decreases significantly in scenarios with dense vehicles and pedestrians, and the DLSSCIL method also achieves good results using control prediction. However, the TLSSCIL method has more adequate perceptual information by using surround view images as inputs, and also has a better performance compared to other methods by using trajectory-guide control prediction, so the TLSSCIL method has an average performance improvement of 13% compared to the TCP, LBC, and DLSSCIL in more challenging navigation environments with unknown weather and dense obstacles. The experimental comparison proves that the TLSSCIL method has better performance and generalization ability.

3.4   Ablation Experiment

To demonstrate the effectiveness of trajectory-guide control prediction, the trajectory planning branch and the control prediction branch are trained separately. The model with the trajectory branch alone is denoted as TLSSCIL-trj, and the model with the control prediction branch alone is denoted as TLSSCIL-ctl. The performance of the two branches under the NoCrash benchmark for the Carla version 0.9.10 in unknown Town2 is shown in Table 4.

Table 4 shows the control prediction branch has insignificant performance degradation in simple scenarios, but in complex scenarios with dense obstacle navigation, there is no significant improvement in driving performance relative to other methods. It shows that the TLSSCIL method is important for future trajectory planning. Using the trajectory planning branch alone, the driving performance decreases extremely, by 24% in the most complex navigation scenario

Table 4. Control vs. Trajectory Town2 (Test)

Weather	Known			Unknown		
Task	Empty	Regular	Task	Empty	Regular	Task
LSTCP-trj	93	75	54	76	62	46
LSTCP-ctl	91	87	68	**79**	69	59
LSTCP	**96**	**90**	**73**	78	**78**	**70**

with unknown weather and dense obstacles, because the planned trajectory cannot be perfectly tracked by the simple PID controller. The ablation experiments show that combining the advantages of the control prediction branch and the trajectory planning branch to make the trajectory guide control prediction is effective.

3.5   Study of Weight Factor δ

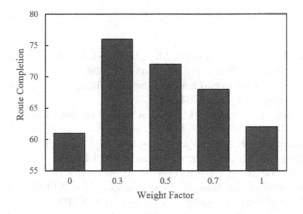

Fig. 5. Test result of different δ.

In order to obtain the optimal values of the weighting factors in Algorithm 1, different values of the weighting factors are tested separately under the NoCrash benchmark in CARLA version 0.9.10 Town2 unknown environment, unknown weather. $\delta \in [0, 0.5]$ indicates that in the case of a turn, since the PID controller does not perform trajectory tracking perfectly, the commands are based on the output of the predicted branch, which is a priori knowledge. To exclude the chance of testing, in addition to taking values in the range of $\delta \in [0, 0.5]$, values in the range of $\delta \in (0.5, 1]$ are also taken, and the results are shown in Fig. 5. $\delta = 0$ and $\delta = 1$ represent the outputs using only the trajectory planning branch or the control prediction branch, which have the lowest route completion, indicating that combining the trajectory planning and control prediction is effective. Among the values within $\delta \in (0, 1)$, the case of $\delta = 0.3$ has the highest completion rate, indicating that considering the respective advantages of different branches fully is necessary for the integration of branching outputs.

4 Conclusion

In this paper, we propose an end-to-end autonomous driving approach based on conditional imitation learning and trajectory guidance. Predicting the semantic bird's eye view centered on the vehicle by estimating the depth of monocular surround view images, provides scale information and interpretable reference for the end-to-end network. It fully combines the advantages of trajectory planning methods and direct control prediction methods, avoiding the problem of PID tracking failure of trajectory planning methods and the problem of control prediction methods relying on the assumption of independent identical distribution. Comparison experiments with TCP, LBC, DLSSCIL, etc., and ablation experiments on the NoCrash benchmark of Carla prove the effectiveness of the proposed approach.

References

1. Argyriou, A., Evgeniou, T., Pontil, M.: Multi-task feature learning. In: Advances in Neural Information Processing Systems 19 (2006)
2. Bojarski, M., et al.: End to end learning for self-driving cars. CoRR abs/1604.07316 (2016). http://arxiv.org/abs/1604.07316
3. Casas, S., Sadat, A., Urtasun, R.: MP3: a unified model to map, perceive, predict and plan. CoRR abs/2101.06806 (2021). https://arxiv.org/abs/2101.06806
4. Chen, D., Zhou, B., Koltun, V., Krähenbühl, P.: Learning by cheating. CoRR abs/1912.12294 (2019). http://arxiv.org/abs/1912.12294
5. Cho, K., et al.: Learning phrase representations using RNN encoder-decoder for statistical machine translation. In: Proceedings of the 2014 Conference on Empirical Methods in Natural Language Processing (EMNLP), Doha, Qatar, pp. 1724–1734. Association for Computational Linguistics, October 2014. https://doi.org/10.3115/v1/D14-1179. https://aclanthology.org/D14-1179
6. Codevilla, F., Müller, M., López, A., Koltun, V., Dosovitskiy, A.: End-to-end driving via conditional imitation learning. In: 2018 IEEE International Conference on Robotics and Automation (ICRA), pp. 4693–4700 (2018). https://doi.org/10.1109/ICRA.2018.8460487
7. Codevilla, F., Santana, E., López, A.M., Gaidon, A.: Exploring the limitations of behavior cloning for autonomous driving. CoRR abs/1904.08980 (2019). http://arxiv.org/abs/1904.08980
8. Dosovitskiy, A., Ros, G., Codevilla, F., López, A.M., Koltun, V.: CARLA: an open urban driving simulator. CoRR abs/1711.03938 (2017). http://arxiv.org/abs/1711.03938
9. He, K., Zhang, X., Ren, S., Sun, J.: Deep residual learning for image recognition. In: Proceedings of the IEEE Conference on Computer Vision and Pattern Recognition (CVPR), June 2016

10. Laskey, M., et al.: Iterative noise injection for scalable imitation learning. CoRR abs/1703.09327 (2017). http://arxiv.org/abs/1703.09327

11. Philion, J., Fidler, S.: Lift, splat, shoot: encoding images from arbitrary camera rigs by implicitly unprojecting to 3D. CoRR abs/2008.05711 (2020). https://arxiv.org/abs/2008.05711

12. Tampuu, A., Matiisen, T., Semikin, M., Fishman, D., Muhammad, N.: A survey of end-to-end driving: architectures and training methods. IEEE Trans. Neural Netw. Learn. Syst. **33**(4), 1364–1384 (2022). https://doi.org/10.1109/TNNLS.2020.3043505

13. Tan, M., Le, Q.V.: EfficientNet: rethinking model scaling for convolutional neural networks. CoRR abs/1905.11946 (2019). http://arxiv.org/abs/1905.11946

14. Wang, T.: Research on motion planning methods based on semantic bird's eye view (2023)

15. Wu, P., Jia, X., Chen, L., Yan, J., Li, H., Qiao, Y.: Trajectory-guided control prediction for end-to-end autonomous driving: a simple yet strong baseline (2022)

16. Yurtsever, E., Lambert, J., Carballo, A., Takeda, K.: A survey of autonomous driving: common practices and emerging technologies. IEEE Access **8**, 58443–58469 (2020). https://doi.org/10.1109/ACCESS.2020.2983149

17. Zhang, J., Ohn-Bar, E.: Learning by watching. CoRR abs/2106.05966 (2021). https://arxiv.org/abs/2106.05966

Trust Management as a Service for RFID Based Applications in Edge Enabled Cloud

Wenjie Sun[1], Guodong Peng[2], Wenchao Pan[1], Junlei Deng[1], Xuemei Cui[1], and Feng Lin[1,2(✉)]

[1] Yunnan Transport Engineering Quality Inspection Co., Ltd., Kunming, China
fenglinytsr@sina.com
[2] Yunnan Traffic Science Research Institute Co., Ltd., Kunming, China

Abstract. The incremental development of the Internet of Things (IoT) technologies and the rapid popularization of IoT based applications has promoted the rapid integration of edge computing and cloud computing into an innovation service paradigm, edge cloud service (ECS). As one of the most important technologies in IoT scenarios, radio frequency identification (RFID) has been widely used for object identification and tracking. Since RFID devices with limited resources are unable to process large amounts of data, it tends to employ ECS to address the issues of insufficient computing and storage resources. However, due to the diversity, uncertainty and dynamic of quality of service (QoS) in edge enabled cloud context, it brings well-known trust concern and has become one of the most challenging issues for RFID based application to adopt ECS. To this end, a novel trust management framework for RFID based application in edge enabled cloud context is proposed. Such a framework can facilitate RFID application developers to evaluate and choose a trustworthy ECS based on their specific QoS requirements. In addition, a trust level evaluation model is proposed, which can effectively evaluates the trust level of various ECSs based on dynamic QoS. The experimental result of a case study using a real-world dataset shows that the proposed framework can effectively evaluate the trust level of various ECSs with multiple and dynamic QoS attributes to improve the availability of RFID applications.

Keywords: Edge cloud service · Trust management · RFID applications

1 Introduction

The emergence of the Internet of Things (IoT) has provided the constant universal connection between people and things (e.g., sensors or mobile devices) [1–3]. Radio frequency identification (RFID) is a key technology for achieving the IoT, which can automatically identify objects through wireless communication [4]. In particularly, RFID based object identification and tracking technology offer a wide variety of conveniences to our daily life and work in many aspects. The smart and end-user devices have been widely applied in RFID scenarios and are the key linker between cloud and edge [5]. However, due to the limited resources of RFID devices, they are unable to process tasks with high computational complexity and large data storage capacity locally [6,7]. For instant, in the interactive scenario of intelligent laboratory, RFID technology can be used to monitor, data collection and process, remote control, device management, and alarm in real time [21]. In this case, RFID devices generate a large amount of end user data, which put huge strains on IoT network. Therefore, the intelligent laboratory can use the edge cloud service (ECS) to process and store the big data generated by RFID devices, which will improve the overall efficiency of IoT network.

However, it is difficult for RFID application developers (RADs) to take full advantage of ECS. They will face many challenges in migrating RFID applications from the traditional centralized data center to the distributed ECS, especially the application of intelligent laboratory with various complex interaction scenarios (e.g., access control system, attendance system, asset management system, security monitoring system, etc.). Such a scenario integrates information technology, sensor technology, communication technology, and network technology to intelligently manage the device, samples and personnel of laboratory, which brings well-known security and trust concerns to RADs. These concerns are usually caused by the specific requirements and characteristics of the existing RFID applications. They are determined by the quality of service (QoS) of ECS provisioned by the edge-cloud service provider (ECP). In addition, with the continuous growth of RFID applications, many ECS with similar functions and characteristics provided by various ECPs have emerged in the commercial market [11]. The ECSs can meet the different QoS requirements of various RFID applications. It is a challenging issue for RADs to select a trustworthy edge cloud service customer (ESC) from a large number of alternatives with similar functions [20]. Consequently, trust management for ECSs becomes a promising solution to address the aforementioned issues [17]. It aims to accurately and objectively evaluate the trustworthiness of ECSs provisioned by various ECPs for RADs decision-making, thereby ensuring the stable operation of their RFID applications.

To this end, we propose a novel framework of trust management as a service in edge cloud context for RFID application. This framework integrates a trust evaluation model to enhance the availability of ECS based RFID applications by assessing the trust level of ESC. It can improve the effectiveness and accuracy of QoS data collection of RFID. In addition, the proposed framework can facilitate RADs to evaluate and choose a trustworthy ECS based on their actual QoS requirements of RFID applications.

The rest of the paper is organized as follows. Section 2 introduces the related work. Section 3 describes the proposed framework in detail. Section 4 presents the experiment of case study and discusses the experimental result. Section 5 presents the conclusions and outlines future work.

2   Related Work

Although the research on the trust management and evaluation has attracted considerable interest of many researchers in recent years, there is little literature on the trust issues of RFID applications in edge enabled cloud context. Most researchers tend to focus on studying cloud service trust management, and a variety of trust evaluation methods and trust models have been proposed by taking QoS attributes as metrics. Kumar et al. [14] proposed a optimal service selection and ranking of cloud services framework. This framework enables customers to compare available service based on QoS. The framework uses the best-worst method to rank and prioritize the QoS criteria. Additionally, it adopts the technique for order preference by similarity to ideal solution (TOPSIS) method to obtain the final rank of cloud services. Similarly, a cloud service selection framework based on fuzzy analytic hierarchy process (AHP) method is proposed in [16]. This framework defines the selection process for cloud service and calculation method for the weights of QoS criteria. The framework applies the calculated weights of QoS criteria to the TOPSIS method to evaluate the final ranking of cloud services. To select the optimal cloud service, Sun et al. [25] proposed a cloud service selection framework with criteria interactions. In this framework, a fuzzy and choquet integral based measure method is proposed to measure and aggregate non-linear relations between QoS criteria (e.g., latency, response time, availability, etc.). Similarly, to addresses a hybrid multi-criteria decision-making issue for the optimal cloud services selection, Jatoth et al. [13] proposed a novel evaluation methodology based on a extended gray technique for order preference by similarity to an ideal solution (TOPSIS) integrated with AHP. This method assigns various ranks to cloud services based on the quantified QoS attributes. Jagpreet and Sarbjeet [22] detailed three multiple criteria decision making (MCDM)-based trust evaluation methods for cloud services. These methods evaluate trust level of cloud services by monitoring compliance provided by cloud providers against the service level agreements (SLAs). The three trust evaluation methods respectively adopt three MCDM methods: AHP, TOPSIS and preference ranking organization methods for enrichment evaluations (PROMETHEE). They enable cloud service customers to evaluate the trust level of a cloud service from different perspectives. Yang et al. [28] proposed a multi-QoS-aware cloud service selection method. It adopts the AHP method to help the customers to select the optimal cloud service.

To determining the most suitable cloud service for customers, a computational framework by integrating the AHP and TOPSIS is proposed in [15]. In this framework, the AHP method is used to calculate the weights of QoS criteria. The TOPSIS method is used to rank the cloud service according to the QoS criteria. In [27], a novel two-way ranking based cloud service framework is proposed to enable cloud customers to select a suitable cloud service. This framework uses the AHP method to evaluate the ranking score of the cloud providers and cloud customers. It takes the QoS attributes value offered by cloud providers into consideration as well as desired by their customers. Somu et al. [24] proposed a three tiers cloud service selection framework. This framework adopts a hypergraph based ranking algorithm integrated computational model (HGCM) and minimum distance-helly property (MDHP) in the service ranking layer. It measures and quantifies the QoS attributes thereby facilitating the customers to rank the cloud services. Moreover, some literature try to evaluate the trustworthiness of cloud service from the perspective of security, but lacking an effective and efficiency method [18,23].

To summarize, most of the existing trust management framework adopt subjective preference weighting and single constant value based trust evaluation model to evaluate trust level of cloud service. Due to the lack of objectivity and flexibility, the existing models are not applicable to the ECS. Therefore, the availability and accuracy of information or data regarding multiple QoS attributes of ECS and the applicability and efficiency of trust evaluation model is still urgent issues to be solved in the edge cloud context (Fig. 1).

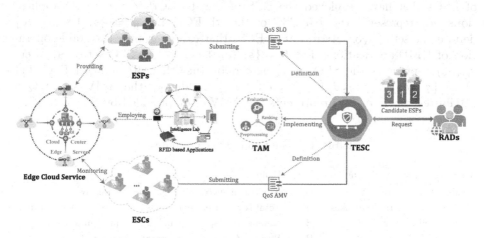

Fig. 1. The proposed trust management framework

3   The Proposed Trust Management Framework

3.1   Framework Architecture

The proposed framework mainly consists of three entities and a trustworthy ECS component (TESC), as shown in Fig. 1. The entities in this framework are ESP, ESC and RADs. The ESP is mainly responsible for signing an SLA [12] with its ESC on the specific SLO [12] of QoS attributes of the cloud service. Besides, ESP focus on operating and maintaining the cloud services to its ESC in accordance with the agreed SLA. The ESC is responsible for monitoring the QoS attributes based on SLA during the use of cloud service, and provides actual monitoring value (AMV) to TESC. The RAD initiates a trust evaluation request to TESC based on its specific QoS requirements and receives the trust evaluation result (i.e., the turst level of cloud service provided by alternative ESPs) from TESC, and chooses an optimal and trustworthy cloud service.

The main role of TESC is to collect QoS attributes information provided by ESP and ESC (i.e., SLO and AMV) and to evaluate cloud services. According to the evaluation request of RAD and the collected QoS attributes information, TESC utilizes the trust evaluation model (TAM) to evaluate the trust level of cloud service of the alternative ECPs. TESC offers the trust evaluation results to RAD so that it can choose a trustworthy cloud service. Next, we will elaborate on the TAM in detail.

3.2   Problem Definition

Let $C = \{c_i | 1 \leq i \leq I\}$ denote the set of ECPs. Let $U = \{u_{ij} \in U_i | 1 \leq i \leq I, 1 \leq j \leq J\}$ denote the set of ESCs, where U_i represents a set of ESCs that have employed the ECS of the ith ECP to run RFID applications, u_{ij} represents the jth ESC of the ith ECP. Let $A = \{a_k | 1 \leq k \leq K\}$ denote the set of QoS attributes of ECSs that can guarantee the stable operation of RFID applications. Let $S = \{s_{ij}(a_k)\}$ denote the SLO set of A, where $s_{ij}(a_k)$ represents the SLO of a_k agreed by the ith ECP and jth ESC. Let $Q = \{q_{ij}(a_k) | 1 \leq i \leq I, 1 \leq j \leq J, 1 \leq k \leq K\}$ denote the AMV set of QoS attributes, where $q_{ij}(a_k)$ represents the AMV of a_k provided by the jth ESC.

Table 1. Definition of QoS attributes [9]

QoS attributes	Abb	Unit	Type	Definition
Availability	av	%	B	Number of successful invocations/total invocations
Throughput	th	invokes/s	B	Total Number of invocations for a given period of time
Successability	su	%	B	Number of response/number of request messages
Reliability	re	%	B	Ratio of the number of error messages to total messages
Latency	la	ms	C	Time taken for the server to process a given request
Response Time	res	ms	C	Time taken to send a request and receive a response

3.3   The Trust Evaluation Model

TAM mainly consists of two stages: normalization preprocessing and trust level evaluation, which will be described in details as follows.

Normalization Preprocessing. For convenience, we take an ECS as an example to detail the normalization preprocessing of QoS data.

Normalize Monitoring Value of QoS. For a given ECP (denoted as c_i, $c_i \in C$), it provides the ECS with the QoS attributes (denoted as A) to its ESCs (denote as U_i). The ECP c_i and its ESCs (denote as u_{ij}, $u_{ij} \in U_i$) respectively submits the SLO (denoted as $s_{ij}(a_k)$, $s_{ij}(a_k) \in S$) and the AMV (denoted as

$q_{ij}(a_k), q_{ij}(a_k) \in Q$) about each QoS attribute $a_k, a_k \in A$ to TESC. The TESC can obtain the AMV set (denoted as $\boldsymbol{S_i(A)}$) and SLO set (denoted as $\boldsymbol{Q_i(A)}$) about A of the ECS provided by c_i. It can be represented as the following matrix.

$$\boldsymbol{S_i(A)} = \begin{bmatrix} s_{i1}(a_1) & s_{i1}(a_2) & \cdots & s_{i1}(a_k) \\ s_{i2}(a_1) & s_{i2}(a_2) & \cdots & s_{i2}(a_k1) \\ \vdots & \vdots & \ddots & \vdots \\ s_{ij}(a_1) & s_{ij}(a_2) & \cdots & s_{ij}(a_k) \end{bmatrix} \tag{1}$$

where, $s_{ij}(a_k)(s_{ij}(a_k) \in S)$ denotes the SLO of the QoS attribute a_k agreed by c_i and u_{ij}.

$$\boldsymbol{Q_i(A)} = \begin{bmatrix} q_{i1}(a_1) & q_{i1}(a_2) & \cdots & q_{i1}(a_k) \\ q_{i2}(a_1) & q_{i2}(a_2) & \cdots & q_{i2}(a_k) \\ \vdots & \vdots & \ddots & \vdots \\ q_{ij}(a_1) & q_{ij}(a_2) & \cdots & q_{ij}(a_k) \end{bmatrix} \tag{2}$$

where, $q_{ij}(a_k)$ $(q_{ij}(a_k) \in Q)$ denotes the average AMV of a_k provided by u_{ij}.

To unify measurement benchmark of QoS attributes, the average monitoring value about each QoS attribute submitted by ESCs is taken as the unique AMV. For a given u_{ij}, it submitted N monitoring values about a_k, then the average AMV $q_{ij}(a_k)$ of a_k can be obtained as follows.

$$q_{ij}(a_k) = \frac{\sum_{n=1}^{N} q_{ij}(a_k)^n}{n} \tag{3}$$

where, $q_{ij}(a_k)^n$ represents the nth monitoring value on a_k submitted by u_{ij}.

We assume that if $s_{ij}(a_k)$ of c_i is not less than $q_{ij}(a_k)$ of u_{ij}, it is considered to be consistency. Therefore, let $N_i(a_k)$ represent the number of times that a_k complies with consistency. Let $|U_i(a_k)|$ represent the number of AMV regarding a_k submitted by its ESCs U_i. Accordingly, the consistency rate of a_k can be obtained as follows, which is denoted as λ_{ik}.

$$\lambda_{ik} = \frac{N_i(a_k)}{|U_i(a_k)|} \tag{4}$$

It should be noted that the QoS attributes of ECS can be divided into two types according to their features: benefit and cost. The benefit QoS attribute means that the higher the value of attribute is, the higher its performance is (e.g., throughput and availability). The cost QoS attribute means that the higher the value of attribute is, the lower its performance is (e.g., latency and response time).

Determine SLO Range of QoS. Let $s_i(a_k)^l$ and $s_i(a_k)^u$ respectively denote the minimum and maximum SLO about each of QoS attributes submitted by c_i, which can be obtained from $S_i(A)$. Then, the SLO of each QoS attribute of c_i can be represented as an interval value: $s_i(\tilde{a}_k) = [s_i(a_k)^l, s_i(a_k)^u]$. Let $\tilde{b}_{ik}$ be the

SLO range of a_k of the ECS provided by c_i, which indicates the QoS guarantee capability on a_k of c_i. It can be obtained by the following equation.

$$\tilde{b}_{ik} = \lambda_{ik} \times s_i(\tilde{a}_k) = \left[\lambda_{ik} s_i(a_k)^l, \lambda_{ik} s_i(a_k)^u \right] = \left[b_{ik}^l, b_{ik}^u \right] \tag{5}$$

where, b_{ik}^l and b_{ik}^u respectively denote the actual minimum and maximum SLO of the QoS attribute a_k.

Therefore, the actual SLO range (i.e., interval value) about A of the ECS of c_i can be denoted as $\tilde{\boldsymbol{b}}_i = \left[\tilde{b}_{i1}, \tilde{b}_{i2}, \cdots, \tilde{b}_{ik} \right]$.

Trust Level Evaluation. The trust level evaluation process consists of four steps, which will be described in details as follows.

Normalized Decision Matrix. According to Equations (1)-(4), the actual SLO interval value $\tilde{\boldsymbol{b}}_i$ about the K QoS attributes submitted by the I ECPs can be obtained. Then, the decision matrix composed of $\tilde{\boldsymbol{b}}_i$, denoted as $\boldsymbol{B} = (\tilde{b}_{ik})_{I \times K}$, which can be represented as follows.

$$\boldsymbol{B} = \left[\tilde{\boldsymbol{b}_1}, \tilde{\boldsymbol{b}_2}, \cdots, \tilde{\boldsymbol{b}_i}, \cdots, \tilde{\boldsymbol{b}_I} \right]^T = \begin{bmatrix} \tilde{b}_{11} & \tilde{b}_{12} & \cdots & \tilde{b}_{1k} \\ \tilde{b}_{21} & \tilde{b}_{22} & \cdots & \tilde{b}_{2k} \\ \vdots & \vdots & \ddots & \vdots \\ \tilde{b}_{i1} & \tilde{b}_{i2} & \cdots & \tilde{b}_{ik} \\ \vdots & \vdots & \ddots & \vdots \\ \tilde{b}_{I1} & \tilde{b}_{I2} & \cdots & \tilde{b}_{IK} \end{bmatrix} \tag{6}$$

To eliminate the impact of different QoS attributes types on the trust evaluation results, the decision matrix $\boldsymbol{B}$ needs to be normalized. The normalized decision matrix $\boldsymbol{B}$ can be denoted as $\boldsymbol{R} = (r_{ik})_{I \times K}$, where $r_{ik} = \left[r_{ik}^l, r_{ik}^u \right]$.

Determine Objective Weights. Most of the existing trust evaluation models employ the subjective preference based weighting methods [8,10,19,20]. Such a method are static constants, which cannot well adapt to the dynamic QoS attributes in the real cloud context. Hence, we adopt an objective weighting method based on deviation maximization to determine the weights of different QoS attributes. The rationale of this method is that if the difference of values about a QoS attribute provided by all ESCs is smaller, it indicates that the impact of this QoS attribute on trust evaluation is smaller [26]. Accordingly, the weights of QoS attributes can be calculated, which can be denoted as $\boldsymbol{\omega} = (\omega_1, \omega_2, \cdots, \omega_k, \cdots, \omega_K)$, where $\omega_k \geq 0$ and conforms to the following constraint.

$$\sum_{k=1}^{K} \omega_k = 1 \tag{7}$$

Calculate Trust Level. For a given ECS with K QoS attributes provided by ECP c_i, let z_i represent the trust value of c_i. It can be calculated by the following equation.

$$z_i = \sum_{k=1}^{K} \omega_k r_{ik} = \sum_{k=1}^{K} \hat{r_{ik}} \tag{8}$$

where $\hat{r_{ik}} = \left[\omega_k r_{ik}^l, \omega_k r_{ik}^u\right]$ denotes the weighted r_{ik}. Then, $\hat{r_{ik}}$ can be represented as $\hat{r_{ik}} = \left[\hat{r_{ik}}^l, \hat{r_{ik}}^u\right]$.

Apparently, the z_i is still an interval, it is difficult to compare ECSs of C directly. Hence, the trust value of ECS need to be normalized to a unified dimension. Let tl_i denote the trust level of c_i, which can be calculated by normalizing its trust value as follows:

$$tl_i = \frac{e^{\lambda_i}}{\sum_{i=1}^{I} e^{\lambda_i}} \tag{9}$$

where $\lambda_i = \sum_{k=1}^{K} (\hat{r_{ik}}^l + \hat{r_{ik}}^u)$ and $\lambda_i \in [0,1]$.

It can be concluded that the closer λ_i is to 1, the more trustworthy the ECS provided by c_i, and vice verse.

$$R = \begin{pmatrix} [0.196, 0.274] & [0.0465, 0.742] & [0.208, 0.273] & [0.159, 0.245] & [0.0452, 0.725] & [0.101, 0.407] \\ [0.14, 0.276] & [0.0698, 0.968] & [0.138, 0.276] & [0.153, 0.279] & [0.0514, 0.644] & [0.0834, 0.371] \\ [0.137, 0.262] & [0.031, 0.839] & [0.131, 0.259] & [0.169, 0.232] & [0.0552, 0.828] & [0.0955, 0.417] \\ [0.16, 0.222] & [0.0388, 0.936] & [0.158, 0.237] & [0.161, 0.225] & [0.0481, 0.966] & [0.104, 0.338] \\ [0.158, 0.231] & [0.0543, 0.677] & [0.151, 0.228] & [0.172, 0.284] & [0.0574, 0.725] & [0.106, 0.456] \end{pmatrix}$$

Fig. 2. The normalized decision matrix R

Rank Cloud Services. The priority of ECSs provided by the alternative ECPs that satisfy the QoS requirements of RAD can be obtained by ranking tl_i. Then, the RAD can choose a trustworthy ECS from the alternative ECPs based on the ranking results.

4   Experiment and Analysis

4.1   Experiment Setup

The purpose of the experiment is to demonstrate the feasibility of the proposed framework in solving trust concern of RADs. For convenience, we consider an interactive scenario in the intelligent laboratory. In this scenario, we assume that a RAD wants to use the framework to choose a trustworthy ECS to deploy its sample, personnel, and devices management system based on RFID for the intelligent laboratory. Therefore, we conduct a case study by using an open source dataset to evaluate the effectiveness of TAM. The dataset (named as QWS) consists of 2,507 pieces of real data produced by hundreds of Web services

Table 2. The SLO interval value of ECPs and ECC on the QoS attributes

	QoS Attributes					
	av	th	su	re	la	res
ECP_1	[87, 96]	[6, 23]	[95, 98]	[58, 73]	[8, 33]	[103, 204]
ECP_2	[62, 97]	[9, 32]	[63, 99]	[56, 83]	[9, 29]	[113, 246]
ECP_3	[61, 92]	[4, 26]	[60, 93]	[62, 69]	[7, 27]	[89, 215]
ECP_4	[71, 78]	[5, 30]	[72, 85]	[59, 67]	[6, 31]	[124, 198]
ECP_5	[70, 81]	[7, 21]	[69, 82]	[63, 74]	[8, 26]	[92, 193]
RAD	[50, 100]	[1, 35]	[50, 100]	[50, 100]	[1, 100]	[50, 300]

Table 3. Trust level evaluation of ECSs

Alternative ECPs	ECP_1	ECP_2	ECP_3	ECP_4	ECP_5
Trust Value	[0.109, 0.474]	[0.0953, 0.477]	[0.0968, 0.525]	[0.102, 0.526]	[0.107, 0.473]
Trust Level	0.197	0.195	0.205	0.206	0.196
Priority	3	5	2	1	4

on the 6 QoS attributes [9]. The definitions of QoS attributes contained in QWS are shown in Table 1, where B (i.e., benefit type) and C (i.e., cost type) in the type column represent the benefit and cost respectively.

4.2   Case Study

To focus on the details of trust level evaluation process for ECS, this experiment simplifies the data preprocessing process of QoS attributes in TAM (i.e., as described in 3.3).

We consider a scenario that a RAD initiates an trust evaluation request to the TESC and specifies the SLOs interval about QoS attributes according to the requirements of its RFID application, as shown in Table 2. The QWS dataset is used as the AMV on the QoS attributes of ECS submitted by the ESCs. In addition, we assume that TESC matched 5 alternative ECPs satisfy the SLO requirements of the RAD according to its evaluation request, denoted as ECP_1, ECP_2, ECP_3, ECP_4 and ECP_5. The maximum and minimum values of these alternative CPSs on each QoS attribute are taken as their actual SLO interval, as shown in Table 2.

The trust level of each ECS of alternative ECPs can be obtained by TAM according to the actual SLO interval. The specific process are described as follows. Firstly, the normalized decision matrix of alternative ECPs R are constructed according the data of Table 2, as shown in Fig. 2. Then, the weights of QoS attributes are calculated by R. The weight vector ω can be calculated as follows.

$$\omega = (0.0295 \quad 0.118 \quad 0.150 \quad 0.167 \quad 0.247 \quad 0.288).$$

Next, the trust value of ECS of alternative ECPs can be aggregated by multiplying R and ω. The trust level of ECSs of alternative ECPs can be calculated according to their trust value. Finally, the priority of ECPs can be obtained by ranking the trust level of their ECSs, as shown in Table 3. It can be seen from the table that ECP_4 is the optimal.

5 Conclusion

In this paper, we propose a novel trust management as a service framework for RFID based applications in edge enabled cloud. This framework facilitates RADs to evaluate and select a trustworthy ECS based on the actual QoS requirements of RFID applications. Moreover, in order to efficiently evaluate the trust level of ECSs, a trust evaluation model is proposed. Such a model adopts an interval attribute based evaluation method and an objective weighting method based on deviation maximization to calculate the trust level of ECSs. The effectiveness of the framework is validated by the case study experiment using a real-world dataset. The experimental result indicates that the proposed trust level evaluation model can effectively utilize multiple QoS attributes to accurately evaluate the trust level of various ECSs. In addition, the proposed framework can effectively tackle the security and trust issues of RFID applications in the scenario interactions of intelligent laboratory. As future work, we aim to develop a prototype for our framework and implement it in the real environment to further verify its effectiveness.

Acknowledgments. This work is supported by Research on the construction of intelligent full-process laboratory system based on RFID (Radio frequency identification) technology (No. YCIC-YF-2022-05), and Research on dynamic monitoring and remote control technology of environmental laboratory (No. JKYZLX-2021-23).

References

1. Chen, D., Mao, X., Qin, Z., Wang, W., Li, X.-Y., Qin, Z.: Wireless device authentication using acoustic hardware fingerprints. In: Wang, Yu., Xiong, H., Argamon, S., Li, X.Y., Li, J.Z. (eds.) BigCom 2015. LNCS, vol. 9196, pp. 193–204. Springer, Cham (2015). https://doi.org/10.1007/978-3-319-22047-5_16
2. Lee, S.K., Bae, M., Kim, H.: Future of IoT networks: a survey. Appl. Sci. **7**(10), 1072 (2017)
3. Chen, D., Zhang, N., Wu, H., Zhang, K., Lu, R., Guizani, M.: Audio-based security techniques for secure device-to-device communications. IEEE Netw. **36**(6), 54–59 (2022)
4. Su, J., Sheng, Z., Leung, V.C.M., Chen, Y.: Energy efficient tag identification algorithms for RFID: survey, motivation and new design. IEEE Wirel. Commun. **26**(3), 118–124 (2019)
5. Khalid, N., Mirzavand, R., Iyer, A.K.: A survey on battery-less RFID-based wireless sensors. Micromachines **12**, 819 (2021)

6. Li, X., Jin, X., Wang, Q., Cao, M., Chen, X.: SCCAF: a secure and compliant continuous assessment framework in cloud-based IoT context. In: Wireless Communications and Mobile Computing 2018 (2018)
7. Chen, D., Wang, H., Zhang, N., et al.: Privacy-preserving encrypted traffic inspection with symmetric cryptographic techniques in IoT. IEEE Internet Things J. 9(18), 17265–17279 (2022)
8. Habib, S.M., Hauke, S., Ries, S., Mühlhäuser, M.: Trust as a facilitator in cloud computing: a survey. J. Cloud Comput. Adv. Syst. App. 1(1), 1–18 (2012)
9. Al-Masri, E., H., Q.M.: Discovering the best web service: a neural network-based solution. In: 2009 IEEE International Conference on Systems, Man and Cybernetics, pp. 4250–4255 (2009)
10. Alabool, H., Kamil, A., Arshad, N., Alarabiat, D.: Cloud service evaluation method-based multi-criteria decision-making: a systematic literature review. J. Syst. Softw. 139, 161–188 (2018)
11. Hayyolalam, V., Kazem, A.A.P.: A systematic literature review on QOS-aware service composition and selection in cloud environment. J. Netw. Comput. Appl. 110, 52–74 (2018)
12. ISO/IEC-JTC-1/SC-38: Information technology-Cloud computing-Service level agreement (SLA) framework-Part 1: Overview and Concepts (2016). https://www.iso.org/standard/67545.html 2021
13. Jatoth, C., Gangadharan, G.R., Fiore, U., Buyya, R.: Selcloud: a hybrid multi-criteria decision-making model for selection of cloud services. Soft. Comput. 23(13), 4701–4715 (2019)
14. Kumar, R.R., Kumari, B., Kumar, C.: CCS-OSSR: a framework based on hybrid MCDM for optimal service selection and ranking of cloud computing services. Clust. Comput. 24(2), 867–883 (2021)
15. Kumar, R.R., Mishra, S., Kumar, C.: A novel framework for cloud service evaluation and selection using hybrid MCDM methods. Arab. J. Sci. Eng. 43(12), 7015–7030 (2018)
16. Kumar, R.R., Shameem, M., Kumar, C.: A computational framework for ranking prediction of cloud services under fuzzy environment. Enterprise Inf. Syst. 16, 1–21 (2021)
17. Li, X., Wang, Q., Lan, X., Chen, X., Zhang, N., Chen, D.: Enhancing cloud-based IoT security through trustworthy cloud service: an integration of security and reputation approach. IEEE Access 7, 9368–9383 (2019)
18. Li, X., Yang, R., Chen, X., Liu, Y., Wang, Q.: Assessment model of cloud service security level based on standardized security metric hierarchy. Adv. Eng. Sci. 52, 159–167 (2020). https://doi.org/10.15961/j.jsuese.201900429
19. Mahmud, K., Usman, M.: Trust establishment and estimation in cloud services: a systematic literature review. J. Netw. Syst. Manage. 27(2), 489–540 (2019). https://doi.org/10.1007/s10922-018-9475-y
20. Noor, T.H., Sheng, Q.Z., Maamar, Z., Zeadally, S.: Managing trust in the cloud: state of the art and research challenges. Computer 49(2), 34–45 (2016). https://doi.org/10.1109/MC.2016.57
21. Patre, S.R.: Passive chipless RFID sensors: concept to applications-a review. IEEE J. Radio Freq. Identif. 6, 64–76 (2022). https://doi.org/10.1109/JRFID.2021.3114104
22. Sidhu, J., Singh, S.: Design and comparative analysis of MCDM-based multi-dimensional trust evaluation schemes for determining trustworthiness of cloud service providers. J. Grid Comput. 15(2), 197–218 (2017)

23. Silva, A., Silva, K., Rocha, A., Queiroz, F.: Calculating the trust of providers through the construction weighted SEC-SLA. Future Gener. Comput. Syst. **97**, 873–886 (2019)
24. Somu, N., Kirthivasan, K., Shankar Sriram, V.S.: A computational model for ranking cloud service providers using hypergraph based techniques. Future Gener. Comput. Syst. **68**, 14–30 (2017)
25. Sun, L., Dong, H., Hussain, O.K., Hussain, F.K., Liu, A.X.: A framework of cloud service selection with criteria interactions. Future Gener. Comput. Syst. **94**, 749–764 (2019)
26. Xu, Z.: A deviation-based approach to intuitionistic fuzzy multiple attribute group decision making. Group Decis. Negot. **19**(1), 57–76 (2010)
27. Yadav, N., Goraya, M.S.: Two-way ranking based service mapping in cloud environment. Future Gener. Comput. Syst. **81**, 53–66 (2018)
28. Yang, Y., Peng, X., Fu, D.: A framework of cloud service selection based on trust mechanism. Int. J. Ad Hoc Ubiquit. Comput. **25**(3), 109–119 (2017)

Distributed Intelligent Collaborative Scheduling Mechanism for Cloud-Edge-End Resources in IoT

Liqiang Wang, Wenchen He, Ruilin liu, and Faqiang Liu[✉]

National Computer Network Emergency Response Technical Team/Coordination Center of China, Beijing 100029, China
libin_bjsg@163.com

Abstract. With the integration and development of the Internet of Things (IoT), the service model of IoT has gradually evolved from traditional "data collection and business processing" to "providing ubiquitous and universal services through collaboration between cloud, edge, and end resources". However, traditional methods rely on centralized servers, which pose challenges in terms of trust, cost, and single points of failure. To overcome these challenges, we propose a distributed intelligent collaborative scheduling mechanism. First, we establish a distributed collaborative environment based on blockchain and construct a resource scheduling model. Second, we propose an intelligent collaborative mechanism for workload balancing based on reinforcement learning. This mechanism ensures the efficient allocation of resources across the network. Lastly, we design simulation experiments to evaluate the effectiveness of the proposed mechanism, and the results demonstrate its efficiency.

Keywords: Resource Scheduling · Cloud-edge-end resource · Blockchain · Reinforcement learning

1 Introduction

With the integration and development of the Internet of Things [1, 2] and the digital economy, the service model of the Internet of Things has evolved from traditional "data collection and business processing" to "providing users with ubiquitous and universal services in collaboration with cloud, edge, and end heterogeneous resources" [3–5]. Increasingly, public and private computing, communication, storage, and other multi-dimensional resources are joining the interconnection collaboration, propelling the Internet of Things towards a more extensive "Internet of Everything."

Currently, most schemes propose relying on a trusted centralized resource agent to facilitate the sharing and collaboration of multi-dimensional resources. However, poor third-party management can result in manipulated managed/requested resources, information leakage, or collaboration task failures. Additionally, unified heterogeneous resource scheduling management places more scheduling pressure on the third-party and faces the risk of a single point of failure.

© The Author(s), under exclusive license to Springer Nature Singapore Pte Ltd. 2024
H. Jin et al. (Eds.): IAIC 2023, CCIS 2060, pp. 56–68, 2024.
https://doi.org/10.1007/978-981-97-1332-5_5

Currently, most schemes propose relying on a trusted centralized resource agent to facilitate the sharing and collaboration of multi-dimensional resources. However, poor third-party management can result in manipulated managed/requested resources, information leakage, or collaboration task failures. Additionally, unified heterogeneous resource scheduling management places more scheduling pressure on the third-party and faces the risk of a single point of failure.

To address these problems, a distributed heterogeneous resource sharing and collaboration scheme based on blockchain has been considered an effective solution [6–9]. This approach utilizes blockchain to support resource scheduling architectures and employs smart contracts to achieve trusted scheduling and resource allocation. Blockchain technology, derived from digital currency (Bitcoin), offers technical characteristics such as anonymous participation, data tamper-proofing, and transaction traceability. Its distributed advantages are unmatched by other technologies.

Traditional resource scheduling problems in cloud-edge collaboration often employ solutions based on game theory, branch constraint, dynamic programming, convex optimization, and heuristic algorithms [10]. However, these methods are usually slow in decision-making and struggle to adapt to dynamic environments. Consequently, they are not well-suited for directly solving the resource cooperative scheduling problem in blockchain-based distributed IoT. Recently, researchers have proposed that deep reinforcement learning (DRL) technology has great potential for solving resource scheduling problems [11]. As an artificial intelligence algorithm with self-learning and online learning capabilities, DRL can dynamically allocate and adjust heterogeneous resources based on past experiences. It provides users with customized compound connection services, effectively improves service quality, and enhances user satisfaction [12].

Therefore, we propose a blockchain-based distributed intelligent resource scheduling mechanism for cloud-edge-end resource collaboration in the IoT, which combines reinforcement learning. Based on the proposed blockchain-based distributed architecture for resource scheduling, we present a system model comprising a resource sub-model, scheduling decision sub-model, user satisfaction sub-model, and load balancing sub-model. We then design a reinforcement learning-based solution to address this problem and conduct simulation experiments to verify the efficiency of the proposed mechanism.

2 System Model and Problem Description

2.1 Blockchain-Based Distributed Architecture for Resource Scheduling

The distributed intelligent architecture for resource scheduling, as illustrated in Fig. 1, comprises two layers: the Internet of Things (IoT) resource layer and the intelligent shared resource scheduling and decision layer.

The resource layer operates based on a collaborative computing approach known as "cloud-edge-end". Resources are categorized into four types: end resources, edge resources, core network resources, and cloud resources, based on their location. These resources can further be classified into computing resources, storage resources, and communication resources. To enable IoT services, resource nodes are equipped with blockchain components, which grant them a blockchain identity. This identity allows the nodes to join the blockchain network, share resources anonymously with external parties, and facilitate resource transactions.

The intelligent shared resource scheduling layer is built upon the blockchain network. All nodes in this layer are equipped with both blockchain components and AI engines. These nodes function independently and can be configured by high-performance resource nodes (such as cloud resource nodes) in practical applications. These nodes serve two main purposes. Firstly, as the primary nodes of the blockchain network, they facilitate the access of various resource nodes, establish the foundational blockchain network, store network status information and relevant node information, and record resource interaction transactions. Secondly, they perform comprehensive network analysis, facilitate centralized online training, and send training results to the edge server.

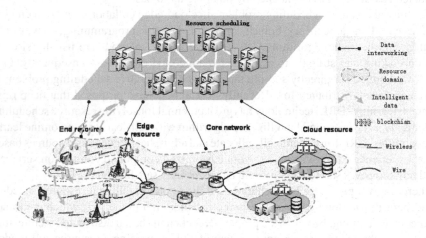

Fig. 1. The distributed architecture for resource scheduling

2.2 Resource Model

We define H Internet of Things terminals that share resources externally, the set is defined as H = $\{1, 2, \cdots, H\}$. The set of networked terminals with computing tasks is defined as N=$\{1, 2, \cdots, N\}$, in the dynamic task offloading scenario, the computing task of terminal i is defined as a triplet $T_i = (F_i, D_i, R_i)\forall i \in$ N, where F_i describes the total number of CPU cycles required by task T_i and T_i represents the amount of data that needs to be transferred to other resource nodes during task offloading, R_i is the expected service quality requirement of task execution.

It is considered that similar tasks have the same R_i, $R_i = (rt_i, re_i, rc_i)$. Edge server is deployed near the base station or router on the edge of the network and can provide services for terminal nodes in a close distance. The resource model defines K edge servers K = $\{1, 2, \cdots, K\}$ and J cloud servers J = $\{1, 2, \cdots, J\}$ to provide resource sharing services. Compared to cloud servers, edge servers have lower service capacity, but they offer the advantage of providing fast services in close proximity to users. K

edge servers are configured at K edge communication nodes, and the overall channel bandwidth of each communication node is defined as B. Users pass Orthogonal Frequency Division Multiplexing (OFDM) channel offloads its tasks to the edge server of the edge communication node, which effectively ensures that there is no interference between communications. Edge server and cloud server are resource nodes, and each resource node has class d resources, $C = (c_1, c_2, \cdots, c_d)$. In order to simplify the model, in this paper, only resources directly related to the execution of tasks are considered, namely CPU capacity and memory size, and the service capability of resource nodes is defined as $C_i = (s_i, m_i)$. Resource nodes provide efficient services through rapid deployment containers. In addition, considering the location relationship among nodes, the geographic location coordinate sets of Internet of Things terminal, edge server and cloud server are defined as $L_d=\{(x_i^d, y_i^d)|\forall i \in N\}$, $L_e=\{(x_i^e, y_i^e)|\forall i \in K\}$, $L_c=\{(x_i^c, y_i^c)|\forall i \in J\}$ respectively.

2.3 Scheduling Decision Model

Assume that time is divided into discrete time slots. In time slots t, the scheduling decision of task T_i is defined as quintuple $D_i(t) = (a_i, num_i, X_i, enum_i, B_i)\forall i \in N$, where $a_i = \{0, 1, 2\}$ stands for the optional position of task T_i, num_i stands for the serial number of the resource node executing task T_i, and the value range is affected by the value of a_i. $X_i = (x_i^1, x_i^2, \cdots, x_i^d)$ stands for the amount of different types of resources allocated to task T_i by the corresponding resource node with serial number num_i. To simplify calculation, $d = 2$ is defined in this model, $enum_i$ represents the edge communication node that allocates communication resources to task T_i and B_i represents the communication bandwidth allocated by the edge communication node to terminals. $a_i = 0$ indicates that the Internet of Things terminal executes task T_i locally, that is, offloads the task to other Internet of Things terminal resource nodes locally or in Device to Device (D2D) communication mode. At this point, $num_i \in N$; $a_i = 1$ indicates that the offloading task T_i is executed to the edge server; at this point, $num_i \in K$; $a_i = 2$ indicates that the offloading task T_i is executed to the cloud server; at this point, $num_i \in J$.

1) Local implementation
Assuming that the D2D communication restriction distance of the terminal is dis_{max}^{D2D}. In the case of local execution, the container resource allocated by the terminal i is $C_l = (s_i, m_i)$. There is no transmission cost for task execution delay, but only calculation delay cost:

$$t_i^{local} = \begin{cases} \dfrac{F_i}{s_i}, D_i \le m_i \\ +\infty, otherwise. \end{cases} \tag{1}$$

The energy consumption of Internet of Things terminal i is $e_i = P_i^d t_i^{local}$, where P_i^d is the calculated power. At this time, the cost of renting resources for task T_i is $pay_i = 0$. When terminal j provides services to complete the execution of task T_i, the task calculation delay t_i^c is.

$$t_i^c = \begin{cases} \dfrac{F_i}{s_j}, D_i \le m_j \\ +\infty, otherwise. \end{cases} \tag{2}$$

The geographical distance between terminals i and j is dis_{ij}, and the path loss is inversely proportional to the $\alpha(\alpha > 2)$-power of the distance. If in the D2D model, all links experience independent and identically distributed sharp fading, and the fading coefficient obeys exponential distribution with mean value of $E[h] = 1$, the power of the receiving end j can be calculated as $P_j = P_i h dis_{ij}^{-\alpha}$, where P_i is the transmitting power of i. The signal noise ratio (SNR) of terminal j receiving signals interfered by cell base station can be calculated as.

$$SINR = \frac{P_j}{I_{BD} + \sigma^2} \tag{3}$$

where I_{BD} is the interference of cell base station to the receiving end, and σ^2 is the link-additive white noise. In D2D mode, the transfer rate between nodes is.

$$r_{ij}^{local} = B_{ij}^{local} \log_2(1 + SINR) \tag{4}$$

B_{ij}^{local} is the link bandwidth allocated to the D2D pair, $SINR$ is the Signal to interference plus noise ratio. The data transmission time can be calculated as.

$$t_i^{tr} = \begin{cases} \dfrac{D_i}{r_{ij}^{local}}, dis_{ij} \le dis_{\max} \\ +\infty, otherwise. \end{cases} \tag{5}$$

Therefore, the computation time cost of completing the task in D2D mode can be calculated as $t_i^{local} = t_i^c + t_i^{tr}$, where t_i^c is calculation delay, t_i^{tr} is data transmission delay. Meanwhile, the energy consumption of Internet of Things terminal i can be calculated as $e_i = P_i t_i^{tr}$. If the unit resource price of terminal j is p_j, the cost for terminal i to rent resources is $pay_i = p_j s_j$.

2) Edge server execution
In time slot t, $a_i = 1$ means that task T_i is unloaded to edge server j to perform $num_i = j$. Task T_i calculates the delay in the edge server as.

$$t_{ij}^c = \begin{cases} \dfrac{F_i}{s_{ij}^{edge}}, \quad D_i \le m_j \\ +\infty, \ otherwise. \end{cases} \tag{6}$$

where s_{ij}^{edge} is the computing resource assigned by the edge server to the container to perform task T_i. In the edge wireless communication network, the data rate from terminal to edge server can be obtained as $r_{ij}^{edge} = B_{ij}^{edge} \log_2(1 + SINR)$, where $SINR =$

$P_j/I_{DB} + \sigma^2$ and I_{DB} are the influence of D2D communication on the base station. Since D2D communication multiplexes the base station downlink, the influence on the uplink link can be ignored, and $I_{DB} = 0$ can be set.

Similarly, $P_j = P_i h dis_{ij}^{-\alpha}$ is the received power of the edge server, and dis_{ij} is the distance between the edge server j and terminal i. The maximum access range of the base station is defined as $dis_{\max}^{edge}$. B_{ij}^{edge} is the communication bandwidth allocated by task T_i execution.

The data transmission delay of task T_i is

$$
t_{ij}^{tr} = \begin{cases} \dfrac{D_i}{r_{ij}^{edge}}, \ dis_{ij} \leq dis_{\max}^{edge} \\ +\infty, \ otherwise. \end{cases} \tag{7}
$$

The overall execution delay of task T_i is

$$
t_i^{edge} = t_{ij}^{tr} + t_{ij}^{c} \tag{8}
$$

When task T_i is executed on edge server, the power consumption of terminal i is

$$
e_i = P_i t_{ij}^{tr} \tag{9}
$$

If the set of tasks performed at edge server j is W_j, there are resource constraints

$$
\sum_{i \in W_j} B_{ij}^{edge} \leq B_j, \ \sum_{i \in W_j} s_{ij}^{edge} \leq s_j, \ \sum_{i \in W_j} D_i \leq m_j \tag{10}
$$

If the unit resource price of edge server j is p_j, the cost of terminal i renting resources is $pay_i = p_j s_{ij}^{edge}$.

3) Cloud Server Execution

In time slot t, $a_i = 2$, $num_i = j$, which means task T_i is unloaded to cloud server j for execution. Based on edge communication, the communication delay between edge server κ and cloud server ι can be calculated as

$$
t_{\kappa\iota}^{tr} = \frac{dis_{\kappa\iota}}{C} \tag{(11)}
$$

where $dis_{\kappa\iota}$ is the geographical distance between κ and ι, C is the signal propagation speed, approximately $C = 3 \times 10^8$ m/s. In the cloud server execution mode, the execution time of task T_i is

$$
t_i^{cloud} = t_{i\kappa}^{tr} + t_{\kappa\iota}^{tr} + \frac{F_i}{s_{i\iota}^{cloud}} \tag{12}
$$

$s_{i\iota}^{cloud}$ refers to the computing resources that cloud server ι allocates to task T_i. If the number of tasks executed on the cloud server ι is W_ι:

$$
\sum_{i \in W_\iota} s_{i\iota}^{cloud} \leq s_\iota, \ \sum_{i \in W_\iota} D_i \leq m_\iota \tag{13}
$$

The terminal power consumption is generated during data transmission. Therefore, the terminal power consumption in cloud server execution scenarios is the same as that in edge server execution scenarios. If the unit resource price of cloud server ι is p_ι, the cost of renting resources for terminal i is $pay_i = p_\iota s_{i\iota}^{cloud}$.

2.4 User Satisfaction Model

In terms of user satisfaction, the smaller the value of delay, cost and energy consumption, the higher the user satisfaction. Therefore, the single factor satisfaction quantification function is designed as follows:

$$S(x_i) = \frac{\arctan\left(\frac{r_i}{x_i}\right)}{\pi/2} \tag{14}$$

where x_i is the actual execution of the task, $r_i(r_i \in R_i)$ is the expected completion constraint. The smaller the value of r_i/x_i is, the lower the user satisfaction is. The user's satisfaction with the execution time, energy consumption and cost of task T_i is

$$S(t_i) = \frac{\arctan\left(\frac{r_i}{t_i}\right)}{\pi/2} \quad S(e_i) = \frac{\arctan\left(\frac{re_i}{e_i}\right)}{\pi/2} \quad S(pay_i) = \frac{\arctan\left(\frac{rc_i}{pay_i}\right)}{\pi/2} \tag{15}$$

The higher the value of satisfaction tendency index is, the more attention users attach to this index, and the greater the influence it has on overall satisfaction. The following constraints must be met to ensure user satisfaction with tasks:

$$S(t_i, e_i, pay_i) \geq 0.5 \tag{16}$$

2.5 Load Balancing Model

According to the resource $C_i = (s_i, m_i)$ defined in the resource model, the current load rate of computing resources and memory resources in single node i is defined as $u_i^s = (c_i^{olds} + \sum_{cp\in EC} c_{cp}^s)/c_i^s$, $u_i^m = (c_i^{oldm} + \sum_{cp\in EC} c_{cp}^m)/c_i^m$, where c_i^{oldm}, c_i^{olds} is the currently used resources of the node, EC is the collection of containers used by resource node i to run assigned tasks, $c_{cp}^s = \sum_{j\in W_i} s_{ji}$, $c_{cp}^m = \sum_{j\in W_i} m_{ji}$ is the amount of computing and storage resources required by the container cp deployed to execute tasks, and W_i is the collection of tasks allocated to resource node i.

Define the overload threshold and underload threshold of resource $c_i^r(c_i^r \in C_i)$ as Thu_r^o and Thu_r^l. The variance of load rates among resources on compute node i is

$$V_i = \frac{1}{d}\sum_{k=1}^{d}\left(u_i^k - \frac{1}{d}\sum_{k-1}^{d} u_i^k\right)^2 = \frac{1}{2}\left(\left(u_i^s - \frac{u_i^s + u_i^m}{2}\right)^2 + \left(u_i^m - \frac{u_i^s + u_i^m}{2}\right)^2\right) \tag{17}$$

In a resource node, the overuse of a single resource may affect the overall running performance of the resource node. Therefore, the load ratio of a single resource node must meet the following constraints.

$$Thu_d^l < u_i^d < Thu_d^o \tag{18}$$

If the node load rate is $u_i^d > Thu_d^o$ or $u_i^d < Thu_d^l$, a request needs to be sent to the blockchain network to trigger the resource scheduling smart contract to reschedule resources and feedback the result. The variance threshold of load rate is defined as Thv_i^d. When the variance of load rate among nodes is greater than the threshold, that is, $V_i > Thv_i^d$, container migration is required.

Define the load of a single resource node i. The weight of resource $c_i^r (c_i^r \in C_i)$ in node i is defined as δ_i^r, and the weighted resource load ratio is

$$\Upsilon_i^r = \delta_i^r u_i^r \tag{19}$$

The weighted load rate of computing resource and memory resource in node i is $\delta_i^s u_i^s$ and $\delta_i^m u_i^m$ respectively. The resource load rate of a single node can be expressed as

$$u_i = \delta_i^s u_i^s + \delta_i^m u_i^m \tag{20}$$

The total computing resource node set is $RS = H \cup K \cup J$, and the variance function of the overall load rate of the network is

$$V = \frac{1}{H+K+J} \sum_{j=1}^{H+K+J} \left(u_j - \frac{1}{H+K+J} \sum_{t=1}^{H+K+J} u_j \right)^2 \tag{21}$$

In the resource model, K edge servers are described, corresponding to K edge communication nodes (base station is taken as an example). There are two cases in which tasks are transmitted through edge communication node i, one is when tasks are scheduled to edge server i for execution; the other is when nodes are scheduled to cloud for execution and need to be relayed through edge communication node. Assume that the maximum communication resource of the edge communication node i is expressed as B_i in bandwidth, assume that the task set performed by the edge server i is W_i, and the task set that forwards data to the cloud server through the edge communication node i is W_{cloud}^i, and the load rate of the edge communication node i can be described as follows

$$u_i^{com} = \frac{\sum_{j \in W_i} B_{ji}^{edge} + \sum_{j \in W_{cloud}^i} B_{ji}^{cloud}}{B_i} \tag{22}$$

where B_{ji}^{edge} and B_{ji}^{cloud} respectively represent the bandwidth allocated by edge server and cloud server to perform tasks. According to the load rate of edge communication nodes, the variance function of load rate among edge communication nodes is described as

$$V_{com} = \frac{1}{K} \sum_{i=1}^{K} \left(u_i^{com} - \frac{1}{K} \sum_{i=1}^{k} u_i^{com} \right)^2 \tag{23}$$

2.6 Problem Description

The decision variable under time slot t is defined as $X = \{D_1(t), D_2(t), \cdots, D_N(t)\}$, and the optimization problem of Internet of Things cloud edge resource scheduling aiming at network load balancing is described as

$$
\begin{aligned}
& P_1: \\
& \min_{X}(V + V_{com}) \\
& s.t. \quad (10), (13), (17), (19) \\
& \qquad a_i = \{0, 1, 2\}, \\
& \qquad num_i \in \mathrm{H} \ or \ \mathrm{J} \ or \ \mathrm{K}
\end{aligned} \tag{24}
$$

Problem analysis: in the decision variables $X = \{D_1(t), D_2(t), \cdots, D_N(t)\}$, $D_i(t) = (a_i, num_i, X_i, enum_i, B_i)$ determine the service task T_i in the corresponding resource scheduling scheme, a_i, num_i are integer variables, decide which resource node that task T_i should be assigned to. $X_i = (s_i, m_i)$ and B_i respectively for the node T_i the allocation of resources, in order to optimize the design, s_i, m_i, B_i are integer variables, that is, node resources are divided by unit.

3 Dynamic Resource Scheduling Mechanism Based on DRL

3.1 Dynamic Resource Cooperative Scheduling Framework Based on Multi-agent

The framework of dynamic resource cooperative scheduling based on multi-agent mainly includes three parts: imitation learning, distributed online decision making and centralized model training. As shown in Fig. 2, imitation learning is completed in the initial stage of the system, training is carried out based on expert data, and parameter θ is generated and sent to agents with decentralized edges. The Agent makes online decisions based on the obtained parameter θ, and continuously uploads the decision result tuple

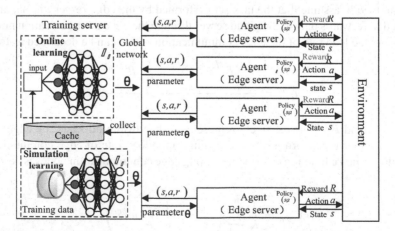

Fig. 2. Dynamic resource cooperative scheduling framework based on multi-agent

(s, a, r) to the decentralized training server. Based on these training samples, the decentralized training server periodically trains parameter θ and issues the resource scheduling decision strategy to update the edge server. The distributed online decision process and centralized model training are iterated alternately, and the scheduling strategy is updated dynamically continuously.

3.2 Simulation Learning

In the initialization stage, the problem P_1 is solved based on the conventional optimization algorithm, and a batch of decision trajectory data $\{\Gamma_1^i, \Gamma_2^i, \Gamma_3^i, \cdots, \Gamma_m^i\}$ is obtained. Each data contains the environment state and the action executed, which can be expressed as $\Gamma_j^i = \langle S_1^i, a_1^i, S_2^i, a_2^i, \cdots, S_{n+1}^i \rangle$, where n represents the number of state changes in trajectory Γ_j^i, S_1^i is the environment state, mainly including the data of each resource node. These "state-action" pairs are extracted to construct a new data set $D = \{(S_1^i, a_1^i), (S_2^i, a_2^i), \cdots\}$, and the model training can be completed by using the classification algorithm.

3.3 Online Resource Scheduling Algorithm Based on DRL

The collaborative scheduling of Internet of Things resources is modeled as markov decision process, including environment, state space S, action space a_s, reward $Reward$ and state transition probability P. The reinforcement learning task corresponds to the quad $MDP = \langle S, a_s, \pi, Reward \rangle$, where π is the strategy obtained through continuous attempts in the environment. According to this strategy, we can know the strategy $a_{s_t} = \pi(S_t)$ that needs to be implemented in the state S_t.

Action space: Action represents Agent's scheduling of Internet of Things resources and task allocation in resource scheduling. Actions in resource scheduling scenarios mainly include task unloading decision a_i, num_i, $cnum_i$, computing resource allocation X_i, communication resource allocation B_i, and communication resource and computing resource are discrete quantities.

Reward: The global reward is defined as the reciprocal of the variance value of the whole network load rate. When the load rate variance value is smaller, the global reward is larger, which promotes the network load balancing.

$$Reward_t = \frac{1}{V_t + V_{com,t}} \tag{(25)}$$

Strategy π: The goal of markov decision model is to find a strategy $\pi_\theta(s, a)$ that can optimize reward from t to t_Δ. The optimal strategy is $\pi^*(S) = \arg^{\max}_{a_t} Q^*_{\pi_\theta}(S_t, a_{t+1})$. Therefore, the approximate optimal strategy can be obtained by deep learning to train the approximate optimal function model of $Q_{\pi_\theta}(S, a)$. Table 1 shows the online resource scheduling algorithms based on DRL.

Table 1. Dynamic resource scheduling algorithm based on DRL

Algorithm 1 Dynamic resource scheduling mechanism based on DRL

Input: network state, task set;

Output: Resource scheduling decision $X = \{D_1(t), D_2(t), \cdots, D_N(t)\}$;

1: Initializes the iteration data times Δ, $it = 0$;

2: Initialize the K Agent, Generate random policy parameters θ ;

3: Select decentralized training server based on blockchain smart contract;

/* Imitation training stage */

4: Based on the conventional optimization algorithm, a batch of decision trajectory data $\{\Gamma_1^i, \Gamma_2^i, \Gamma_3^i, \cdots, \Gamma_m^i\}$ was generated, and the classification algorithm was called for training to obtain all Agent parameters θ^i.

5: Store relevant data in the blockchain network;

/* Distributed online decision making */

6: while True do:

7:     for $it++$; Agent k in K do:

8:         Interact with the environment to obtain node load status information;

9:         Make resource scheduling decisions independently;

10:        Upload the resource scheduling decision tuple (s,a,r) to the decentralized training server

11:    end for

/* Concentrated model training */

12:    if $it = 0$ then

13:        The decentralized training server updates the parameters of the edge Agent ;

14:    end if

15: end while

---

4   Simulation and Performance Analysis

In this paper, we conducted simulation experiments using Remix, Ganache, and TensorFlow. The experimental parameters are shown in Table 2. The goal was to validate the effectiveness of the dynamic resource scheduling mechanism based on DRL in a collaborative scheduling scenario involving edge and end resources in the cloud-based Internet of Things (IoT) environment. We applied the Proximal Policy Optimization (PPO) and Deep Q-Network (DQN) algorithms to observe the average network load variation during each update of the decision model. As depicted in Fig. 3, as the number of reinforcement learning iterations increased, the variance of the network load decreased gradually. This variance represents the average load of the network, indicating that the level of network load balance improved with the increasing number of iterations in the current simulation environment. The dynamic resource scheduling strategy based on DRL demonstrated its ability to adapt well to the distributed network environment through continuous self-learning. It also optimized the decision model while providing fast reasoning and decision-making capabilities.

Table 2. Experimental parameters

Parameters	Value
Number of nodes	10
Number of tasks	30
Learning rate	0.002
Reward discount factor	0.9
Minimum number of samples	64

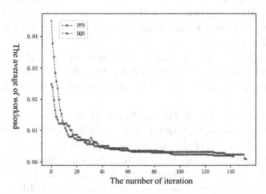

Fig. 3. Impact of the increase in the number of iterations on the average network load

5 Conclusion

Based on blockchain technology and DRL, we have developed a distributed intelligent resource sharing model for IoT services. This model incorporates a cloud-edge-end collaborative approach to resource scheduling. To ensure user satisfaction and improve network stability, we have constructed a user satisfaction model and a load balancing model. The resource scheduling problem, with the goal of network load balancing and constraints of user satisfaction and node resource availability, is formulated as an optimization problem. However, this problem is proven to be NP-hard and challenging to solve using conventional optimization methods. In order to address this, we approach it as a Markov decision problem and design a dynamic resource scheduling mechanism based on DRL. This mechanism provides a heuristic approximation solution. Simulation experiments have been conducted to validate the effectiveness of the dynamic resource scheduling mechanism based on DRL.

References

1. Al-Fuqaha, A., Guizani, M., Mohammadi, M., et al.: Internet of things: a survey on enabling technologies, protocols, and applications. IEEE Commun. Surv. Tutorials 17(4), 2347–2376 (2015)
2. Rui, J., Sun, D.: Architecture design of the internet of things based on cloud computing. In: Seventh International Conference on Measuring Technology & Mechatronics Automation. IEEE (2015)
3. Singh, J., Pasquier, T., Bacon, J., et al.: Twenty security considerations for cloud-supported internet of things. IEEE Internet Things J. 3(3), 269–284 (2017)
4. Yi, X., Helal, A.: Scalable cloud-sensor architecture for the internet of things. IEEE Internet Things J. 3(3), 285–298 (2016)
5. Lin, J., Yu, W., Zhang, N., et al.: A survey on internet of things: architecture, enabling technologies, security and privacy, and applications. IEEE Internet Things J. 4(5), 1125–1142 (2017)
6. Nguyen, D.C., Pathirana, P.N., Ding, M., et al.: Integration of blockchain and cloud of things: architecture, applications and challenges. IEEE Commun. Surv. Tutorials 22(4), 1 (2020)
7. Dai, H.N., Zheng, Z., Zhang, Y.: Blockchain for internet of things: a survey. IEEE Internet Things J. 6(5), 8076–8094 (2019)
8. Jiang, H.B., Li, J., Zhao, P., et al.: Location privacy-preserving mechanisms in location-based services: a comprehensive survey. ACM Comput. Surv. 54(1), 1–36 (2021)
9. Nakamoto, S.: Bitcoin: A peer-to-peer electronic cash system (2008). https://bitcoin.org/bitcoin.pdf
10. Huang, H., Xue, Y., Wu, J., Tao, Y., Hu, M.: Temporal computing resource allocation scheme with end device assistance. IEEE Internet Things J. 9(18), 16884–16896 (2022)
11. Wu, H., Zhang, Z., Guan, C., et al.: Collaborate edge and cloud computing with distributed deep learning for smart city internet of things. IEEE Internet Things J. 7(9), 8099–8110 (2019)
12. Liu, C.H., Chen, Z., Tang, J., et al.: Energy-efficient UAV control for effective and fair communication coverage: a deep reinforcement learning approach. IEEE J. Sel. Areas Commun. 36(9), 2059–2070 (2018)

Large-Scale Network Adaptive Situation Awareness Method in Spatio-Temporal Dimension

Hongbin Zhang[1,2] (ORCID), Ying Xu[1] (ORCID), Bin Liu[3,4] (ORCID), Dongmei Zhao[2(✉)] (ORCID),
and Yikang Bai[1] (ORCID)

[1] School of Information Science and Engineering, Hebei University of Science and Technology,
Shijiazhuang 050000, China
[2] Hebei Key Laboratory of Network and Information Security, Hebei Normal University,
Shijiazhuang 050024, Hebei, China
zhaodongmei666@126.com
[3] School of Economics and Management, Hebei University of Science and Technology,
Shijiazhuang 050000, China
[4] Research Center of Big Data and Social Computing, Hebei University of Science,
Shijiazhuang, China

Abstract. In large-scale networks, the state space is exploding and changing dynamically. This leads to difficulties in collecting and analyzing situational awareness data, so we construct an adaptive situational awareness model in spatio-temporal dimensions. In the spatial dimension, vulnerabilities's threats are assessed through attack graphs combined with Shapley values. At the same time, vulnerability threats are dynamically quantified by updating the status node reachability probability in real time. In the temporal dimension, a game model is established by analyzing vulnerability attack graph nodes to dynamically adjust the observation frequency of high-risk vulnerabilities, focusing on the safety status characteristics of high-risk assets. Experimental results show that our proposal integrates the security features of both space and time dimensions. This method can better focus on high-risk vulnerabilities and accurately reflect the dynamic changes in the network security situation, ensuring timeliness and accuracy in network security detection.

Keywords: Situation awareness · vulnerability assessment · time game

1 Introduction

With the development and popularization of computer networks and communication technology, network security issues are becoming increasingly prominent. Computer viruses, malicious attacks, information leaks, and other network attacks are becoming more severe. Multifaceted network security threats and risks are also increasing [1]. In this context, network security situational awareness technology has become an essential means to understand network situations and protect network security [2]. However, as the network expands, the number of analyzed data sources grows exponentially. Furthermore, the surge in various types of network assets, the complexities of changing states, and cross-domain relationships make state collection extremely challenging [3]. How to timely and accurately detect the security status of large-scale networks and accurately grasp the network security situation is an urgent problem that needs to be solved.

Existing network security situational assessment methods mostly rely on traffic analysis and employ machine learning techniques to construct assessment models [4]. However, with the huge amount of traffic data, the processing process has some problems, such as resource waste and processing delays. The root cause of threat traffic and attacks is that vulnerabilities in the network are exploited by attackers[5]. Therefore, this paper establishes an assessment model based on vulnerability information and assesses vulnerability threats from the perspective of attackers. By focusing on the observation of high-risk vulnerabilities to facilitates the rapid discovery of problems, locate risks, and understand the network security situation in a more timely and accurate manner.

2 Related Work

Regarding the acquisition of network situation data, Zhang Yongzheng et al. [6] studied a dynamic polling scanning technology based on network intrusion characteristics, to improve the system's non-frequent detection mode in high-traffic networks under fixed data storage capacity. In order to improve the quality and efficiency of network security situation element extraction, Zhao Dongmei et al. [7] proposed a parallel reduction algorithm based on attribute importance matrix. They deleted redundant attributes according to the reduction rules to achieve efficient extraction of network security situation elements. Sun Qian et al. [8] also considered that when the detection of the whole node will cause a large load and low efficiency, they propose a prediction-based dynamic polling scanning optimization strategy for network security detection.

In terms of situation assessment, Tang Zanyu et al. [9], based on information fusion, identified the attack stage, calculated the probability of a successful network attack and the realization probability of the network attack stage, and finally evaluated the network security situation. Wang Jinheng et al. [10] developed a probabilistic neural network-based network security situation assessment model, taking into account the characteristics of the network security situation and common assessment levels. They optimized the correction factors to improve the accuracy of the situational assessment. Sun Pengyu et al. [11] addressed the issue of opportunity strategy selection in network security and proposed a network security defense decision-making method that enables real-time measurement of the system's security state. Qiu Mingyang et al. [12] studied the network security assessment method based on a time-probability attack graph, considering the impact of vulnerability utilization time and scanning time on network offensive and defensive confrontation. In order to accurately obtain the security situation of the whole network, Chang Liwei et al. [13] designed a network security situation awareness model based on the basic characteristics of traffic, which includes five core elements: traffic detection, attribute extraction, decision engine, multi-source fusion, and situation assessment. With the help of convolutional neural networks, the network security situation was objectively and efficiently assessed.

Although situation acquisition and situation assessment have been studied in large-scale networks, they are still based on traditional intrusion detection ideas to collect status information and obtain situation elements. In addition, the above studies mainly adopt a full-range and fixed-period acquisition strategy in situation acquisition, resulting in a long detection cycle and no timely risk warning. To address the research gaps mentioned, we propose a large-scale network adaptive situation awareness method in the spatio-temporal dimension. This method aims to enhance the efficiency and accuracy of acquiring situational information, promptly identify potential security vulnerabilities and threats, and evaluate the overall network security situation.

3   Adaptive Situational Awareness Model for Large-Scale Networks

3.1   Space Optimization Strategy Based on Vulnerability Threat Assessment

Firstly, a vulnerability attack graph is constructed in the spatial dimension based on network topology and vulnerability information. Vulnerability threats are dynamically assessed by calculating the Shapley value of each node combined with the method of state transition probabilities in the attack graph [14]. Using the method in this article not only considers the importance of the vulnerability but also the probability of the vulnerability being exploited in specific scenarios, resulting in more accurate assessment results.

Vulnerability Attack Graph. This paper defines the attack graph VAG based on the vulnerability node[15], which is described as a five-tuple, and the specific meaning of each tuple is as follows:

1) V represents the set of nodes in the attack graph, $V = \{V_i | i = 1, 2..., N\}$ represents the vulnerability at the corresponding location, and N is the number of nodes in the attack graph.
2) H represents a host collection, $H = \{H_i | i = 1, 2, ..., m\}$ represents each host set in the attack diagram;
3) E represents the set of directed edges between state nodes in the attack graph; for example, $e_{i,j}$ is the directed edge connecting nodes V_i and V_j.
4) EXP represents the exploitability score of the vulnerability node, where $EXP(V_i)$ represents the exploitability score of vulnerability node V_i in the figure.
5) $Impact$ represents the system influence of the vulnerable node after being exploited by the attacker, where $Impact(V_i)$ represents the influence score of the vulnerable node V_i in the figure.

Shapley Value Based on Vulnerability Attack Graph. When evaluating vulnerability threats, the idea of a cooperative game is introduced; each vulnerability node is regarded as an agent in the game, and an attack by an attacker is regarded as a cooperative behavior among these agents. Use Shapley values as solutions to cooperative games. Define π as an arrangement of all agents, X_i^π represents all the predecessor nodes of node v_i in the arrangement, and use $\pi(j)$ to represent the position of node j in the arrangement. In the vulnerability attack graph, an attack path can be regarded as a cooperative behavior between nodes, for all paths $l \in L$, define the average marginal contribution of each node v_i to cooperation X_i^π as the Shapley value, represented by $SV_i(F)$.

$$SV_i(F) = \frac{1}{N!} * \sum_{c \in L} [F(X_i^c \cap v_i) - F(X_i^c)] \tag{1}$$

where F represents the characteristic function, which is used to numerically evaluate the performance of each cooperation, and $F(X_i^c \cap v_i) - F(X_i^c)$ represents the difference of the characteristic function change before and after joining node V_i, which is the marginal contribution.

In this paper, the cooperative behavior between the vulnerability nodes means that the attacker launched the attack, so the performance of cooperation can also be called the benefit of the attack. So consider the ultimate payoff an attacker might get for launching an attack.

$$F(c) = \sum_{v \in c} Imapct(v) * length(v) \tag{2}$$

where $Imapct(v)$ is the influence score of the vulnerability on the network system, that is, the income obtained by the attacker after exploiting the vulnerability, the indicators are shown in Table 1, and $length(v)$ represents the shortest path length in the attack graph.

Table 1. Attack benefit indicators

Benefit index	Parameter values	The numerical
Confidentiality impact (C)	N/L/H	0.0/0.22/0.56
Integrity impact (I)	N/L/H	0.0/0.22/0.56
Availability impact (A)	N/L/H	0.0/0.22/0.56
Remediation level (RL)	O/T/W/U	0.87/0.90/0.95/1.0

The attack benefit score is:

$$Imapct(V) = 1 - (1 - C)(1 - I)(1 - A) \times RL \tag{3}$$

Calculation of State Transition Probability. When assessing the threat of vulnerabilities based on attack graphs, it is necessary to calculate the probabilities of attackers reaching various state nodes. The calculation of vulnerability exploitability often relies on the static scoring results of the CVSS [16]. However, vulnerabilities have lifecycles, and therefore, it is important to describe their dynamic characteristics in order to dynamically calculate their exploitability probability [17].

$$EXP(V) = 20 \times AV \times AC \times AU \times RC \times E \tag{4}$$

Scoring index as shown in Table 2.

Table 2. Vulnerability exploitability score index

Availability index	Parameter values	The numerical
Attack path (AV)	N/A/L	1.00/0.65/0.40
Attack complexity (AC)	L/M/H	0.71/0.61/0.35
Identity authentication (AU)	M/S/N	0.45/0.56/0.74
Report credibility (RC)	X/U/R/C	0.90/0.95/1.00/1.00
Exploitability (E)	X/U/P/F/H	1/0.85/0.90/0.95/1

Therefore, the formula for calculating the state transition probability is

$$P_{ij} = f \times EXP(V_i) \qquad (5)$$

where, f is the proportional coefficient, and P_{ij} represents the probability that an attacker will use vulnerability V_i to reach the successor node when the predecessor node is satisfied.

Vulnerability Threat Assessment. Based on the consideration of state transition probabilities, this paper calculates the average marginal contribution of each vulnerability node under all possible cooperation scenarios to determine the threat value of each vulnerability node. The formula is as follows:

$$Th_i = P_i \times SV_i(F) \qquad (6)$$

Sort vulnerability nodes based on their threat values and select high-risk vulnerabilities for observation.

Considering the impact of attackers on state transitions, the paper analyzes threat intelligence to understand attacker intentions. The threat severity value of the attacker in the threat intelligence is used as the weight to update the state transition probabilities corresponding to the vulnerabilities[18]. The state transition probability matrix in the vulnerability attack graph is dynamically updated in real time, allowing for dynamic quantification of vulnerability threats.

Algorithm 1. Vulnerability threat assessment dynamic update algorithm

---

Input: vulnerability attack graph (VAG), threat intelligence, state transition probability matrix P

Output: Dynamic updated results of vulnerability threat assessment

---

BEGIN

1): Initialize:

   Initialize the state transition probability matrix;

   Create *Thetas* to store the dynamic vulnerability threat assessment results.

2): Calculate the weight factor $\varpi = f(TL)$ based on the threat severity value TL of the attacker intentions in the threat intelligence;

3): For all S_i, Update the state transition probability, $P'(S_i) = \varpi \times P(S_i)$, Update the state transition probability matrix;

4): Utilize the Shapley value of the vulnerability node and update the vulnerability threat value using equation (6).Add the updated vulnerability threat value to *Thetas*.

END

3.2   Situation Assessment Based on the Time Game Model

In large-scale networks, network security monitoring parties usually do not have the authority to directly repair vulnerabilities [19]. So the scenario in this paper is that the security monitoring party selects the vulnerability scanning period, finds the vulnerability, and generates an alarm to remind the user to reduce the risk. The attacker exploits the vulnerabilities and causes damage to the system. This process conforms to the game idea. Therefore, a time-game model is established based on the attack behavior and risk observations of both parties regarding the vulnerable nodes [20]. It enables the security monitoring party to optimize their observation strategy, enhancing observation efficiency under the premise of limited cost. Finally, by introducing CVSS, we can quantify the utility of both parties in the game and assess the network security situation.

Game Model Definition.   Network Security Awareness Model-Timing Game (NSAM-TG): including the participants of the game, the security status of the target network, the strategy set of both sides of the game, the time variable, and the utility function, that is, NSAM-TG = (N, S, P(t), t, U).

The meaning of each NSAM-TG tuple is expressed as follows:

1) N: participate in the game; N_a on behalf of the attacker, N_d on behalf of the security monitoring; $N = (N_a, N_d)$.

2) S: The set of security statuses of the target network, $S = (S_0, S_1, S_2, \cdots, S_n)$, and S_i represents the security status of the target network at time i, generated by the vulnerability threat assessment model.

3) $P(t)$: the strategy space of both sides of the game. The strategy of the attack side is defined as the set of illegal attack actions taken on the target network. The monitoring strategy is the set of time intervals between two consecutive actions, $P(t) = (P_a(t), P_d(t))$.

4) t: The time variable in the attack and defense game. The actual network attack and defense scenario is a continuous and real-time confrontation. Changes in network security status, gains and decisions in the attack and defense game are all related to time.

5) U: Represents the set of utility functions of both parties in the game. $U = (U_a, U_d)$ represents the utility function of the attacker and the utility function of the security monitoring party, respectively.

Situation Quantification. The utility of both network attackers and defenders depends on the benefits and costs brought by the strategies they choose.

The attacker's benefit refers to the actual harm caused by the attacker. We posit that the attack benefit is related to the vulnerability exploitation probability and the impact of vulnerabilities on the network [15]. Therefore, the attack benefit is defined as follows:

$$benefit_a = P \times Imapct(v) \tag{7}$$

where P is the state transition probability in the vulnerability attack graph, and $Imapct(v)$ is calculated as shown in formula (3).

Because the situation element used in this paper is vulnerability information, the strategic cost for the attacker is defined as the difficulty of exploiting the vulnerability. Additionally, the attack cost is related to the number of vulnerabilities. As the number of vulnerabilities increases, the attacker has more attack paths to choose from, resulting in lower costs. Therefore, the attack cost function is defined as follows:

$$cost_a = \mu \times EXP(v) \tag{8}$$

Among them, μ is the correction factor, which is the combined function value of the percentage weight of the policy vulnerability type in the overall and the increase rate, and the quantification of $EXP(v)$ is shown in formula (4).

The monitoring party's benefit. The benefit of the monitoring party is directly proportional to the reduction in risk achieved by implementing their strategies. Which is defined as follows:

$$benefit_d = R_t - R_{t+1} \tag{9}$$

R_t represents the system risk value at time t. The risk value is dependent on the vulnerability threat value, the losses incurred after vulnerability exploitation, and the intensity of vulnerability discovery. It is defined as follows:

$$R = \sum_{i=1}^{n} \frac{Th_i * \alpha Impact(v_i)}{\frac{T}{t} * \frac{n}{N}} \tag{10}$$

α represents the value level of assets, which reflects the varying degrees of their importance. t denotes the time interval between consecutive vulnerability scans by the monitoring party. T represents the attacker's time for exploiting vulnerabilities, i.e., the time required from vulnerability discovery to exploitation. N represents the total number of vulnerabilities, while n indicates the number of detected vulnerabilities.

Cost of the monitoring party. Similar to the cost of attacker strategies, the cost of the monitoring party represents the cost incurred by adopting a strategy. It is directly proportional to the scanning intensity, where a higher scanning intensity results in increased costs. The cost to the security monitoring party is defined as follows:

$$\cos t_d = \frac{1}{t} \times \frac{n}{N} \tag{11}$$

Based on the above calculations, it can be concluded that the utility functions of both parties in the game are as follows:

$$U_a = benefit_a - \cos t_a = p \times Imapct(v) - \frac{1}{T} \times EXP(v) \tag{12}$$

$$U_d = benefit_d - \cos t_d = (R_t - R_{t+1}) - \frac{1}{t} \times \frac{n}{N} \tag{13}$$

According to the utility values of both parties in the game, the security situation value S of the target network is calculated as:

$$S = U_d - U_a \tag{14}$$

The value of $|S|$ represents the degree of safety or danger of the target system. When $S > 0$, the network system is safe. The greater $|S|$, the greater the degree of network security. When $S < 0$, the situation is reversed.

4   Experiment Analysis

4.1   Description of the Experiment

In order to verify the effectiveness and feasibility of our proposal, a small local area network was built to carry out related experiments, and the following network structure was adopted. The network contains a total of six devices, namely DNS server, Web server, Database server, FTP server, PC1, and PC2. The topology is shown in Fig. 1.

In this experiment, the attacker attempted to attack a server located on the organization's internal network. It ultimately wants to obtain root permissions on the FTP server and illegally read sensitive data on it.

First, collect the information elements in the current network using information element collection tools. Supplement and refine the relevant vulnerability indicators by utilizing vulnerability management platforms such as CVD and CNVD, and conducting extensive research. Calculate the vulnerability information based on the indicator scores provided in Tables 1 and 2 using formulas (2) and (6), resulting in the following vulnerability details (Table 4):

Due to the FTP server being the target of the attacker in this network, it is considered the absorptive state. The relevant information elements, including topological information, vulnerability information, port information, and others, are inputted into the attack graph generation tool, Mulval. The resulting attack graph is shown below.

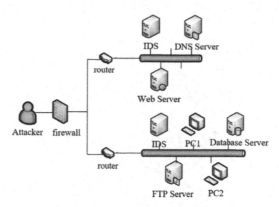

Fig. 1. Experimental network topology

Table 3. Vulnerability information

Devices	CVE-ID	EXP	Impact
DNS server(H1)	CVE-2013–0722(V1)	5.9	0.926
Web server(H2)	CVE-2018–0075(V2)	3.3	0.926
	CVE-2018–0074(V3)	3.8	0.926
PC1(H3)	CVE-2018–2219(V4)	9.5	0.926
PC2(H4)	CVE-2017–5904(V5)	4.7	0.587
FTP server(H5)	CVE-2016–6171(V6)	3.5	0.587
Database server(H6)	CVE-2017–2583(V7)	4.7	0.130

Table 4. Shapley values of vulnerability nodes

Vulnerability	V1	V2	V3	V4	V5	V6	V7
Shapley value	15.372	7.711	10.825	12.717	12.534	9.783	9.649

4.2 Experimental Results and Analysis

Vulnerability Threat Assessment. To verify the effectiveness of this study, the threat assessment of each vulnerability is first conducted using the vulnerability assessment method, which considers the marginal contribution proposed in this paper. Based on the EXP and Impact of vulnerabilities provided in Table 3, combined with the calculation rules of formulas (3) and (1), calculate the marginal contribution value and Shapley value of vulnerability nodes, and the results are shown in the following table:

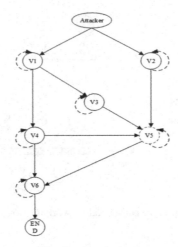

Fig. 2. Attack graph.

Based on the information presented in Table 3 and Fig. 2, the corresponding state transition probability matrix is calculated using formula (5).

$$P = \begin{bmatrix} 0 & 0.640 & 0.360 & 0 & 0 & 0 & 0 \\ 0 & 0.332 & 0 & 0.216 & 0.452 & 0 & 0 \\ 0 & 0 & 0.513 & 0 & 0 & 0.487 & 0 \\ 0 & 0 & 0 & 0.462 & 0 & 0.538 & 0 \\ 0 & 0 & 0 & 0 & 0.313 & 0.287 & 0.400 \\ 0 & 0 & 0 & 0 & 0 & 0.552 & 0.448 \\ 0 & 0 & 0 & 0 & 0 & 0 & 1 \end{bmatrix}$$

The threat value vector $Th = (8.412, 3.275, 6.126, 8.0579, 7.3194, 5.891)$ for the vulnerabilities is calculated using the formula (6). According to the evaluation results, V1, V4, and V5 are the three vulnerabilities with the highest threat values. The main reason is that V1 serves as a starting point for the entire network, making it susceptible to being exploited as an entry point for attacks. If an attacker successfully exploits this vulnerability, multiple attack paths become available, increasing its potential harm. Since V2 only has one subsequent exploitation node, its impact is relatively smaller. V4 and V5 appear in multiple paths, resulting in higher threat values.

The attack behavior information in the attack scenarios can be obtained from the IDS detection systems deployed in each region. The attack behavior information targeting vulnerable hosts during the experimental process is presented in Table 5 below.

From the attack sequence analysis, it can be seen that the vulnerabilities exploited during the attack are basically consistent with the high-threat vulnerabilities evaluated previously, which confirms the accuracy of the vulnerability threat assessment in this article (Table 6).

CVSS provides a method to capture the primary characteristics of vulnerabilities and assign a numerical score to reflect their severity. However, a major drawback of this standard is the lack of consideration for the specific network environment in which the

Table 5. Attack sequence list

Attack number	Attack time	Attack source	Attack target	Exploited vulnerability
1	08:44–08:46	H0	H1	V1
2	08:46–08:50	H1	H2	V3
3	08:50–08:56	H2	H4	V5
4	08:56–09:15	H4	H5	V6

Table 6. Comparison of evaluation based on Shapley values and CVSS

Vulnerability threat ranking	Shapley value	CVSS score
1	V1	V4
2	V4	V1
3	V5	V2
4	V3	V5
5	V6	V3
6	V2	V6

vulnerabilities exist. The comparison between the evaluation based on Shapley values calculated in this study and CVSS is presented as follows:

It is evident that V2 has a higher threat value in the CVSS score. However, from the attack graph, it can be observed that V2 only exists in one attack path, and its exploitability score is low, indicating a high cost for vulnerability exploitation. In this case, the threat value of V2 is relatively lower. Therefore, the Shapley value assessment provides a more objective evaluation, yielding more reasonable results that align with the actual circumstances.

Security Situational Assessment. Perform attack-defense experiments on the network environment depicted in Fig. 1, and collect secure data samples. The weights of each host during the attack process are shown in Table 7.

Table 7. Host importance weight

Host	H1	H2	H3	H4	H5	H6
Weight	0.34	0.18	0.05	0.05	0.2	0.18

According to the defined offensive and defensive revenue function formulas in this article, we calculate the revenue for both offense and defense in the NSAM-TG model.

By analyzing the equilibrium revenue at each sampling time point, we can obtain the revenue values for both parties during that time period (Fig. 3).

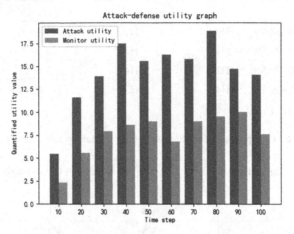

Fig. 3. Attack-defense utility graph

Under the network environment illustrated in Fig. 1, the security situation values are computed using the proposed method in this paper, as well as the HMM method [21] and the Analytic Hierarchy Process [22]. A comparative experiment is conducted, and the resulting network's security situation change curve is depicted in Fig. 4.

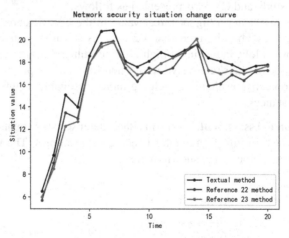

Fig. 4. Network security situation change curve

According to the above figure, it can be observed that the security situation values assessed using the method proposed in this paper are generally consistent with the results obtained from the HMM method and the AHP. However, due to the frequent observation of high-risk vulnerabilities in this paper's method, it highlights the emphasis on high attack gains, resulting in a higher security situation value compared to other evaluated methods. The experiment demonstrates that this paper accurately reflects the change in the network security situation while focusing on high-risk vulnerabilities.

To achieve a comparison of the reliability of the evaluation results, we compare the results with the actual network security situation and calculate the root mean square error (RMSE) for each of the three evaluation methods. We also employ the Mean Absolute Error (MAE) algorithm to determine the reliability of the three evaluation methods. The calculation process is as follows:

$$RMSE = \sqrt{\sum_{i=1}^{n} \frac{(S_i' - S_i)^2}{n}} \tag{15}$$

$$MAE = \sum_{i=1}^{n} \frac{|S_i' - S_i|}{n} \tag{16}$$

Among them, S_i' is the real situation value, and S_i is the predicted situation value. The prediction accuracy results of the three prediction methods are shown in Table 8.

Table 8. Performance comparison of three evaluation methods

Method	RMSE	MAE
Textual method	0.0375	0.215
Reference 22 method	0.0352	0.274
Reference 23 method	0.0425	0.315

It can be seen from the experimental data in Table 8 that the indicators of the experimental evaluation results in this paper are comprehensively lower than those in the other two groups of experiments. Therefore, the method in this paper is more stable in situation assessment, which proves the reliability of experimental assessment methods.

5 Conclusion

This paper introduces an adaptive situational awareness approach for large-scale networks. First, the vulnerability threats were dynamically assessed, and vulnerabilities with major threat risks were screened out in the spatial dimension. Secondly, a game model is utilized to assess the network security posture, allowing the security monitoring part to adjust the frequency of actions in the time dimension, considering cost constraints. Experimental results show that this method can accurately assess changes in security situations, providing a basis for timely security responses. Future research will consider applying this method to large-scale networks with a significant amount of vulnerability information to further optimize the evaluation model.

Acknowledgement. This research was supported by the National Natural Science Foundation of China under Grant No. 61672206, No. 61572170, Central Guide Local Science and Technology Development Fund Project (216Z0701G), S&T Program of Hebei under Grant No. 18210109D, No. 20310701D, No. 20310802D, No. 21310101D, National cultural and tourism science and technology innovation project (2020).

References

1. Speech by Xi Jinping at the Symposium on Network Security and Informatization. China Inf. Secur. (05), 23–31 (2016)
2. Jia, Y., Han, W., Yang, X.: Research status and development trends of network security situation awareness. J. Guangzhou Univ. (Nat. Sci. Ed.) **18**(03), 1–10 (2019)
3. Feng, P., Tao, L.: Research progress on network security situation awareness in the big data environment. Secrecy Sci. Technol. **04**, 27–33 (2016)
4. Li, Y., Wang, C., Huang, G., Zhao, X., Zhang, B., Li, Y.: Comparison of network security situation awareness analysis framework and implementation methods. Acta Electronica Sin. **47**(04), 927–945 (2019)
5. Huang, Z.: Analysis of the impact of web system vulnerabilities in internet enterprises on cybersecurity. Netw. Secur. Technol. Appl. **05**, 10–12 (2023)
6. Zhang, Y., Xiao, J., Yun, X., et al.: DDoS attack detection and control methods. J. Softw. **23**(08), 2058–2072 (2012)
7. Zhao, D., Li, H.: Network security situation element extraction method based on parallel reduction. Comput. Appl. **37**(04), 1008–1013 (2017)
8. Sun, Q.: Key Technologies for Adaptive Network Security Detection and Defense Strategies in Large-scale Networks. Northwestern University (2019)
9. Tang, Z., Liu, H.: Research on network security situation assessment method under multi-stage large-scale network attacks. Comput. Sci. **45**(01), 245–248 (2018)
10. Wang, J., Shan, Z., Tan, H., et al.: Network security situation assessment based on genetic optimization PNN neural network. Comp. Sci. **48**(06), 338–342 (2021)
11. Sun, P., Zhang, H., Tan, J., et al.: Network security defense decision-making method based on game theory. Comput. Eng. **48**(11), 145–151 (2022). https://doi.org/10.19678/j.issn.1000-3428.0063866
12. Qiu, M., Sai, Y., Wang, G., et al.: Network security assessment method based on time-probability attack graph. Fire Control Command Control **47**(01): 145–149+155 (2022)
13. Chang, L., Liu, X., Qian, Y., et al.: Network security situation awareness model based on convolutional neural network and multi-source fusion. Computer Science **50**(05), 382–389 (2023)
14. Duan, C.: Research on Vulnerability Assessment and Defense Mechanism Selection of Network Systems Based on Game Theory. Hangzhou Dianzi University (2020). https://doi.org/10.27075/d.cnki.ghzdc.2020.000188
15. Zhang, K., Liu, J.: Network intrusion path analysis method based on dynamic exploitability of vulnerabilities. Inf. Netw. Secur. **21**(04), 62–72 (2021)
16. Common Vulnerability Scoring System (CVSS), 30 May 2019. http://www.first.org/cvss/
17. Gao, H., Wang, S., Zhang, H., Liu, B., Zhao, D., Liu, Z. : Network security situation assessment method based on absorbing Markov chain. In: Proceedings of the 2022 International Conference on Networking and Network Applications (NaNA), Urumqi, China, pp. 556–561 (2022). https://doi.org/10.1109/NaNA56854.2022.00102
18. Zhang, H., Yin, Y., Zhao, D., et al.: Network security situation awareness model based on threat intelligence. J. Commun. **42**(06), 182–194 (2021)

19. Liu, Y.: Overview of global network security situation in 2022. Secrecy Sci. Technol. **03**, 61–64 (2023)
20. Sun, P., Tan, J., Li, C., et al.: Network security defense decision-making method based on time differential game. Inf. Netw. Secur. **22**(05), 64–74 (2022)
21. Wang, Z., Lu, Y., Zhao, D.: Network security risk assessment method based on hidden Markov model. J. Air Force Eng. Univ. (Nat. Sci. Ed.) **20**(03), 71–76 (2019)
22. Zhao, X., Xu, H., Wang, T., Jiang, X., Zhao, J.: Research on multidimensional system security assessment based on AHP and gray correlation. In: Han, W., Zhu, L., Yan, F. (eds.) CTCIS 2019. CCIS, vol. 1149, pp. 177–192. Springer, Singapore (2020). https://doi.org/10.1007/978-981-15-3418-8_13

Real-Time Detection of Human Heart Rate and Blood Pressure During Exercise

Litao Guang, Jiancheng Zou(⊠), and Zibo Wen

College of Science, North China University of Technology, Beijing 100041, China
zjc@ncut.edu.cn

Abstract. Regular moderate exercise has great benefits for both our physical and mental well-being. Nowadays, more and more people attach great attention to exercise, and among the population that needs to engage in moderate exercise, individuals with hypertension have a high proportion. Although there has been great progress in non-contact detection of heart rate and blood pressure, little research has been done to achieve detection during exercise. Different from the traditional contact measurement and IPPG (Imaging Photo Plethysmography)-based static heart rate and blood pressure detection, this paper designs a non-contact system that can monitor heart rate and blood pressure in real time during exercise and count some common exercises. This paper realizes real-time ROI(Region of Interest) detection and tracking through human pose estimation and hand key points detection model, which is used to calculate heart rate and blood pressure, and uses the estimated maximum heart rate to give the limit of exercise intensity. In addition, human posture estimation is also used to count movements such as squats and exercise bikes. Compared with the heart rate measured by the smart bracelet, the error is 3.4% ~ 6.7%. Under static, compared with the Omron electronic sphygmomanometer, the systolic blood pressure error is -7.39% ~ 3.43%, and the diastolic blood pressure is -6.67 ~ 4.53%. This system has low equipment requirements and low environmental requirements. It is easy to operate and does not cause any discomfort. It is suitable for daily life and will not interfere with normal activities.

Keywords: YOLOv8-pose · Mediapipe · Heart rate Detection · blood pressure detection · Exercise control

1 Introduction

In recent years, with the rapid development of society, the richness of material life has not only improved people's living standards, but also led to changes in their living habits. Unhealthy eating, rest, and lack of physical activity habits have also brought higher disease risks. By persisting in some simple physical activities, our heart and lung function can be enhanced to a certain extent. It can also promote blood circulation, metabolism, improve body composition, relax the mind, and benefit physical and mental health [1–4]. However, exercise seems to be good for the body, but it should also be moderate. If the amount of exercise is too large, it increases the burden on the body and is more likely to

H. Jin et al. (Eds.): IAIC 2023, CCIS 2060, pp. 84–97, 2024.
https://doi.org/10.1007/978-981-97-1332-5_7

cause joint and muscle damage, as well as cardiovascular accidents. Monitoring heart rate is a simple and quite effective method for controlling amount of exercise [5]. In addition, among the population that needs to engage in moderate exercise, individuals with hypertension have a high proportion. A large number of studies have confirmed that the blood pressure problems of hypertensive patients can be improved through moderate exercise, which is also applicable to elderly hypertensive patients [6, 7].

But after exercise, it is also important to pay attention to the blood pressure situation, as the latest research has found that the increase in blood pressure after exercise is related to the risk of death [8]. Therefore, in addition to heart rate detection, it is also necessary to consider blood pressure detection. Previously, there were also examples of detecting heart rate during exercise [9]. However, the approach they adopted is minimally invasive, and although it is minimally invasive, it will inevitably have an impact on the subjects, leading to an impact on the experimental results.

For a better detection experience, based on deep learning human detection and pose estimation (YOLOv8), hand detection (Mediapipe) and IPPG-based non-contact blood pressure and heart rate measurement, this paper designs a camera to capture images to continuously detect human heart rate and blood pressure in real time during exercise, and realizes the counting of part of the exercise to control the amount of exercise and find the appropriate exercise intensity.

2 Principle

2.1 Imaging Photo Plethysmography (IPPG)

The measurement of heart rate and blood pressure can generally be divided into two types: contact and non-contact. Apart from the discomfort caused by wearing contact devices, these devices can also become interference in most sports, so a non-contact detection method is necessary.

In 2000, Schmitt et al. first proposed an IPPG improved from PPG [10], which can obtain rich physiological signals using common cameras only under natural light. Its generation cannot be separated from the basic law of spectrophotometry (Beer-Lambert law), which describes the relationship between the strength of a substance's absorption of a certain wavelength of light and the concentration of the absorbing substance and its liquid layer thickness. Here, it is used to detect changes in blood's reflection of light.

In a sufficiently long time compared to the detection time, when light (can be natural light) irradiates the human body, human skin, bones, muscles, fat and other tissues or parts can be considered to be a fixed value for the reflection of light, but the volume of blood can lead to changes in the reflected light: as blood volume increases, more light is absorbed, and less is captured by the device, and vice versa. And the beating of the heart drives changes in blood volume, which ultimately leads to changes in the light captured by the device.

In this way, the device captures the changes in reflected light over a period of time, and then processes the obtained image signal. The two-dimensional image signal is transformed into a one-dimensional wave signal, which reflects the changes in heart beat and blood volume, and is further used to detect heart rate and blood pressure.

As early as 2005, Wieringa et al. successfully extracted heart rate signals at different wavelengths using photo capacitance pulse technology [11].

2.2 Detection of Heart Rate

Selection of Region of Interest (ROI). When detecting heart rate, we already know that it uses the heart rate to drive changes in blood volume, which in turn affects changes in reflected light to reflect the rhythm of the heart rate. However, how to select appropriate areas for processing still needs to be addressed.

Firstly, it should be easy to obtain. Choosing a non-contact detection method is to make it easier to obtain data, reduce discomfort and tension caused by contact devices, and even to avoid affecting movement, making it suitable for most scenarios. Due to the significant impact of clothing on the penetration of light when collecting information, the skin surface is naturally the best choice because it can be directly photographed and has minimal interference. Additionally, the depth of human tissue that light can penetrate is limited, so it is necessary to frequently expose shallow skin, which is generally limited to the head and hands.

Secondly, considering the degree of changes in blood volume, it is natural to think of collecting information in areas with dense blood vessels. The distribution of blood vessels in the human body is quite complex, among which the facial blood vessels are densely packed and the most abundant. Changes in blood volume cause more significant changes in reflected light, which is beneficial for reducing pulse wave noise and improving experimental accuracy. Based on the experiment of Verkruysse W. et al. [12], a large area of the forehead with minimal skin interference was selected as the area of interest. However, considering the possibility of hair or hat occlusion, the cheek with dense blood vessel distribution was also selected as the ROI.

Obtaining and processing of wave signals. On the one hand, the changes in reflected light caused by heart beating are captured by the camera, and the specific changes are the changes in pixel values within the ROI. The R, G, and B channels within the obtained ROI are separated, and after averaging the grayscale values, each person's three channel continuous IPPG signal can be obtained.

On the other hand, due to the fact that most of the light contained in natural light is either absorbed by the body's melanin or water, while green light is absorbed by red blood cells in large quantities, red light is more likely to pass through skin tissue. These facts tell us that green light and red light are more suitable for detecting changes in blood volume and thus reflecting the beating of the heart.

By selecting multiple individuals to detect and extract the three channel IPPG signal, we can easily observe that the curve noise of the green channel is relatively low. Therefore, in this article, the green channel will be selected to extract the video heart rate signal.

At this point, the obtained wave signal still has a lot of noise interference due to various factors. Before calculating the heart rate, we still need to perform filtering and denoising operations. The IPPG signal of this green channel is a discrete distribution of points generated as the number of video frames increases. Periodic discrete Fourier transform is used to filter it [13], converting the signal from time domain to frequency domain. At this point, high-frequency signals are filtered out, and then the signal is restored to the time domain using inverse Fourier transform.

Overall, the obtained IPPG signal of the green channel is filtered through forward and inverse Fourier transform to obtain the final heart rate value.

2.3 Detection of Blood Pressure

There are many similarities between blood pressure detection and heart rate detection, but they also pose new requirements and challenges. In addition to the basic Beer Lambert law, another important driving discovery in measuring blood pressure is the relationship between blood pressure BP and pulse wave conduction time PTT [14].

Pulse Transit Time (PTT), refers to the conduction time of pulse waves generated during cardiac contraction from the heart to a certain location. PTT is commonly used in blood pressure measurement.

After extensive research and analysis, it has been found that there is a relatively stable relationship between blood pressure BP and pulse wave conduction time PTT [15]. In a sufficiently long and short period of time, the interference of most other factors, such as vascular elasticity, can be ignored. Therefore, the relationship between the two can be simplified and approximated as [16]:

$$BP = a + bPTT \tag{1}$$

Selection of Region of Interest (ROI). Due to the fact that both measurements are based on IPPG, the ROI selection requirements for measuring blood pressure are similar to those for measuring heart rate. However, due to the need for PTT, an ROI with a significant difference in distance from the head to the heart is also required. In this case, due to the less interference factors in the forehead region during heart rate measurement, this article chooses to use the forehead part as the "proximal end" and choose a hand that is easy to collect and has a dense distribution of blood vessels as the "distal end".

Because the forehead is closer to the heart than the hand, the pulse wave signal of a heart beat will first reach the forehead and then the palm. After obtaining the pulse wave signal of the forehead and palm, it is necessary to continue to obtain the time difference between them.

Further processing. Every time our heart beats, it actually undergoes a contraction and a relaxation. When the heart contracts, it causes the ventricle to eject blood outward, increasing the volume of blood in the blood vessels and increasing the absorption of light, resulting in a decrease in the detectable reflected light. This is reflected in the video as a decrease in the grayscale value of pixels within the ROI in each frame of the image. When the heart relaxes, the opposite is true. This tells us that in the obtained wave signal graph, the time difference obtained by using two signal peaks is related to human diastolic blood pressure, also known as low pressure, in the relationship between blood pressure and pulse wave conduction time. Similarly, the time difference obtained from the trough is related to systolic blood pressure, also known as high pressure.

In this way, after obtaining two pulse wave signals, the peaks and troughs generated by the two pulse wave signals can be found, and then the difference between the peaks and troughs can be calculated. The obtained data can be recorded for filtering and sorting. Due to the fact that BP and PTT are not completely linear, they are only approximated through linear fitting. Considering that the fitting degree of the first-order equation to the relationship between pulse conduction time and blood pressure is limited, the obtained PTT information is combined with the actual blood pressure values collected as the dataset, and a neural network is used for fitting training. With the help of the nonlinear activation function ReLU for processing, it can be used to fit stronger relationships or even non-linear relationships, which can better improve the model $BP = a + bPTT$ and achieve non-contact blood pressure estimation.

However, in [16], it is necessary to remain stationary in order to measure blood pressure, which clearly cannot be applied to real-time detection during exercise.

2.4 YOLOv8 and Attitude Estimation

Unlike the method of facial recognition and then locating and tracking facial ROI regions in [13], this article will use pose estimation to select.

The You Only Look Once (YOLO) series has been one of the most popular models in the field of computer vision since its first release in 2015. The one used in this article is YOLOv8, which is the latest version of YOLO. Like the previously highly anticipated YOLOv5, it was published by the publisher Ultralytics. Currently, it supports object detection, instance segmentation, image classification, object tracking, and attitude estimation tasks. The attitude estimation section is used in this article.

Pose estimation is a technology in the field of computer vision that detects the image of a person in an image (or video). Through pose estimation, specific joints of the human body are located in the image to determine the position of the person in the image. By locating several key points and connecting them with their physical meaning and actual connection, the pose displayed by the person in the image can even be constructed. Simply put, the purpose of posture estimation is to reconstruct several key points of human information.

Using YOLOv8-pose, theoretically, we can obtain an output for each person in each frame of the video, including a set of key points in the image, as well as a bounding box for this person.

The key point information of the human body is used in two parts: to locate and track the ROI of head heart rate and blood pressure detection, and to count movements.

The information of the character bounding box is used to slice each frame of the image, slicing each person's image separately for subsequent operations.

The model used in this article was trained on the coco key point dataset, which has a total of 17 key points. The numbers represent the index, starting from 0 to 16, representing the nose to the right ankle (Fig. 1):

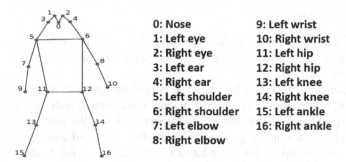

0: Nose	9: Left wrist
1: Left eye	10: Right wrist
2: Right eye	11: Left hip
3: Left ear	12: Right hip
4: Right ear	13: Left knee
5: Left shoulder	14: Right knee
6: Right shoulder	15: Left ankle
7: Left elbow	16: Right ankle
8: Right elbow	

Fig. 1. 17 key points of the COCO dataset

2.5 Mediapipe and Hand Estimation [17]

Mediapipe is a multimedia processing framework open-sourced by Google that provides a series of pre trained models and tools that allow users to directly handle rich multimedia tasks such as video, audio, pose estimation, gesture recognition, and more.

Although Mediapipe can also perform human pose detection, it only supports one person and only works on the CPU. Although Mediapipe has more key points, due to factors such as insufficient detection accuracy, this paper chooses to use the more feature-rich and equally efficient YOLOv8-pose for detection in terms of human pose estimation, while hand detection still needs to rely on Mediapipe.

Mediapipe uses deep learning models and computer vision technology to detect and track hand key points. Firstly, Mediapipe has trained a large amount of hand image data. When conducting detection, only the trained model needs to be loaded, and then the input image (in this article, a small image of a person after slicing) needs to be passed to the model. The model will detect the hand area in the image and locate the positions of various key points in the hand. After detecting the position of key points in the hand, Mediapipe also utilizes computer vision technology to track these key points. This processing helps to maintain consistency of key points between consecutive frames, resulting in better recognition accuracy.

The hand detection provided by it will provide key point positioning for 21 hand joints (Fig. 2):

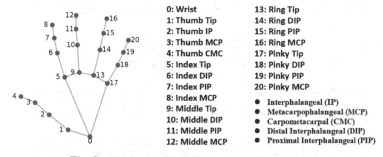

0: Wrist	13: Ring Tip
1: Thumb Tip	14: Ring DIP
2: Thumb IP	15: Ring PIP
3: Thumb MCP	16: Ring MCP
4: Thumb CMC	17: Pinky Tip
5: Index Tip	18: Pinky DIP
6: Index DIP	19: Pinky PIP
7: Index PIP	20: Pinky MCP
8: Index MCP	
9: Middle Tip	• Interphalangeal (IP)
10: Middle DIP	• Metacarpophalangeal (MCP)
11: Middle PIP	• Carpometacarpal (CMC)
12: Middle MCP	• Distal Interphalangeal (DIP)
	• Proximal Interphalangeal (PIP)

Fig. 2. Mediapipe gives 21 key points for the hand

2.6 Exercise Control

A simple and quite effective method for controlling exercise intensity is to estimate it through heart rate, specifically by using the maximum heart rate. There are many formulas to choose [18].

The maximum heart rate refers to when people engage in exercise, as the amount of exercise increases, the heart rate and oxygen consumption also increase. However, there is an upper limit to the increase in heart rate. Once the heart rate reaches this maximum value, it will no longer increase with the increase of exercise intensity. And as age increases, this upper limit, also known as the maximum heart rate, will gradually decrease.

Many sports enthusiasts like to roughly estimate their maximum heart rate by using a formula (220 minus age), but in reality, the maximum heart rate cannot be directly obtained by applying such a simple formula. For individuals, the true maximum heart rate value needs to be obtained through professional measurement. For example, is it suitable for patients with heart or lung diseases? The answer is no [19].

Fortunately, for most people, they do not need a very accurate maximum heart rate value to strictly control their exercise intensity, but just provide basic reference value is enough. For professional athletes, if they have already measured their closest maximum heart rate, it can naturally be used for more precise control.

In summary, we still need to have a certain estimate of the maximum heart rate. Therefore, this article will choose the following formula to estimate the maximum heart rate [20]:

$$HRmax = 207 - 0.7 \times age. \tag{2}$$

3   Algorithms and Experiments

In order to achieve the experimental objectives, the system designed in this article mainly consists of five parts: human posture estimation part, hand detection part, ROI selection part, heart rate and blood pressure detection part, and motion counting part (Fig. 3).

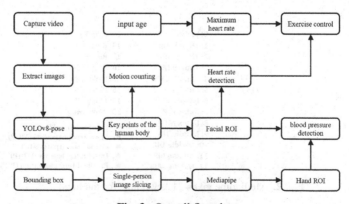

Fig. 3. Overall flowchart

The entire experiment was conducted under natural light, requiring the tester to expose their face and hands. Common mobile phone cameras and computer cameras were used to record videos, which were either recorded as videos or directly transmitted in real-time to the program for processing.

3.1 Human Pose Estimation Part

The video stream enters the human pose estimation section frame by frame in chronological order. In this section, character bounding box information and human key point information will be output based on YOLOv8 pose. As shown in the following figure (Fig. 4):

Fig. 4. YOLOv8-pose detection results

3.2 Hand Detection Part

Using the character bounding box information generated in the previous step, slice the characters in the initial image separately, so that Mediapipe can be used to capture the corresponding character's hand key point information faster and more accurately on these small images (Fig. 5):

Fig. 5. Hand key point detection results

3.3 ROI Selection Section

Based on the output of YOLOv8 pose's human key points, the cheek ROI is given using key points with indices 1, 2 (for both eyes). Unlike commonly used facial recognition methods, there are many fewer key points related to the face. But fortunately, the identifying key points can still be detected with sufficient accuracy. Due to the lack of key points that can be directly used to select ROI, through practical testing with multiple people, the forehead ROI is ultimately determined using key point information as follows:

$$left = min(x(1), x(2)),$$
$$right = max(x(1), x(2)),$$
$$top = min(y(1), y(2)) - (right - left)/2, \tag{3}$$
$$bottom = top - (right - left)/2.$$

Among them, $x(i)$ represents the horizontal coordinate of the key point with index, and $y(i)$ represents the vertical coordinate, the same below (Fig. 6).

Fig. 6. Facial key points and forehead ROI

At the same time as the ROI of the forehead, there is also the ROI of the cheeks area, which is given using key points with indices 0,3,4 (nose and ears). The ROI of the cheeks area is given as follows (Fig. 7):

$$a = min(x(3), x(4)),$$
$$b = max(x(3), x(4)),$$
$$left = 2a/3 + x(0)/3,$$
$$right = 2b/3 + x(0)/3, \tag{4}$$
$$top = y(0) - (b - a)/16,$$
$$bottom = top + (b - a)/3.$$

Fig. 7. Cheeks ROI

Through Mediapipe hand recognition, we will obtain information on 21 key points of the hand of each small image. The ROI of the hand is easier to determine than that of the face, as Mediapipe provides sufficient information on hand key points. Obviously, a good solution is to use four points with indexes 0, 1, 5, and 17 to provide the ROI of the hand used for detecting blood pressure:

$$left = min(x(0), x(1), x(5), x(17)),$$
$$right = max(x(0), x(1), x(5), x(17)),$$
$$top = min(y(0), y(1), y(5), y(17)),$$
$$bottom = max(y(0), y(1), y(5), y(17)).$$

(5)

Whether it's the front and back sides of the hand or the gesture, this choice ensures that there is enough hand skin in the box and minimal external interference (Fig. 8).

Fig. 8. Hand ROI

3.4 Heart Rate and Blood Pressure Detection Section

Due to the lack of data collected at the beginning, the calculation of heart rate will give many abnormal values. This article will calculate the heart rate after obtaining 100 frames of video data. The results will be displayed outside the top left corner of the character bounding box, and blood pressure will be displayed inside the top left corner (Fig. 9):

Fig. 9. Heart rate and blood pressure detection results

In addition, considering that heart rate has a more universal reference value for controlling exercise volume, a line chart will be provided for the estimated heart rate in the order of image frame processing. After inputting the age of the tester, the system will automatically estimate the maximum heart rate of the tester and take 70% of the maximum heart rate as the warning line (see pink dashed line in the figure). When the tester's heart rate does not exceed the warning line, the line chart representing the heart rate will be displayed in blue. When the current heart rate of the tester exceeds this limit, the broken line representing the heart rate will be converted into a bright red color as a warning (Fig. 10):

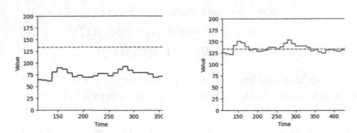

Fig. 10. Heart rate statistics chart and warning line

3.5 Motion Counting Section

There are many common sports, such as squatting, running, bench press, walking, exercise bike, skipping rope, etc. The 17 key human body points provided by YOLOv8-pose are sufficient to simply count most movements. The purpose of motion counting is to standardize movements to a certain extent (such as squatting, pulling up, etc., those that do not meet the standards will not be counted) and to provide a new reference index for controlling one's own exercise volume.

Squat counting: For squatting, the main consideration is the bending angle of the legs. This article will calculate the changes in the angle between the legs. Using the key points provided by YOLOv8-pose, we select the key points with indexes 11, 13, 15 and 12, 14, 16 to calculate the angle between the legs.

After obtaining the corresponding results, take the average of them. When this value is less than 75°, increment the count and introduce a Boolean value that allows starting the next counting process only when the angle exceeds 160°.

Pull up counting: Similar to squatting, consider selecting key points with indices 5, 7, 9 and 6, 8, 10 to calculate the angle between the arms, and take the average of the two. When this value is less than 75°, increment the count and introduce a Boolean value that allows starting the next counting process only when the angle exceeds 160°.

Exercise bike counting: when two legs can be detected, let

$$m = \frac{|y(13) - y(15)|}{|y(11) - y(13)|}, n = \frac{|y(14) - y(16)|}{|y(12) - y(14)|} \tag{6}$$

be the ratio of the difference in the heights of the two legs of the upper and lower legs respectively. When the size of m, n is interchanged, the number of times is increased by one. When only one leg can be detected to get m for example, take $m < 0.4$, the number of times plus one, and then introduce a Boolean value, so that $m > 0.6$ when the next counting process can begin.

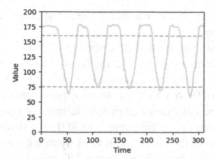

Fig. 11. Legs angle change chart and counting line

Taking squatting as an example, provide a graph of leg angle changes (Fig. 11):

By combining all the above results, the final output is obtained. Due to the similar vertical coordinate range between the line chart of heart rate and the leg angle change chart, the angle change chart and heart rate change chart can be merged together to provide a comprehensive view of the entire detection result (without displaying key point information of the human body and hands) (Fig. 12):

Fig. 12. The final result output plot

After experiments, using only the CPU (12th Gen Intel(R) Core (TM) i5-12500H 2.50 GHz) can achieve a processing speed of 7.8 frames per second for 1920 * 1080 videos while using GPU (1080Ti) can be quicker. However, it is still not fast enough to achieve processing of 30 frames per second by using GPU (1080Ti). Therefore, in the experiment, a multi-threaded approach was used to obtain the latest frame of image captured by the camera in real-time to ensure real-time performance.

4 Result Analysis and Outlook

Through multiple tests conducted on different weather conditions, different venues, and different testers, it was found that the intensity of light in the environment has a significant impact on the detection results. With sufficient light, the detection results will be more accurate. Compared with the smart bracelet, the error of heart rate varies from 3.4% to 6.7% in different scenarios.

Using the error calculation method described in [16], the comparison between the Omron electronic sphygmomanometer and the blood pressure measurements provided in this article, under static conditions, showed an error range of -7.39% to 3.43% for systolic pressure and -6.67% to 4.53% for diastolic pressure. However, due to the lack of reliable blood pressure detection equipment during exercise, accurate comparison data cannot be provided. In the motion state, the ROI area is retrieved for each frame. Because of the movement of the character, it is likely that the area of the ROI area will change, which will cause significant interference. This may lead to greater error in blood pressure detection based on [16] the improved motion state. However, fortunately, the overall trend of change is the same, so it still has good reference value.

Overall, this article achieves heart rate detection under pose estimation rather than facial recognition. Different from the static fixed IPPG blood pressure measurement in [16], it achieves a more adaptable automatic tracking non-contact blood pressure detection. Unlike [9] minimally invasive detection of heart rate during running, non-contact detection has a smaller impact on subjects.

At present, this article only implements single person detection. But by using character bounding box information, multiple individuals can be transformed into multiple individuals, which provides a way to promote single person detection to real-time detection of multiple individuals in the future.

Acknowledgment. We sincerely appreciate the support provided by the Organized Research - Collaborative Urban Slow Travel Environment Planning and Design System Based on the Concept of Smart Accessibility for People with Disabilities and the Elderly in the Capital City project.

References

1. Connolly, M.L., Bowden, S.C., Pascoe, M.C., et al.: Development and psychometric validation of the mental health-related barriers and benefits to exercise (MEX) scale in healthy adults. Sports Med. Open **9**, 18 (2023)
2. Sick, T.: The role of exercise intensity in physical education for improving body composition. J. Phys. Educ. Recreation Dance **93**(8), 50–57 (2022)
3. Steineck, I.I.K., Ranjan, A.G., Schmidt, S., Norgaard, K.: Time spent in hypoglycemia is comparable when the same amount of exercise is performed 5 or 2 days weekly: a randomized crossover study in people with type 1 diabetes. BMJ Open Diabetes Res. Care **9**(1), e001919 (2021)
4. Balasekaran, G., Mayo, M., Ng, Y.C.: Effects of large exercise-induced weight loss on insulin sensitivity and metabolic risk factors in young males with obesity. J. Sports Med. Phys. Fitness (2023)
5. Yiiong, S.P., Ting, H., Tan, D.Y.W., Chia, R.: Investigation of relation between sport's motion and heart rate variability (HRV) based on biometric parameters. IOP Conf. Ser. Mater. Sci. Eng. **495**, 012015 (2019)
6. Oliveira, J., Mesquita-Bastos, J., de Melo, C.A., Ribeiro, F.: Postaerobic exercise blood pressure reduction in very old persons with hypertension. J. Geriatr. Phys. Ther. **39**(1), 8–13 (2016)

7. Gasparini-Neto, V.H., Caldas, L.C., de Lira, C.A.B., et al.: Profile of blood pressure and glycemic responses after interval exercise in older women attending (in) a public health physical activity program. J. Bodyw. Mov. Ther. **25**, 119–125 (2021)
8. Lee, J., Vasan, R.S., Xanthakis, V.: Association of blood pressure responses to submaximal exercise in midlife with the incidence of cardiovascular outcomes and all-cause mortality: the Framingham heart study. J. Am. Heart Assoc. **9**(11), e015554 (2020)
9. Schimpchen, J., Correia, P.F., Meyer, T.: Minimally invasive ways to monitor changes in cardiocirculatory fitness in running-based sports: a systematic review. Int. J. Sports Med. **44**(2), 95–107 (2023)
10. Wu, T., Blazek, V., Schmitt, H.J.: Photoplethysmography imaging: a new noninvasive and noncontact method for mapping of the dermal perfusion changes. In: European Conference on Biomedical Optics (2000)
11. Wieringa, F.P., Mastik, F., van der Steen, A.F.: Contactless multiple wavelength photoplethysmographic imaging: a first step toward "SpO2 camera" technology. Ann. Biomed. Eng. **33**(8), 1034–1041 (2005)
12. Verkruysse, W., Svaasand, L.O., Nelson, J.S.: Remote plethysmographic imaging using ambient light. Opt. Express **16**(26), 21434–21445 (2008)
13. Wei, J., Zou, J., Li, J., Li, Z., Yang, X.: Non-contact heart rate detection based on fusion method of visible images and infrared images, pp. 62–75. Springer, Heidelberg (2022). https://doi.org/10.1007/978-3-031-06788-4_6
14. Sola, J., Proenca, M., Ferrario, D., Porchet, J.A., Falhi, A., et al.: Noninvasive and nonocclusive blood pressure estimation via a chest sensor. IEEE Trans. Biomed. Eng. **60**(12), 3505–3513 (2013)
15. Xia, J.S.: Design of blood pressure detection system based on pulse wave. M.S. Dissertation, Xidian University (2017)
16. Zou, J., Zhou, S., Ge, B., Yang, X.: Non-contact blood pressure measurement based on IPPG. J. New Media **3**(2), 41–51 (2021)
17. Zhang, F., et al.: MediaPipe hands: on-device real-time hand tracking. arXiv, abs/2006.10214 (2020)
18. Reis, D., Ferreira, M.T., et al.: Are age-predicted equations valid in predicting maximum heart rate in individuals after stroke? Disabil. Rehabil., 1–7 (2023)
19. Han, S.H., et al.: Is age-predicted maximal heart rate applicable in patients with heart or lung disease? Ann. Rehabil. Med. **46**(3), 133–141 (2022)
20. Gellish, R.L., Goslin, B.R., Olson, R.E., McDonald, A., Russi, G.D., Moudgil, V.K.: Longitudinal modeling of the relationship between age and maximal heart rate. Med. Sci. Sports Exerc. **39**(5), 822–829 (2007)

Basketball Shooting and Goaling Detection Based on DWC-YOLOv8

Zibo Wen, Jiancheng Zou$^{(\boxtimes)}$, and Litao Guang

College of Science, North China University of Technology, Beijing 100041, China
Zjc@ncut.edu.cn

Abstract. Smart sports provide intelligent and digital sports services and experiences through the application of artificial intelligence technology, promoting the development and progress of the sports field. This article introduces a basketball shooting and goaling detection algorithm based on DWC-YOLOv8, aiming to promote the intelligent development of campus basketball games. Firstly, we use the self-made dataset to train the YOLOv8 object detection model based on depth-wise separable convolution with the of 98.9% and FPS of 52. The training results show that the model has high detection accuracy for basketball and basket, and meet the requirements of real-time detection. Then, a judgment method for detecting shots and goals is proposed. Four sets of basketball game shooting clips are selected for experiments and the recognition rate for shots and goals is 95.92% and 95.83%, respectively. The experimental results show that the algorithm effectively realizes the detection of basketball shots and goals. This study is of great significance for promoting the intelligent and digital development of campus basketball games.

Keywords: Basketball shooting and goaling · Object detection · Deep learning · Depth-wise separable convolution

1 Introduction

In basketball games, scoring goals is the only way to score. An automated goal recognition system can provide real-time statistical data, accurately record players' shooting and goaling times as well as their hit rate and other information, thereby evaluating players' performance, tactical effectiveness and the overall strength of the team. This provides valuable technical statistics for teams and coaches to guide training and tactical adjustments. Therefore, basketball goal recognition has important significance and application value in basketball matches, training and sports competitions.

However, in basketball games, basketball usually appears in a high-speed motion state, which may cause the basketball in the image to appear blurry or ghostly. In addition, factors such as insufficient lighting, background interference and occlusion can also increase the difficulties of detecting basketball targets, thereby affecting basketball goal recognition. Therefore, object detection algorithms need to have strong feature extraction capabilities to accurately detect basketball and baskets, and further identify shooting and goal events.

H. Jin et al. (Eds.): IAIC 2023, CCIS 2060, pp. 98–107, 2024.
https://doi.org/10.1007/978-981-97-1332-5_8

There have been some related studies on the recognition of basketball shooting and goaling, and the traditional method is to use sensor devices to detect whether the basketball passes through the basket or net. Lu et al. designed an automatic testing system for basketball shooting based on a single-chip microcomputer [1]. The sensor is fixed on the net, and when the ball enters the basket, it collides with the net. The sensor detects the presence of the basketball and sends a signal to the system. The system judges whether to score based on changes in the sensor signal. This method is not susceptible to external interference, such as light conditions or other visual obstacles, and therefore has high accuracy. However, installing sensors in baskets or nets requires high installation and maintenance costs, and disassembly is also relatively difficult, which limits the popularity of this method in practical applications.

With the popularization of computer vision technology, researchers have begun to use object detection algorithms to detect basketballs and baskets in shooting videos, and apply relevant algorithms to determine whether a goal has been scored. Liu et al. proposed an improved image intelligent detection method based on the Internet of Things, combining background differencing and inter-frame differencing to detect the basketball in videos [2]. Finally, an improved Cam-shift target tracking algorithm was proposed to detect and recognize basketball goaling, with a goal recognition rate of 92.52%. Lan proposed a combination of background differencing and three-frame differencing method, as well as Hough transform, to detect the basketball and basket in the video [3]. The basketball feature parameter data was collected by a camera installed directly above the basketball stands, and compared with the system configuration parameters to determine whether to score. The goal recognition rate reached 93.33%, and the missed detection rate was 6.67%. Both background differencing and inter-frame differencing use pixel level differences to detect targets [4, 5]. Their advantages lie in their simplicity and high computational efficiency. However, their feature extraction ability for targets is limited, and they cannot provide a deeper understanding and description of the targets. For target detection tasks in complex scenes, there is a high misjudgment rate and missed judgment rate.

Compared to traditional object detection methods, object detection algorithms based on deep learning train on large-scale datasets by constructing deep neural networks, automatically learning and extracting advanced semantic features of images, and have stronger representation capabilities for targets. They effectively overcome the limitations of traditional methods and demonstrate significant advantages in dealing with small targets and detection tasks in complex scenarios. Therefore, in this article, we adopt an object detection model based on deep learning to more accurately identify the basketball and basket in game videos.

At present, the mainstream target detection models include the two-stage RCNN series and the single stage SSD series, YOLO series [6–12]. The YOLO series achieve a good balance between recognition accuracy and detection speed, so this study uses the latest YOLOv8 model from the YOLO series as our object detector [13]. Firstly, the Bottleneck module in YOLOv8 has been modified by replacing the conventional convolution operation with depth-wise separable convolution, making the model more lightweight and improving inference speed to meet the needs of real-time detection [16]. In addition, the DWC-YOLOv8 model is trained using the self-made dataset to better adapt to practical scenarios of campus basketball matches. Then, use the trained object detector to detect the basketball and basket in the game video, and obtain their position and size information. Finally, a method for identifying basketball shots and goals is proposed, which analyzes this information to determine whether shooting and goaling behavior has occurred.

2 DWC-YOLOv8 Network Model

2.1 YOLOv8 Network Structure

Based on the different depth and width of the network, YOLOv8 can be divided into five levels: n, s, l, m and x. This article uses YOLOv8n as the benchmark model, and its network structure is shown in Fig. 1, mainly composed of Backbone, Neck, and Head. Backbone is based on CSPDarknet53 and consists of a series of convolutional layers and residual blocks, used to extract features of different scales from images. Neck adopts a PAN-FPN structure, utilizing features extracted by Backbone to perform cross layer connections between up-sampling and down-sampling, fusing feature maps of different scales, which can help the model detect targets of different scales. In addition, Backbone and Neck both use a C2f module, which has more residual connections compared to the C3 module in YOLOv5, achieving richer gradient flow to accelerate model convergence. Head has three sets of detection heads, and use Decoupled Head to separate the regression branch and prediction branch.

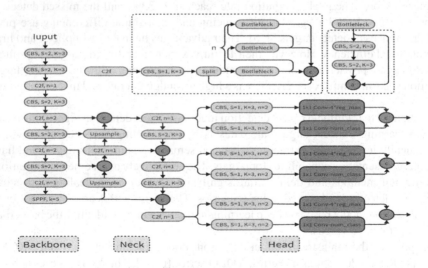

Fig. 1. YOLOv8n model structure diagram

2.2 Depth-Wise Separable Convolution

Convolution is an important component in CNN. By stacking multiple convolutional layers, different levels of feature maps are extracted from input data. As the network deepens, higher-level semantic features are gradually extracted, thereby improving the model's expression ability and performance. However, due to the increase in network complexity, the parameters and computational complexity of the model have also significantly increased, facing limitations in memory overhead and computational resources.

In order to reduce the parameters and computational complexity of convolution operations, Alex proposes a group convolution method, which divides the number of input data channels into g groups, performs routine convolution on each group, and finally merges the output channels of all groups to form the final output feature map [14, 15]. Due to the reduction in the number of channels for each group of input data after grouping, the number of channels for each convolutional kernel is also reduced, reducing the parameters to 1/g of that of conventional convolution, but isolating information exchange between channels in different groups. Depth-wise separable convolution divides convolution operations into deep convolution and point by point convolution [17]. Deep convolution is an extreme group convolution that performs convolution operations on each channel of input data, with the same number of output channels as the input channels. To enhance information exchange between channels, using a 1x1 convolutional kernel (point by point convolution, typically found in lightweight networks such as Mobile-Net) performs convolution operations on the output feature maps of deep convolution. By combining deep convolution and point by point convolution, the parameters and computation can be effectively reduced while maintaining good feature extraction capabilities. From Fig. 1, we can see that the Bottleneck module in C2f contains two convolutional layers. All the convolution operations of Bottleneck are replaced with depth-wise separable convolutions, and comparative experiments are conducted to observe the performance changes before and after the model changes.

3   Model Training and Detection Algorithm

3.1   Acquisition and Production of the Dataset

The quality of the dataset has a significant impact on the design and training results of object detection algorithms. This experiment uses a self-made dataset. We collect several videos of campus basketball matches from the internet and use Python's OpenCV library to extract frames from the videos. Due to the high similarity between adjacent video frames, in order to reduce the similarity of the dataset and avoid adverse effects on model training, a frame is extracted every 3 frames to convert the video dataset into an image dataset. All images have a resolution of 1920x1080 and are saved in JPG format. Use the annotation tool to annotate the image, including the position and size of the basketball and hoop. During the annotation process, we manually filter and remove blurry images to ensure the quality of the dataset, and ultimately produce a total of 6058 images and 11704 labels. Figure 2 shows an example of partially annotated data.

Fig. 2. Basketball and hoop examples of the dataset

3.2 Training Platform and Parameter Settings

We use the Windows 10 operating system and programming software PyCharm to build an experimental platform, with an Intel (R) Core (TM) i7-7800X @ 3.50GHz 3.50 GHz CPU and a graphics card configured with two Nvidia 2080 @ 32GB GPUs. We also use CUDA 11.8 and CUDNN to accelerate parallel computing of GPUs.

Divide the dataset into training, validation and testing sets in a 7:2:1 ratio. Set the input image size of the model to 640 x 640, the training epochs to 300, and the batch size to 16.

3.3 Evaluation Indicators

In order to evaluate the training performance of the model on the dataset, the parameters, precision, recall, mAP@0.5 and the FPS are used as evaluation indicators for the model.

3.4 Results and Analysis for Training

We trained the YOLOv8n model and DWC-YOLOv8n model on the dataset, and the training results are shown in Table 1.

Table 1. Comparison of performance of YOLOV8n before and after improvement

Algorithm	Parameter / (10^6) Precision	Recall map@0.5 FPS/(frame $\cdot$ s^{-1})
YOLOv8n	3.00 99.10%	97.5% 98.4% 40
DWC-YOLOv8n	2.07 98.20%	98.3% 98.9% 52

From Table 1, it can be seen that compared to the original model, although the Precision of the DWC-YOLOv8n model decreases by 0.9%, the Recall increases by

0.8%, and the map@0.5 improves by 0.5%, reaching 98.9%, with a slight improvement in accuracy, and a reduction of approximately 31% in the parameters, effectively improving the detection speed of the model.

3.5 Detection Algorithm for Shooting

We remember that the bounding box of basketball is A, where the center point is a, and the bounding box of the basket is B, where the center point is b, and the distance between a and b is D. The minimum outer rectangle of A and B is C, and the diagonal length of C is L. If the following two conditions are met:

- The lower boundary of B is located above the upper boundary of A.
- The ratio of D to L is less than the preset threshold.

This indicates that the basketball has reached the vicinity of the basket and can be considered a shot. After multiple experiments, when the threshold is set to 0.4, the accuracy of identifying shots is the highest.

3.6 Detection Algorithm for Goaling

If shooting occurs, mark it as f1 starting from the current frame. When the upper boundary of A is smaller than the lower boundary of B, it is marked as the end of the shot, ending at the current frame and marked as f2.

Between f1 and f2, A and B will have overlapping areas. We record the area of A as X, and the overlapping area of A and B as Y. The ratio of Y to X represents the degree to which the basketball enters the basket. The higher the ratio, the greater the likelihood of scoring a goal. In addition, the trajectory of basketball can be approximated as a parabola, and when the basketball touches the basket, rebound, or net, its speed and direction will change. Specifically, if a goal is scored, there will be friction with the net when the basketball enters the basket, resulting in a significant change in its trajectory. Therefore, the abscissa of a is x1, and the abscissa of b is x2. The smaller the offset of x1 relative to x2, the greater the likelihood of a basketball goal. Based on the analysis of two situations, if the following two points are met simultaneously:

- The ratio of Y to X exceeds the preset threshold t1.
- The ratio of the offset of x1 relative to x2 to the width of B is lower than the preset threshold t2.

It can be determined as a goal. Through multiple experimental observations, when we set t1 to 0.6 and t2 to 0.1, the algorithm has the highest goaling recognition rate.

4 Result and Analysis

To test the effectiveness of this scheme in identifying shots and goals, we intercept shot clips with a duration of approximately 1 min and 50 s from four basketball game videos and conduct four sets of experiments, as shown in Fig. 3. Then, the number of shots and goals scored in each game (shooting clips) is counted and compared with the number of times recognized by the algorithm. The statistical results are summarized in Tables 2 and 3. "Actual Numbers" refers to the actual number of shots and goals taken, "Detect Numbers" refers to the number of shots and goals recognized by the algorithm, and "True Detect Numbers" refers to the number of correctly recognized goals.

Fig. 3. Four groups of basketball games

Table 2. Statistics of Shooting Times

Groups	Actual Numbers	Detect Numbers	Precision
G1	59	57	96.61%
G2	67	64	95.52%
G3	59	56	94.92%
G4	59	57	96.61%

According to the results in Table 2, it can be seen that the shooting recognition rates of the four experiments are 96.61%, 95.52%, 94.92% and 96.61%, respectively, with an average accuracy of 95.92%. From Table 3, it can be seen that the goaling recognition rates of the four experiments are 95.65%, 100.00%, 91.67% and 96.00%, respectively. The goal recognition rate of the third group is slightly lower, with an overall average accuracy of 95.83%. The results indicate that the shooting and goaling detection algorithm proposed in this study effectively recognizes shooting and goal behavior.

Table 3. Statistics of Goaling Times

Groups	Actual Numbers	Detect Numbers	True Detect Numbers	Precision
G1	25	23	22	95.65%
G2	22	21	21	100.00%
G3	26	24	22	91.67%
G4	26	25	24	96.00%

Compared to the method of using signals sent by sensors to computers in reference [1] to detect shots and goals, the method based on video analysis adopted in this study not only accurately identifies shots and goal events, but also saves additional equipment installation and maintenance costs. The research in references [2] and [3] mainly focuses on the area near the basket, using background differencing and three-frame differencing method to distinguish the basketball in this area from the background, and detecting the basketball in motion through the difference between frames, which is suitable for goal recognition tasks in simple scenes. Compared to this, the YOLOv8 based object detection model used in this article can more accurately identify basketball and baskets in complex scenes and the designed shooting and goaling detection algorithm can adapt to the needs of basketball shooting and goal detection in different scenarios.

5   Discussion

In the goaling detection experiments, only the second group achieves a goal recognition rate of 100%, while the other groups have false positives, as shown in Fig. 4. These false positives are due to basketball hitting the backboard or basket, which may pass through the front or back of the net and be misjudged as a goal. After analysis, it is found that the DWC-YOLOv8 model used in this study is a target detection algorithm based on two-dimensional planar images, which is difficult to obtain depth information and spatial relationships between objects. Therefore, it is difficult to process targets with pre and post relationships.

In basketball games, the goaling detection precision of 100% is required. In order to achieve this goal, in future research work, we can consider using binocular cameras to obtain image information from different perspectives. By analyzing the images from these two perspectives and using disparity information to infer the depth relationship between objects in the scene, combined with the basketball goal detection algorithm proposed in this article, it can more accurately determine whether the basketball has passed through the net, further improving the accuracy of basketball goal detection.

Fig. 4. False Positive

6 Conclusion

This study achieves high recognition accuracy and detection rate using the DWC-YOLOv8 model on the self-made dataset with the mAP@0.5 of 98.9% and FPS of 52. Compared with the original model, the mAP@0.5 improves by 0.5% and the parameters reduce by approximately 31%. On the basis of detecting basketball and the basket, the basketball shooting and goaling detection algorithm proposed in this study achieves high recognition rates in experiments with the shooting recognition rate of 95.92% and the goaling recognition rate of 95.83%, which provides experimental basis for intelligent analysis of campus basketball matches and strong support for teachers' basketball teaching and guiding students' training.

Although this study has achieved certain results, there are still some shortcomings. For example, the shooting detection algorithm proposed in this paper only determines whether shooting behavior has occurred when the basketball reaches the vicinity of the basket, while in reality, shooting behavior already occurs when the player throws the ball out of the hand. In subsequent research, specific body postures and movements taken by basketball players during shooting can be utilized to predict their shooting behavior using posture estimation models based on deep learning, thereby shortening the algorithm's latency.

Acknowledgments. We sincerely appreciate the support provided by the Organized Research - Collaborative Urban Slow Travel Environment Planning and Design System Based on the Concept of Smart Accessibility for People with Disabilities and the Elderly in the Capital City project.

References

1. Lu, H., Zhao, N., Huan, Y.: Design and realization of the automatic test system of basketball shooting in sports college entrance examination. Foreign Electron. meas. Technol., **26**(4), 34–36+43 (2007)
2. Liu, N., Liu, P.: Goaling recognition based on intelligent analysis of real-time basketball image of Internet of things. J. Supercomput. **78**(1), 123–143 (2022)
3. Lan, K.: Basketball goal recognition based on image analysis. J. Huai Hua Univ. **40**(5), 83–87 (2021)
4. Chien, S.Y., Ma, S.Y., Chen, L.G.: Efficient moving object segmentation algorithm using background registration technique. IEEE Trans. Circuits Syst. Video Technol. **12**(7), 577–586 (2002)
5. Zhang, Y., Wang, X., Qu, B.: Three-frame difference algorithm research based on mathematical morphology. Procedia Eng. **29**, 2705–2709 (2012)
6. G, R., Donahue, J., Darrell, T., et al: Rich feature hierarchies for accurate object detection and semantic segmentation. In: Proceedings of the IEEE Conference on Computer Vision and Pattern Recognition, 580–587 (2014)
7. G R: Fast r-CNN. Proceedings of the IEEE international Conference on Computer Vision, 1440–1448 (2015)
8. Ren S, He K, G R, et al: Faster r-CNN: Towards real-time object detection with region proposal networks. Advances in neural information processing systems 28, (2015)
9. Liu, W., Anguelov, D., Erhan, D., Szegedy, C., Reed, S., Fu, C.-Y., Berg, A.C.: SSD: Single Shot MultiBox Detector. In: Leibe, B., Matas, J., Sebe, N., Welling, M. (eds.) ECCV 2016. LNCS, vol. 9905, pp. 21–37. Springer, Cham (2016). https://doi.org/10.1007/978-3-319-464 48-0_2
10. Fu, C.Y, Liu, W., Ranga, A., et al.: DSSD: Deconvolutional single shot detector. arXiv preprint arXiv (2017)
11. Li Z., Zhou F.: FSSD: feature fusion single shot multi-box detector. arXiv preprint arXiv, (2017)
12. Redmon, J., D, S., G, R., et al: You only look once: Unified, real-time object detection. In: Proceedings of the IEEE Conference on Computer Vision and Pattern recognition, 779–788 (2016)
13. ULTRALYTICS Homepage. https://github.com/ultralytics/ultralytics
14. K, A., S, I., Hinton, G.E.: Image-net classification with deep convolutional neural networks. Adv. neural Inf. Proces. Syst., **25** (2012)
15. Su, Z., Fang, L., Kang, W., Hu, D., Pietikäinen, M., Liu, Li.: Dynamic Group Convolution for Accelerating Convolutional Neural Networks. In: Vedaldi, A., Bischof, H., Brox, T, Frahm, J.-M. (eds.) ECCV 2020. LNCS, vol. 12351, pp. 138–155. Springer, Cham (2020). https://doi.org/10.1007/978-3-030-58539-6_9
16. Chollet, F.: XCEPTION: deep learning with depth-wise separable convolutions. In: Proceedings of the IEEE Conference on Computer Vision and Pattern Recognition, 1251–1258 (2017)
17. Hua, B.S, Tran, M.K, Yeung, S.K.: Pointwise convolutional neural networks. In: Proceedings of the IEEE Conference on Computer Vision and Pattern Recognition, 984–993 (2018)

Design of High-Power Power Amplifiers for BDS-3 Terminal Based on InGaP/GaAs HBT MMIC and LGA Technology

Zhenbing Li[1], Shilin Jia[1], Jinrong Zhang[1], Jialong Fu[1], Junjie Huang[1],
Xiaochuan Fang[2], Gang Li[3], and Guangjun Wen[1(✉)]

[1] School of Information and Communication Engineering, University of Electronic Science and Technology of China, Chengdu, Sichuan, People's Republic of China
wgj@uestc.ed
[2] Antennas and Electromagnetics Research Group, School of Electronic Engineering and Computer Science, Queen Mary University of London, London E1 4NS, UK
[3] School of Information and Software Engineering, University of Electronic Science and Technology of China, Chengdu, Sichuan, People's Republic of China

Abstract. With the development and popularization of Beidou-3 navigation satellite system (BDS-3), to ensure its unique short message function it is necessary to integrate a radio frequency (RF) power amplifier (PA) with high performance. Therefore, in this paper, an L-band high-power 10 W PA chip is designed in InGaP/GaAs heterojunction bipolar transistor (HBT) technology combined with temperature-insensitive adaptive bias technology, class-F harmonic suppression technology, and LGA packaging technology. The linear gain of the PA chip reaches 37.1 dB and P_{1dB} reaches 40.6 dBm. The results show that the PA chip has high power, high gain, and high linearity, which has obvious advantages over similar PA chip designs and can meet the short message functions of the BDS-3 terminals in various application scenarios.

Keywords: BDS-3 Terminal · High-power · Class-F Power Amplifier · temperature-insensitive adaptive bias · short message function

1 Introduction

The most notable difference between China's Beidou Satellite Navigation System (BDS) [4] and other three satellite navigation systems in the world [1–3] is its unique short message communication function [5], which means that the end users of the BDS can send short messages through satellites, giving ordinary communication terminals the possibility of satellite communication. Therefore, BDS plays an important role in disaster relief, battlefield positioning, field rescue, and other fields. In BDS terminals, the PA chip is the core component responsible for implementing short message functionality, and its performance directly affects the communication quality of this feature.

BDS terminals often operate in the L-band, and due to the short message function requirements, the PA chips in BDS terminals are required to reach at least 3 W, and

© The Author(s), under exclusive license to Springer Nature Singapore Pte Ltd. 2024
H. Jin et al. (Eds.): IAIC 2023, CCIS 2060, pp. 108–118, 2024.
https://doi.org/10.1007/978-981-97-1332-5_9

5 W in some special application scenarios (such as mountainous areas, canyons, and other special areas) [7]. If the terminal is integrated into a high-quality satellite phone, the PA chip output power is required to reach 10 W. On the other hand, while the PA chip is miniaturized and fully functional, its other performance parameters should also be at a high level in the industry. At present, there are some L-band PA chip products in the industry [8–11], but their output power does not reach 10 W, and some are not even suitable for BDS-3. Meanwhile, although many universities have carried out corresponding research works on L-band PA chips [12–14], they also have the problem that output power is low, making it difficult for high-quality communication of BDS satellite phones.

Therefore, this paper introduces a 10 W high-power PA chip design operating at 1.6–1.65 GHz based on InGaP/GaAs HBT technology for BDS-3 terminals, by adopting temperature-insensitive adaptive bias technology, class-F harmonic suppression technology, and LGA packaging technology with double-link power synthesis circuit architecture. The simulation results show that the output power and linearity of the designed PA chip are at the highest level in the industry (P_{1dB} reaches 10 W), which can meet the short message function of BDS-3 terminals and the requirements of different application scenarios.

2 InGaP/GaAs HBT Technology

The first generation of semiconductor materials, represented by silicon materials, replaced bulky electronic tubes, leading to the development of the microelectronics industry centered on integrated circuits and a leap in the entire IT industry. However, due to the low breakdown voltage, the transistor cannot withstand high power output [15]. The second-generation semiconductor materials are compound semiconductors, including various III-V group compounds semiconductors, such as gallium arsenide, indium phosphide, gallium arsenide phosphate (GaAsP), gallium arsenide aluminum oxide (GaAlAs), and indium gallium phosphate (InGaP). The second-generation semiconductor materials have higher electron mobility and breakdown voltage, which can be used in high-frequency and high-power application environments. Therefore, they are widely used in fields such as satellite communication, mobile communication, optical communication, GPS navigation, etc. [16]. With the continuous development of semiconductor technology, the third generation of broadband bandgap semiconductor materials [17] and the fourth generation of ultra-wide bandgap semiconductor materials have emerged [18]. Considering both cost and performance, this article ultimately chose InGaP/GaAs HBT technology to design PA chips for BDS terminals. The research results indicate that when GaAs is doped with InGaP, the lattice matching is good, the bandgap of the second-generation semiconductor material is improved, and the current gain increases due to the small bandgap. On the other hand, high-concentration doped emitter materials passivate the substrate of HBT, reducing the composite current effect. Compared with pHEMT, HBT is more suitable for high-yield and low-cost L-band high-power applications and is suitable for high-frequency satellite communication. The schematic diagram of the InGaP/GaAs HBT stack structure used in this article is shown in Fig. 1.

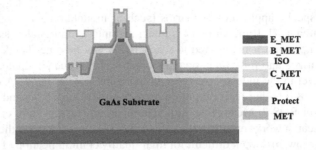

▮	E_MET
▮	B_MET
▮	ISO
▮	C_MET
▮	VIA
▮	Protect
▮	MET

Fig. 1. Stacked structure of InGaP/GaAs HBT

Figure 2 shows the Volt-Ampere characteristic curve of HBT (emitter area equals the emitter finger width × finger length × finger numbers: $3 \times 40 \times 3$ um^2), and the V_{Knee} of the HBT is 0.5 V.

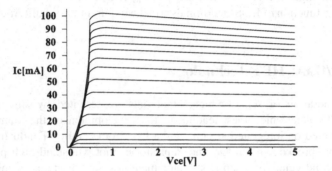

Fig. 2. V-I curve of HBT (emitter area $3 \times 40 \times 3$ um^2)

3 W BDS-3 Terminal PA Chip Design

3.1 W PA Chip Circuit Design

In this paper, because the required functions need PA to provide an output power of 10 W in BDS terminals, the overall circuit performance of the PA chip is to achieve 10 W output powers under the supply voltage of 5 V while keeping high efficiency, linearity, and small size unified requirements. According to the design goals, based on the RF characteristics of the HBT device, the 10 W PA uses a double-link power synthesis circuit architecture as shown in Fig. 3, which is composed of two MMIC Dies and an LGA PCB. To realize the power synthesis of the signals with identical phase and amplitude at the output, and minimize the synthesis loss, the power amplifiers on the upper and lower links are completely symmetrical, which avoids the use of 3dB Couplers and Wilkinson power synthesizers. The power is spitted evenly at the input by an LC-matching network and thus enlarged by the amplifiers in each link. Meanwhile, the outputs of the upper and lower links are matched from the optimal impedance to 100 Ω. Therefore, at the synthesis node, upper and lower links are combined to achieve the matching to 50 Ω. The design is to synthesize two 5 W PA channels, whose structure is shown in Fig. 4, using a three-level cascade amplification circuit structure. In each level of the amplification circuit, HBT is connected in series with resistors and capacitors to form a unit for improving the stability of PA. The output stage adopts a structure of 36 HBT units in parallel, with P_{1dB} greater than 37dBm and gain greater than 12dB. The driver stage (intermediate stage) adopts a parallel structure of 8 HBT units to provide linear output power greater than 26dBm, with a gain greater than 13dB; The input stage adopts 2 HBT units in parallel, providing linear output power greater than 14dBm and gain greater than 13dB. After the final synthesis of two 5 W PA circuits, the P_{1dB} of PA should reach 40dBm.

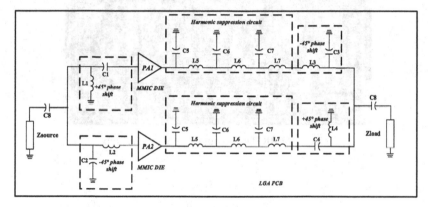

Fig. 3. Double-link power synthesis circuit architecture

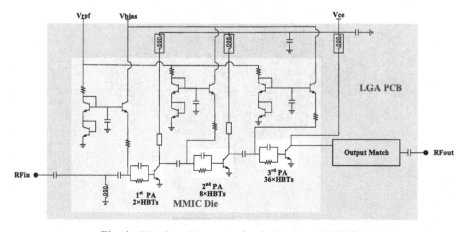

Fig. 4. Circuit architecture of a single channel 5 W PA

One of the MMIC Dies of the 10 W BDS terminal PA chip proposed in this paper is shown in Fig. 5, including a three-level amplification circuit, temperature insensitive adaptive bias circuits at all levels, partial matching circuits (excluding the output matching circuit), and an ESD electrostatic protection circuit. The designed LGA PCB is shown in Fig. 6, including bypass capacitor, RF choke, partial matching network circuit between each stage, power synthesizer, power distributor, and Class-F output matching network circuit with harmonic suppression function.

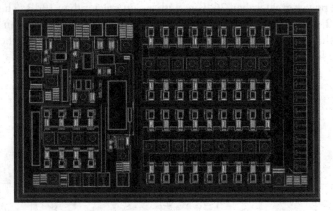

Fig. 5. MMIC Die layout of the 10 W PA chip

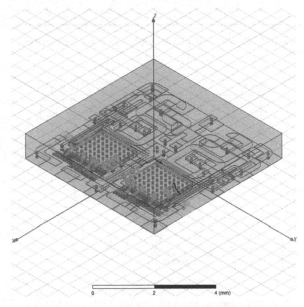

Fig. 6. LGA PCB layout of the 10 W PA chip

3.2 Temperature-Insensitive Adaptive Bias Circuit Design

The quiescent point of a power amplifier will shift with the change of input power and temperature, leading to phase distortion of the output signal and premature gain compression, seriously affecting the linearity of power amplifier. To avoid this issue, a temperature-insensitive adaptive bias circuit was designed to provide bias current for each stage of amplification circuit, as shown in Fig. 7. The transistor HBT0 represents the amplification transistors in each stage, and the temperature-insensitive adaptive bias circuit is connected to the base of the HBT in each stage through a ballast resistor R1. The voltage relationship between each node is as follows:

$$V_{be\_HBT0} = V_B - R_1 I_b - V_{be\_HBT1} \tag{1}$$

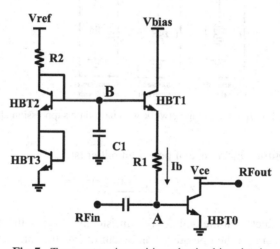

Fig. 7. Temperature-insensitive adaptive bias circuit

Due to the rectification characteristics of the base-emitter diode, the base-emitter voltage V_{be_HBT0} changes according to the input power or temperature. By adopting a temperature-insensitive adaptive bias circuit, RF signal will leak into the bias circuit through R1. HBT1 also has rectification characteristics, therefore V_{be_HBT1} also varies with input power or temperature. From Eq. (1), the changes in V_{be_HBT0} and V_{be_HBT1} cancel out with each other, thus stabilizing the bias point. Furthermore, in bias circuit, the leaked RF signal is short-circuited to ground through bypass capacitor C1 to stabilize the potential at pointB, which ensures the stable compensation of V_{be_HBT1} to V_{be_HBT0}. Meanwhile, the ballast resistor R1 can effectively limit the current from bias circuit to HBT0 base, thereby further improving the temperature-insensitivity and linearity of power amplifier chip.

3.3 Design of Class-F Output Matching Network with Harmonic Suppression Characteristics

PA generates a large amount of harmonic components during operation, with a high proportion of second, third, and fifth harmonics. Therefore, this article proposes a Class-F output matching circuit with harmonic suppression function, which achieves not only fundamental matching, but also second harmonic short circuit, third and fifth harmonic open circuit to improve the energy conversion efficiency of power amplifier chips.

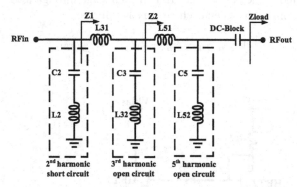

Fig. 8. Class-F output matching network with harmonic suppression characteristics

The optimal output impedance of power amplifier is:

$$Z_1 = \frac{(V_{ce} - V_{Knee})^2}{2P_{out}} \tag{2}$$

where V_{ce} is the collector-emitter voltage of transistor and V_{Knee} is the knee point voltage of transistor. To increase the operating bandwidth of PA, the intermediate impedance should satisfy the following relationship:

$$Z_2 = \sqrt{Z_{Load}Z_1} \tag{3}$$

where Z_{Load} is 50Ω.

The impedance conversion ratio is defined as:

$$m_1 = \frac{Z_{Load}}{Z_2} \tag{4}$$

$$m_2 = \frac{Z_2}{Z_1} \tag{5}$$

The load impedance Z_{Load} is converted to Z_1, through a two-stage LC matching network. To ensure the fundamental impedance conversion while ensuring the open circuit of third and fifth harmonics, the capacitance and inductance of third and fifth harmonics meet the following relationship:

$$L_{51} = \frac{Z_2\sqrt{m_1 - 1}}{\omega} \tag{6}$$

$$L_{31} = \frac{Z_1\sqrt{m_2 - 1}}{\omega} \tag{7}$$

$$C_5 = \frac{24\sqrt{m_1 - 1}}{25Z_{Load}\omega} \tag{8}$$

$$C_3 = \frac{8\sqrt{m_2 - 1}}{9Z_2\omega} \tag{9}$$

$$L_{52} = \frac{Z_{Load}}{24\omega\sqrt{m_1 - 1}} \tag{10}$$

$$L_{32} = \frac{Z_2}{8\omega\sqrt{m_1 - 1}} \tag{11}$$

In the second harmonic tuning branch, the inductance of series resonant network L_2 can be calculated based on the value of C_2:

$$L_2 = \frac{1}{4\omega^2 C_2} \tag{12}$$

The values of each component can be calculated according to Eqs. (2) - (12), and the established circuit is shown in Fig. 8. The simulated S_{11} for this circuit is displayed in the Smith chart, as shown in Fig. 9. It is clear that the fundamental wave is perfectly matched, the second harmonic is located at the short-circuit point, and the impedance of the third and fifth harmonics is relatively high compared to the fundamental wave. Then, the ideal devices are replaced by the SPICE models of Murata and they are rearranged according to the actual layout dimensions before conducting simulation again. The simulation results show that the Class-F output matching network with harmonic suppression characteristics can improve the PAE of the power amplifier chip by about 2%.

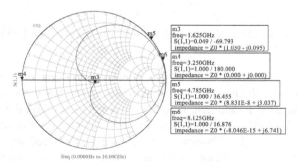

freq (0.0000Hz to 10.00GHz)

Fig. 9. The impedance of fundamental wave, the second, third, and fifth harmonics of the Class-F output matching circuit

4 Simulation Results and Analysis

In this paper, a 10 W BDS-3 terminal power amplifier chip based on InGaP/GaAs HBT technology is designed for wireless communication terminals with limited battery capacity, such as satellites and cellular cell phones. The 10 W terminal power amplifier chip is designed based on ADS platform with an overall size of $8 \times 8 \times 1$ mm^3. The 10 W BDS-3 terminal power amplifier chip has a quiescent operating current of approximately 560 mA at 5 V supply voltage, which is at the leading level in the industry. Meanwhile, the small-signal, power, and large-signal characteristics of the amplifier chip in the frequency band of 1.6 ~ 1.65 GHz are simulated by ADS. The small-signal simulation results are shown in Fig. 10. The S_{11} and S_{22} are less than -15 dB and -20 dB, respectively, which indicates the chip is perfectly matched. Meanwhile, S_{21} at the higher harmonics of the chip are all less than -50 dB, indicating the designed Class-F matching network has excellent harmonic suppression. The simulation results are shown in Fig. 11, which illustrate that the linear gain of the power amplifier reaches 37.1 dB at 1.625 GHz (typical), the gain fluctuation is less than $\pm$ 0.35 dB, $P_{1\,dB}$ reaches 40.6 dBm, Psat reaches 41.7 dBm, and PAE reaches 37.6%; Meanwhile, the higher harmonic rejection ratios are less than -62 dBc, AM-PM distortion is less than $|2.5°|$, and the Third-order InterModulation Distortion (IMD3) is less than -27 dBc.

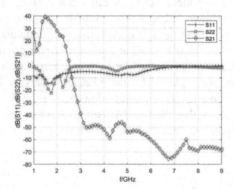

Fig. 10. Small-signal S parameters of the 10W BDS terminal PA chip

When the power supply voltages are 5 V, the simulation results show this 10 W PA achieves all expected RF characteristics requirements, and has the advantages of high gain and high linearity. In addition, temperature simulation of PA during operation is conducted using ANSYS Workbench. Under natural air convection conditions, the average temperature of the 10W BDS terminal PA chip under saturated operating conditions is about 159 °C, with a maximum temperature of about 198 °C, and the temperature is mainly concentrated at the bottom of the transistor. This temperature is within the safe working temperature range of HBT (according to the characteristics of HBT, the maximum safe temperature does not exceed 209 °C).

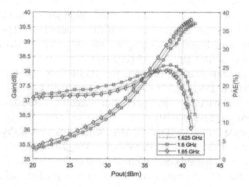

Fig. 11. Power characteristics of the 10W BDS terminal PA chip

5   Conclusions

Based on InGaP/GaAs HBT technology, a 10 W power amplifier chip operating within 1.6–1.65GHz band is proposed by adopting temperature insensitive adaptive bias technology, Class-F harmonic suppression technology, dual-link power synthesis circuit structure, and LGA packaging process. The simulation results demonstrate that when the power supply voltage is 5 V, the static working current of the power amplifier is 560 mA. Meanwhile, its linear gain can reach 37.1dB, P_{1dB} is greater than 40.6 dBm, the saturation power ($Psat$) is 41.7 dBm, and PAE is 37.6%. The higher harmonic suppression ratios are all less than –62 dBc. It is clear that the chip can be applied with high gain, large output power, and strong linearity, and its functions and performance are at the leading level of industry. The chip size is minimized to $8 \times 8 \times 1$ mm^3 and has good thermodynamic characteristics. Additionally, the chip also has an ESD electrostatic protection circuit, which can satisfy the short message function requirements of BDS-3 terminals in different application scenarios. Furthermore, other similar power amplifier chip designs can also take this 10 W BDS-3 terminal PA chip as an example, which gives reference to multiple properties such as high power, large gain, strong linearity, high integration, and thermal stability.

References

1. Seo, J., Walter, T.: Future dual-frequency GPS navigation system for intelligent air transportation under strong ionospheric scintillation. IEEE Trans. Intell. Transp. Syst. **15**(5), 2224–2236 (2014)
2. Bakuła, M., Przestrzelski, P., Kaźmierczak, R.: Reliable technology of centimeter GPS/GLONASS surveying in forest environments. IEEE Trans. Geosci. Remote Sens. **53**(2), 1029–1038 (2015)
3. Benevides, P., Nico, G., Catalão, J., Miranda, P.M.A.: Analysis of galileo and GPS integration for GNSS tomography. IEEE Trans. Geosci. Remote Sens. **55**(4), 1936–1943 (2017)
4. Liu, S., Zhu, S., Zhang, J.: Investigating performance of BDS-2+3 dual-frequency absolute positioning with broadcast ephemerides, RTS and final MGEX products. IEEE Access **11**, 22034–22050 (2023)

5. Guo, S., et al.: Integrated navigation and communication service for LEO satellites based on BDS-3 global short message communication. IEEE Access **11**, 6623–6631 (2023)
6. Liu, B., et al.: Smartphone-based positioning augmented by BDSBAS Ionospheric corrections. IEEE Geosci. Remote Sens. Lett. **20**, 1–5 (2023)
7. Performance requirements and test methods for BDS RDSS unit. http://m.beidou.gov.cn/zt/bdbz/201712/W020171226815455418203.pdf. Accessed 19 Oct 2015
8. RF3183-RF Micro Devices. https://pdf1.alldatasheetcn.com/datasheet-pdf/view/716775/RFMD/RF3183.html. Accessed 2010
9. SKY77354 PA Module for Quad-Band GSM/EDGE. https://www.skyworksinc.com//media/SkyWorks/Documents/Products/17011800/SKY77354_202533a.pdf. Accessed 2013
10. YP163137. http://www.innotion.com.cn/cpxq?product_id=164. Accessed 2013
11. LXK6618 1.6–1.65G PA. http://www.creotechco.com/product_second-6.html. Accessed 2013
12. Chen, S., Zheng, Y., Zhang, G.: Design of HBT power amplifier for Beidou satellite mobile communication. Guti Dianzixue Yanjiu Yu Jinzhan/Res. Prog. Solid State Electron. **35**, 334–339 (2015)
13. Zheng, Y., Chen, S., Zhang, G.: A high-power power amplifier for BeiDou satellite handsets. Micrielectronics **46**, 293–296 (2016)
14. Li, Z., Sun, H., Li, J., Huang, J., Huang, Y., Wen, G.: 5W High-power High-linearity L-band InGaP/GaAs HBT PA MMIC for RDSS Applications. In: 2021 International Conference on UK-China Emerging Technologies (UCET), pp. 185–189. IEEE, Chengdu, China(2021)
15. Kim, D., Krishnamohan, T., Smith, L., Wong, H.S.P., Saraswat, K.C.: Band to band tunneling study in high mobility materials: III-V, Si, Ge and strained SiGe. In: 2007 65th Annual Device Research Conference, pp. 57–58. IEEE, South Bend, IN, USA (2007)
16. Ramella, C., Camarchia, V., Piacibello, A., Pirola, M., Quaglia, R.: Watt-Level 21–25-GHz integrated Doherty power amplifier in GaAs technology. IEEE Microwave Wirel. Compon. Lett. **31**(5), 505–508 (2021)
17. Bao, M., Gustafsson, D., Hou, R., Ouarch, Z., Chang, C., Andersson, K.: A 24–28-GHz Doherty power amplifier with 4-W output power and 32% PAE at 6-dB OPBO in 150-nm GaN technology. IEEE Microwave Wirel. Compon. Lett. **31**(6), 752–755 (2021)
18. Ochs, T.R., et al.: Fourth-generation microstructured semiconductor neutron detector (MSND)-based 3He replacement (HeRep) for high pressure 3He detectors. In: 2016 IEEE Nuclear Science Symposium, Medical Imaging Conference and Room-Temperature Semiconductor Detector Workshop (NSS/MIC/RTSD), pp. 1–5. IEEE, Strasbourg, France (2016)

Design of High-Performance Broadband Power Amplifier Based on Normal Distribution Real Frequency Matching Method

Zhenbing Li[1], Xiangrui Wu[1], Junjie Huang[1], Shilin Jia[1], Jinrong Zhang[1], Xiaochuan Fang[2], Chu Chu[3], and Gang Li[4(✉)]

[1] Centre for RFIC and System, School of Information and Communication Engineering, University of Electronic Science and Technology of China, Chengdu, Sichuan, People's Republic of China
[2] Antennas and Electromagnetics Research Group, School of Electronic Engineering and Computer Science, Queen Mary University of London, London E1 4NS, UK
[3] College of Computer Science, Sichuan Normal University, Chengdu 610066, Sichuan, China
[4] School of Information and Software Engineering, University of Electronic Science and Technology of China, Chengdu, Sichuan, People's Republic of China
ligangpm@uestc.edu.cn

Abstract. In order to improve the working bandwidth of the power amplifier and make it suitable for a variety of application scenarios, a real frequency matching method based on normal distribution generating weights is proposed in this paper. Based on the working frequency of the power amplifier, a normal distribution function is constructed to generate frequency-related function values as the weights of the cost function of the real frequency method (that is, the weights of each frequency point are different), and sideband frequency points are mainly considered according to the actual frequency band. The matching circuit with weak normal distribution is obtained. The simulation results based on MATLAB algorithm show that the average insertion loss of the proposed real frequency matching method in the frequency band is reduced by 9.1%@1.7 GHz compared with the traditional real frequency matching method. The parameters obtained by MATLAB simulation are brought into ADS for circuit simulation. The simulation results show that when the designed power amplifier is less than 1.7 GHz, S21 is greater than 17 dB. It is proved that the proposed real frequency matching method can improve the bandwidth of power amplifier, and can provide theoretical and experimental reference for the design of high broadband power amplifier.

Keywords: Real frequency method · normal distribution weight · power amplifier · broadband

1 Introduction

With the rapid development of modern communication systems such as satellites and 5G, in order to achieve high-speed data transmission and meet the requirements of different application scenarios, the signal bandwidth requirements are increasingly high [1, 2]. As

H. Jin et al. (Eds.): IAIC 2023, CCIS 2060, pp. 119–132, 2024.
https://doi.org/10.1007/978-981-97-1332-5_10

one of the most important devices in the transmitter system of wireless communication system, the performance of power amplifier directly affects the transmission quality of wireless signal [3–7], and the matching circuit is a key part of the power amplifier. Its design performance directly determines the output power, efficiency, linearity, bandwidth and other technical indicators of the amplifier [8, 9].

Since the 1983 International Microwave Symposium (IEEE MTT-S International Microwave Symposium) discussed the topic of matching and designing power microwave amplifiers [10], The matching design technology of microwave power amplifier has been widely concerned by scholars [11–15]. The purpose of the matching network design is to convert the impedance of the load port to the optimal impedance required by each component, so that the impedance of the RF circuit source or load is converted to 50 Ω to achieve the maximum power transmission. However, there are uncertain factors in the circuit, such as component parasitic parameters and package parameters, which increase the difficulty of designing the matching circuit [16]. The existing matching methods are mainly aimed at specific circuits or specific topologies, and the matching circuits designed therefore have great limitations [17–21]. At the same time, compared with lumped parameter components, the equivalent capacitance and inductance of distributed parameter components used in microwave frequency band are more like nonlinear components that change with frequency. Therefore, classical circuit design and network synthesis components are no longer suitable for circuit matching design of microwave power amplifiers.

In view of this, in 1977, Carlin of Cornell University proposed a Real Frequency Technique [22], which firstly calculates the values of the real and imaginary parts of each discrete frequency point, and then obtains the impedance function with the minimum error through rational polynomial optimization approximation. This method is also called the Real Frequency Line Segment Technique (RFLST). Different from the initial Real Frequency technique, Yaman proposed Simplified real frequency Technique in 1982 [23], which uses rational polynomials to represent S parameters and can directly represent the matching network without Hilbert transformation. In order to design the matching circuit of microwave MMIC power amplifier, Perennec et al. further improved the simplified real frequency technique, so that it can simultaneously take into account gain, noise factor and return loss [24]. In 2017, Ramazan proposed a Fast Simple regenerative Real Frequency Technique technology [25]. By assigning a specific initial value to a variable and fluctuating in a small interval, the convergence rate is significantly faster than that of random assignment. However, it is necessary to assign a good initial value to each polynomial in advance, otherwise the optimization effect has great uncertainty. Literature [26] proposes a matching circuit design method based on finite zero real frequency technology. By introducing finite zero to better control the second harmonic impedance of the power amplifier, so as to improve the bandwidth and efficiency of the power amplifier, a broadband continuous class F power amplifier operating at 1.15–2.2 GHz is designed accordingly. The saturation output power is 40.5–43.2 dBm, the saturation drain efficiency is 70%–83%, and the saturation gain is 11.3–13.7 dB. Literature [27] proposes a distributed parameter matching network design method based on real frequency technology. The cost function used in this method can better describe the matching of power amplifier impedance. The Richards transform is also introduced to obtain the impedance function representing the distributed parameter matching network. The proposed method is verified by a broadband power amplifier of 0.9–2.8 GHz. The test results show that in the operating band, the output power is about 39.5 dBm, the additional efficiency is between 52.2%–85.1%, and the gain is in the range of 14.2–16.8 dB, but its efficiency is low.

To sum up, both the existing real frequency technology and the simplified real frequency technology consider the impedance matching problem in the bandwidth equally. Although they can achieve good matching effect, they can only design matching circuits that meet the performance of each frequency point compromise, and it is difficult to further improve the matching effect. Therefore, this paper proposes a real frequency matching method based on normal distribution, which adopts normal distribution to assign different weights to different frequency points in the frequency band. According to the nature of normal distribution, it can be seen that the center frequency in the bandwidth can have the highest weight by properly designing the mean value of the normal distribution, and further designing the appropriate standard deviation can obtain the weights of other frequencies. This method can make use of the normal distribution characteristic of most practical systems to design a better matching network, and further design the RF communication system with high broadband and high conversion efficiency. The main contributions of this paper are as follows:

1. This paper proposes to introduce weight parameters into the cost function of the real frequency method to reduce the insertion loss at the cutoff frequency and increase the bandwidth.
2. This paper proposes a general method for generating weight values based on normal distribution functions.
3. This paper proposes a calculation method for constructing a normal distribution function based on bandwidth and cutoff frequency.

2 Related Research

Existing broadband matching methods often use real frequency technology and simplified real frequency technology, which consider the performance of each frequency point in the bandwidth to optimize the matching network, that is, the existing real frequency technology gives the same weight to each frequency point in the bandwidth. The cost function (also known as the error function) of the existing real frequency technology is generally defined as [28]:

$$\Delta Z = 100\% \cdot \sum_{i=1}^{N} \left| \frac{Z_{in}(\omega_i) - Z_{opt}(\omega_i)}{\left| Z_{opt}(\omega_i) \right|} \right|^2 \tag{1}$$

where, $\omega_i(i = 1, 2...)$ is the discrete frequency points randomly selected in the load operating frequency band, N is the total number of discrete frequency points, $Z_{opt}(\omega_i)$ is the optimal impedance at frequency point ω_i (that is, the impedance value of the actual test), and $Z_{in}(\omega_i)$ is the impedance function of the port driver point of the load matching network. It can be seen from Eq. (1) that the expression of $Z_{in}(\omega_i)$ can be obtained by minimizing ΔZ through the optimization algorithm. Further, the matching network can be obtained through the network synthesis method, and then the matching circuit with broadband and high efficiency characteristics can be designed [29–39].

The cost function of the existing real frequency technology allocates the same weight to each frequency point within the bandwidth, without considering the boundary condition, and can only achieve the compromise performance of each frequency point, but cannot further improve the overall performance of the system. And only the matching circuit of general application scenarios can be designed, and the cost function cannot be adjusted according to the actual application requirements to meet various application requirements.

3   Real Frequency Matching Method Based on Normal Distribution

3.1   Cost Function Based on Normal Distribution

In view of the problems that the performance of frequency points in existing broadband matching methods is difficult to meet the performance of each frequency point and that the cost function cannot be adjusted according to the actual application requirements, this paper first proposes a real frequency matching method that integrates the statistical characteristics of normal distribution, which uses the normal distribution to optimize the performance of each frequency point in the bandwidth. Different normal distribution functions can be designed according to the actual application requirements to set different weights for the optimal performance of each frequency point, which can effectively improve the performance of the broadband matching network. Through normal distribution, different weight values are constructed for different frequencies to change the weight of impedance error at each frequency point in the cost function, so as to focus on optimizing the frequency of concern in the bandwidth according to needs (generally the highest value is set at the cutoff frequency). The formula for calculating the cost function of the real frequency matching method with integrated normal distribution statistical characteristics is as follows:

$$\Delta Z = 100\% \cdot \sum_{i=1}^{N} \left| \frac{k_i \cdot Z_{in}(\omega_i) - Z_{opt}(\omega_i)}{|Z_{opt}(\omega_i)|} \right|^2 \tag{2}$$

Where, the smaller the calculated ΔZ, the larger the realized transmission gain of the matching circuit. Where, $k_i(i = 1, 2...)$ is the weight value of discrete frequency points in the cost function. According to this formula, a universal matching network can be designed for any concern frequency points within the bandwidth.

As can be seen from formula (1), the simplest method is to assign different values to $k_i(i = 1, 2...)$ and directly give a constant value to the concerned frequency points. For example, when each $k_i(i = 1, 2...)$ is equal (for example, all are 1), the cost function proposed is consistent with the traditional cost function and can be fully compatible with the traditional method. However, if $k_i(i = 1, 2...)$ is a constant value, different $k_i(i = 1, 2...)$ values should be selected according to different application scenarios or measured data during design, and iterative optimization should be carried out until the selected value is the optimal value, which has low optimization efficiency and is not universal.

Based on the characteristic that the normal distribution is in accordance with the general physical development law, this paper further proposes to construct a specific normal distribution function through the actual designed bandwidth, and bring different frequency values into the constructed normal distribution function to obtain the output value of the normal distribution function, which is the weight of the frequency point. The method has universality and takes into account the general law of conventional circuit design.

3.2 Construct a Normal Distribution Function

Set the working frequency of the matching circuit as (ω_1, ω_2), then the mean value calculation method of the constructed normal distribution function is as follows:

$$\mu = r(\frac{\omega_1 + \omega_2}{2}) \tag{3}$$

Here, r is an adjustable constant factor, which can be set according to actual needs. The proposed standard deviation calculation method of the real frequency normal distribution function is as follows:

$$\sigma = q\sqrt{\frac{\omega_1}{\omega_2}} \tag{4}$$

Here, q is an adjustable constant factor, which can be set according to actual needs. So, according to μ and σ, the normal distribution function can be uniquely determined as:

$$f(x) = \frac{1}{\sqrt{2\pi}\sigma}e^{-\frac{(x-\mu)^2}{2\sigma^2}} \tag{5}$$

Accordingly, the calculation formula of the real frequency method cost function based on normal distribution is obtained as follows:

$$\Delta Z = 100\% \sum_{i=1}^{N} \left| \frac{f(\omega_i)Z_{in}(\omega_i) - Z_{opt}(\omega_i)}{|Z_{opt}(\omega_i)|} \right|^2 \tag{6}$$

4   Simulation of Real Frequency Matching Method Based on Normal Distribution

Based on the proposed real frequency matching method, polynomial h and g are also generated to obtain the transmission gain of each frequency filter sampled, and the types of components and coefficients of the matching network are calculated. The specific process is as follows:

A) Real frequency technology initialization
The initial value of the polynomial h(p) is generated by the normalized angular frequency Wa of the sampling point load within the input bandwidth, and the real and imaginary parts of the reflected load. The specific steps are as follows:

1. Set $x = \omega_2$ to generate the row vector X_d for linear regression of auxiliary polynomials;
2. Generate the initial denominator polynomial g_0, and calculate $g_0(j\omega) = gR + jgX$
3. Generate array vectors A_d and B_d to generate polynomials $A(x)$ and $B(x)$
4. Polynomial $A(x)$ and $B(x)$ by polynomial curve fitting
5. Determine the initial coefficient h0

B) Nonlinear least squares coefficient optimization polynomial h

By calling **lsqnonlin** function of MATLAB, all optimization coefficients of polynomial $h(p)$ are calculated by several iterations based on the following formula. The specific steps are as follows:

$$\min_x \left\{ \varepsilon_1^2 + \varepsilon_2^2 + \varepsilon_3^2 + ... + \varepsilon_{nopt}^2 \right\} \tag{7}$$

In this case, it is the objective function shown in Eq. (6).

1. Set unknown vector $x = [.]$ and set coefficient vector $h = x$.
2. Generate strict Hurwitz polynomial g.
3. Generate the complex form of load reflection coefficient L11.
4. Calculate the transmission power gain
5. Using the objective function of Eq. (6) under the selected sampling frequency w, the optimization vector x is obtained by minimizing the value of the objective function under the sampling frequency.

C) Generates a strict Hurwitz polynomial g

By setting the optimized $h = x$, the strict Hurwitz polynomial g is calculated using the following formula

1. Set $U = [h_n h_{n-1}...h_0]$ and $V = [(-1)^n h_n (-1)^{n-1} h_{n-1}...h_0]$. The convolution of U and V is calculated

$$H(-p^2) = h(p)h(-p) = H_0 + H_1 p + ... + H_n p^n \tag{8}$$

2. Generate even polynomials

$$G(-p^2) = H(-p^2) + (-1)^k p^{2k} = G_0 + G_1 p^2 + ... + G_n p^2 \tag{9}$$

3. Store row vectors.

$$X = [(-1)^n G_n (-1)^{n-1} G_{n-1} ... G_0] \tag{10}$$

Calculate the root of $G(X)$.

3. Calculate the root pk and pkm of G(-p2) in the right half plane and the left half plane. The calculation is as follows:

$$pk = sqrt(-Xk) \tag{11}$$

$$pkm = -pk_1 \tag{12}$$

1. Generate $\tilde{g}(p)$ single polynomial A, which is C = poly(pkm)
2. The strict Hurwitz polynomial g(p) is obtained by calculating $gc = \sqrt{|G_n|} \times C$. Where gc is a MATLAB row vector of order (n + 1), containing the coefficient $\{g_0, g_1, ..., g_n\}$ of $g(p) = g_0 + g_1 p + ... + g_n p^n$, that is:

$$gc = \{g_n g_{n-1} ... g_0\} \tag{13}$$

D) Calculate the transmission gain T

Through the h and g obtained in steps 2 and 3, the transmission gain of different frequency points w is calculated. The specific steps are as follows:

1. Define complex variables
2. Use h and g to calculate the polynomial values hval and gval.
3. through the input reflection coefficient (gval + hval)/(Gval-HVAL) to obtain transmission gain

E) Network Synthesis

Through the input h and g, the component parameters and the initial component types of the matching network are obtained by long division, and the values of the input matching items are calculated by the network synthesis formula to the normalized values of the input matching devices.

Its calculation formula is as follows:

$$Z_L = \frac{g(s) + h(s)}{g(s) - h(s)} = L_{1s} + \cfrac{1}{C_{2s} + \cfrac{1}{L_{3s} + ... + \frac{1}{X_{ns}+1}}} \tag{13}$$

Where, $C = \frac{C_n}{2\pi f_0 R_0}$, $L = \frac{L_n R_0}{2\pi f_0}$, and the results are stored in CV.

4.1 MATLAB Simulation of Broadband Power Amplifier Based on Real Frequency Method

In this section, a broadband power amplifier design example is used to describe the proposed real-frequency network matching design method based on normal distributed weights. The optimal impedance of the input and output of the power amplifier used in this paper is obtained by simulated load traction. The designed bandwidth of this broadband power amplifier is 0.8–1.7 GHz. Because the designed bandwidth is very wide, only the impedance in the operating frequency band is considered. The result of load traction performed by the power amplifier (only the bandwidth at the 100 MHz step frequency is displayed) is shown in Table 1.

This paper simulates the proposed real frequency matching method based on MAT-LAB. When the traditional real frequency matching method is adopted, the obtained input and output matching simulation results are shown in Fig. 1. As can be seen from the figure, the output matching effect basically reaches the ideal state, but the input matching effect is poor. Although the simulated power transmission efficiency of the input matching reaches 88.78% at 1.7 GHz, the efficiency within the bandwidth shows

Table 1. Results of power amplifier load traction

Frequency (MHz)	Input impedance (Ω)	output impedance (Ω)
800	249.192 + j*494.942	54.708 + j*4.955
900	258.372 + j*242.572	54.708 + j*4.955
100	258.372 + j*242.572	66.132 + j*6.098
1100	258.372 + j*242.572	54.708 + j*4.955
1200	258.372 + j*242.572	66.132 + j*6.098
1300	258.372 + j*242.572	66.132 + j*6.098
1400	258.372 + j*242.572	66.132 + j*6.098
1500	184.018 + j*37.958	63.484 + j*17.853
1600	184.018 + j*37.958	63.484 + j*17.853
1700	3.907 + j*0.654	49.268 + j*23.458

a monotonous decreasing trend, failing to achieve the broadband characteristics, and its out-of-band suppression is poor, which is easy to affect the communication performance of other frequency bands. Therefore, the input broadband design of the power amplifier (the power amplifier parameters provided in this section) cannot be realized by using the traditional real frequency matching method.

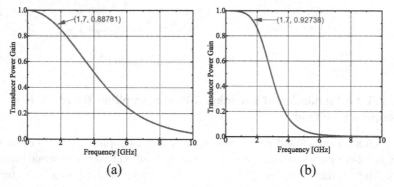

(a)   (b)

Fig. 1. Transmission efficiency of traditional real-frequency matching method, (a) transmission efficiency of input matching circuit; (b) Transmission efficiency of the output matching circuit

In order to realize the input broadband design function of the power amplifier, the proposed real frequency matching method based on normal distribution is adopted to optimize the input matching. The simulation results obtained by MATLAB are shown in Fig. 2. As can be seen from the figure, the power transmission efficiency of the simulation output reaches 97.88% at 1.7 GHz, and the efficiency within the bandwidth tends to be stable, realizing the broadband characteristics, and its out-of-band suppression is very good, reducing the impact on other frequency bands.

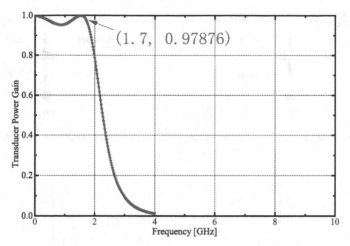

(1. 7, 0. 97876)

Fig. 2. Transmission gain of the real frequency matching method based on normal distribution

By comparing the data in Fig. 1 (a) and Fig. 2, it can be seen that the proposed real frequency matching method based on normal distribution can realize the broadband function, and the transmission efficiency at the cutoff frequency of 1.7 GHz is 9.1% higher than that of the traditional method, and the out-of-band suppression ability is stronger.

Among them, the input matching circuit parameters obtained by simulation of real frequency matching circuit based on normal distribution are as follows: CV = [2.9296 2.1479 1.1934], h = [−0.1360–0.0825–0.5637], g = [0.1360 0.4071 1.1479]; The output matching circuit parameters are: CV = [1.3670 6.6653 1.9869], h = [0.0642 0.0877 0.1570], g = [0.0642 0.3427 1.0122].

4.2 Broadband Power Amplifier Design Based on ADS

In this section, ADS simulation is carried out according to the power amplifier parameters CV, h and g obtained by MATLAB simulation based on the real frequency matching method of normal distribution. The input and output matching circuits are shown in Fig. 3. R, C and L constitute a low-pass filter, in which the resistance of R in the input matching circuit is 11.8 Ω, the capacitance of C is 1.23 pF, and the inductance of L is 2.6 nH. The resistance of R in the output matching circuit is 10.567 Ω, the capacitance of C is 0.957 pF, and the inductance of L is 1.659 nH.

The simulation results obtained through direct simulation are shown in Fig. 4, where S11 is the reflection coefficient of the input matching circuit, S22 is the reflection coefficient of the output matching circuit, and S21 is the transmission coefficient. The simulation results after simple ADS optimization are shown in Fig. 5. As can be seen from the figure, when the designed input and output matching circuits are less than 1.8 GHz, S21 is greater than 17 dB, that is, both have good transmission characteristics within the designed bandwidth.

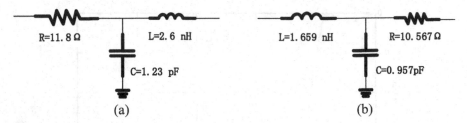

Fig. 3. Input-output matching circuit, (a) input matching circuit; (b) Output matching circuit

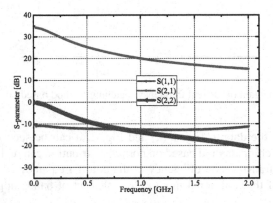

Fig. 4. S parameters of ADS simulation directly based on MATLAB simulation parameters

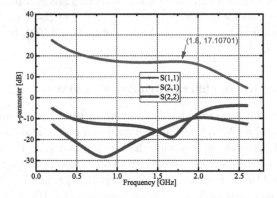

Fig. 5. S-parameter simulation based on ADS simple optimization

The optimized small signal simulation results are shown in Fig. 6. After optimization, gain fluctuation is significantly improved, and the -10 dB bandwidth of S_{11} and S_{22} can still cover the target range.

Further, the simulation results with a large signal platform are shown in Fig. 7. After the optimization of the power characteristics of large signal simulation results, it can be found that the 1dB compression point is about 4, the gain is above 17, and basically can ensure that the gain fluctuation is less than 2, the output power is above S_{21}, S_{11},

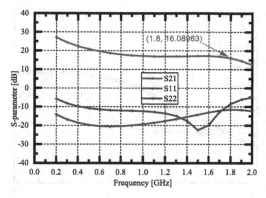

Fig. 6. Small signal simulation results

S_{22} maintain good characteristics in the target frequency, which proves that the real frequency method power amplifier matching design is basically successful.

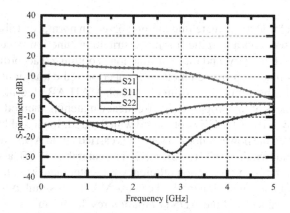

Fig. 7. Simulation results of large signals

Further, after designing the input and output matching circuit of the power amplifier, the conversion efficiency of the power amplifier obtained by simulation is shown in Fig. 8.

It can be seen from Fig. 4 and Fig. 5 that the designed power amplifier has broadband characteristics in the frequency band of 0.7–1.7 GHz, and it can be seen from Fig. 8 that the conversion efficiency can reach 50%. It can be seen from the MATLAB simulation results that the direct calculation results of the real frequency method have made PA show broadband characteristics; The results of ADS large signal simulation of single-tube power amplifier show that the transmission efficiency of broadband circuit designed based on normal distribution real frequency method is greater than 50%. The proposed real frequency method based on normal distribution can be used in power amplifier design.

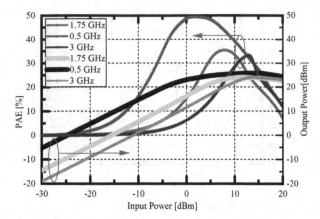

Fig. 8. Shows the power amplifier conversion efficiency based on the proposed normal separate real frequency matching method

5 Conclusion

In this paper, a real frequency matching method based on normal distribution generating weights is proposed, and a specific normal distribution function is constructed based on frequency band boundary, and the cost function is optimized accordingly. MATLAB simulation shows that the proposed method is superior to the traditional real frequency matching method. The simulation results based on MATLAB algorithm show that the average insertion loss of the proposed real frequency matching method in the frequency band is reduced by 9.1%@1.7 GHz compared with the traditional real frequency matching method. Further, the power amplifier design is carried out in ADS. Simulation results show that in the bandwidth of 0.7–1.4 GHz, S21 is greater than 17 dB, and at 0 dBm input power, the power amplifier conversion efficiency is greater than 50%, indicating that the proposed real frequency matching method can use MATLAB simulation to obtain actual circuit parameter values, and the ADS simulation results are in line with the expected effect. This paper shows that the real frequency matching method based on normal distribution can improve the bandwidth of power amplifier, and can provide theoretical and experimental reference for the design of high bandwidth power amplifier.

References

1. Cheng, Z., Zhang, Z., Liu, G., et al.: Design of hybrid EF class power amplifier based on GaN HEMT. J. Microw. **35**(2), 34–37 (2019). https://doi.org/10.14183/j.cnki.1005-6122.201 902008
2. Design of ultra-wideband power amplifier based on modified simplified real frequency technology
3. Palmer, W.D., Abdomerovic, I., Asbeck, P.M., LaRocca, T., Raman, S.: Advancing silicon mm-wave transmitter ICs for satellite communications. In: 2014 IEEE International Microwave and RF Conference (IMaRC), Bangalore, India, pp. 49–52 (2014). https://doi.org/10.1109/IMaRC.2014.7038967

4. Bhavsar, M.L., Srivastava, P., Singh, D.K., Parikh, K.S.: K-band 8-watt power amplifier MMICs using 150nm GaN process for satellite transponder. In: 2021 IEEE MTT-S International Microwave and RF Conference (IMARC), KANPUR, India, pp. 1–4 (2021). https://doi.org/10.1109/IMaRC49196.2021.9714700

5. Kalyan, R., Ghosh, B., Sreekavya, M.K., Harshit, K., Bindu, K.R., Lankapalli, S.: Design of a 25W C-band power amplifier for satellite communication. In: 2021 IEEE MTT-S International Microwave and RF Conference (IMARC), KANPUR, India, pp. 1–4 (2021). https://doi.org/10.1109/IMaRC49196.2021.9714652

6. Piacibello, A., et al.: A 5-W GaN Doherty amplifier for Ka-band satellite downlink with 4-GHz bandwidth and 17-dB NPR. IEEE Microw. Wirel. Compon. Lett. 32(8), 964–967 (2022). https://doi.org/10.1109/LMWC.2022.3160227

7. Shinjo, S., Hangai, M., Yamaguchi, Y., Miyazaki, M.: Advanced GaN HEMT modeling techniques and power amplifiers for millimeter-wave applications. In: 2020 IEEE/MTT-S International Microwave Symposium (IMS), Los Angeles, CA, USA, pp. 566–569 (2020). https://doi.org/10.1109/IMS30576.2020.9223979

8. Muhammad, F.H., You, F., Weimin, S., et al.: Broadband power amplifier using hairpin bandpass filter matching network. IET Electron. Lett. 56(4), 182–184 (2020)

9. Liu, GH., Li, S.D., Cheng, Z.Q., et al.: A power amplifier based on filter matching circuit. In: Cross Strait Quad Regional Radio Science and Wireless Technology Conference, Taiyuan, China, pp. 1–3 (2019)

10. Carlin, H., Yarman, B.: Workshop broadband matching and design of microwave amplifiers. In: IEEE MTT-S International Microwave Symposium Digest, pp. 21–27 (1983)

11. Fettweis, A.: Parametric representation of brune functions. Int. J. Circuit Theory Appl. 7(1), 1979 (1979)

12. Yarman, B., Fettweis, A.: Computer-aided double matching via parametric representation of Brune functions. IEEE Trans. Circuits Syst. 37(2), 212–222 (1990)

13. Yarman, B.: A simplified real frequency technique for broadband matching complex generator to complex loads. RCA Rev. 43, 529–541 (1982)

14. Yarman, B.: Design of Ultrawideband Antenna Matching Networks: Via Simplified Real Frequency Technique. Springer, Dordrecht (2008). https://doi.org/10.1007/978-1-4020-8418-8

15. Yarman, B.: Design of Ultrawideband Power Transfer Networks. Springer, Cham (2010)

16. Fano, R.: Theoretical limitations on the broadband matching of arbitrary impedances. IRE Trans. Circuit Theory 8(2), 165 (1947)

17. Wu, Y., Liu, Y., Li, S., et al.: A generalized dual-frequency transformer for two arbitrary complex frequency-dependent impedances. IEEE Microw. Wirel. Compon. Lett. 19(12), 792–794 (2009)

18. Chuang, M.: Dual-band impedance transformer using two-section shunt stubs. IEEE Trans. Microw. Theory Tech. 58(5), 1257–1263 (2010)

19. Wu, Y., Liu, Y., Li, S., et al.: New coupled-line dual-band dc-block transformer for arbitrary complex frequency-dependent load impedance. Microw. Opt. Technol. Lett. 54(1), 139–142 (2012)

20. Nallam, N., Chatterjee, S.: Multi-band frequency transformations, matching networks and amplifiers. IEEE Trans. Circuits Syst. I Regul. Pap. 60(6), 1635–1647 (2013)

21. Liu, Y., Zhao, Y., Liu, S., et al.: Multi-frequency impedance transformers for frequency-dependent complex loads. IEEE Trans. Microw. Theory Tech. 61(9), 3225–3235 (2013)

22. Carlin, H.: New approach to gain bandwidth problems. IEEE Trans. Circuits Syst. 24(4), 170–175 (1977)

23. Yarman, B., Carlin, H.: A simplified real frequency technique appliable to broadband multistage microwave amplifiers. IEEE Trans. Microw. Theory Tech. 30(12), 2216–2222 (1982)

24. Perennec, A., Olomo, A., Jarry, P., et al.: Optimization of gain, VSWR and noise of the broadband multistage microwave MMIC amplifier by the real frequency method: synthesis in lumped and distributed elements. In: IEEE MTT-S International Microwave Symposium Digest, pp. 363–366 (1990)
25. Kopru, R.: FSRFT-fast simplified real frequency technique via selective target data approach for broadband double matching. IEEE Trans. Circuits Syst. II Express Briefs, 141–145 (2017)
26. Yang, Z., Yao, Y., Liu, Z., et al.: Design of high efficiency broadband continuous class-F power amplifier using real frequency technique with finite transmission zero. IEEE Access 6, 61983–61993 (2019). https://doi.org/10.1109/ACCESS.2018.2875010
27. Dai, Z., et al.: A new distributed parameter broadband matching method for power amplifier via real frequency technique. IEEE Trans. Microw. Theory Tech. 63(2), 449–458 (2015)
28. Dai, Z., He, S., You, F., Peng, J., Chen, P., Dong, L.: A new distributed parameter broadband matching method for power amplifier via real frequency technique. IEEE Trans. Microw. Theory Tech. 63(2), 449–458 (2015). https://doi.org/10.1109/TMTT.2014.2385087
29. Hughes, T., Smith, M.: On the minimality and uniqueness of the Bott-Duffin realization procedure. IEEE Trans. Autom. Control 59(7), 1858–1873 (2014)
30. Koga, T.: Synthesis of a resistively terminated cascade of uniform lossless transmission lines and lumped passive lossless two-ports. IEEE Trans. Circuit Theory 18(4), 444–455 (1971)
31. Youla, D., Rhodes, J., Marston, P.: Driving-point synthesis of resistor-terminated cascades composed of lumped lossless passive 2-ports and commensurate TEM lines. IEEE Trans.Circuit Theory 19(6), 648–664 (1972)
32. Rhodes, J., Marston, P., Youla, D.: Explicit solution for the synthesis of two-variable transmission-line networks. IEEE Trans. Circuit Theory 20(5), 504–511 (1973)
33. Scanlan, S., Baher, H.: Driving point synthesis of a resistor terminated cascade composed of lumped lossless 2-ports and commensurate stubs. IEEE Trans. Circuits Syst. 26(11), 947–955 (1979)
34. Fettweis, A.: Factorization of transfer matrices of lossless two-ports. IEEE Trans. Circuit Theory CT17(1), 86–94 (1970)
35. Basu, S., Fettweis, A.: On the factorization of scattering transfer matrices of multidimensional lossless two-ports. IEEE Trans. Circuits Syst. 32(9), 925–934 (1985)
36. Basu, S., Fettweis, A.: On synthesizable multidimensional lossless two-ports. IEEE Trans. Circuits Syst. 35(12), 1478–1486 (1988)
37. Kilinc, A., Yarman, B.: High precision LC ladder synthesis part I: lowpass ladder synthesis via parametric approach. IEEE Trans. Circuits Syst. I Regul. Pap. 60(8), 2074–2083 (2013)
38. Yarman, B., Kilinc, A.: High precision LC ladder synthesis part II: immittance synthesis with transmission zeros at DC and infinity. IEEE Trans. Circuits Syst. I Regul. Pap. 60(10), 2719–2729 (2013)
39. Yarman, B., Kopru, R., Kumar, N., et al.: High precision synthesis of a Richards immittance via parametric approach. IEEE Trans. Circuits Syst. I Regul. Pap. 61(4), 1055–1067 (2014)

Dynamic Deployment of DNN Inference Tasks Based on Distributed Proximal Policy Optimization

Wenchen He, Yitao Li, and Liqiang Wang[✉]

National Computer Network Emergency Response Technical Team/Coordination Center of
China, Beijing 100029, China
libin_bjsg@163.com

Abstract. With the widespread adoption of edge computing, Deep Neural Networks (DNN) inference tasks are gradually deployed on edge computing nodes. The inference and decision-making process of intelligent services is moved to the edge side, reusing edge resources to provide ubiquitous services. However, during the service process, due to constrained edge resources or factors such as terminal mobility, DNN inference tasks may experience long delays or service interruptions, affecting the timeliness and continuity of the services. To address the problem of deteriorated communication conditions and reduced data transmission efficiency during terminal mobility, which leads to decreased service quality or even interruptions, a dynamic deployment method for DNN inference tasks based on distributed proximal policy optimization (DPPO) is proposed. Building upon an edge-terminal collaborative architecture for dynamic deployment of DNN inference tasks, this method takes into account the terminal's location, communication conditions, and the availability of resources in accessible edge nodes. The process involves DNN model caching, inference computation offloading, as well as communication and computation resource allocation. The experimental results demonstrate that the proposed method can adapt to the dynamic environment of the edge and achieve the integration and on-demand allocation of edge multidimensional resources, effectively ensuring service continuity.

Keywords: Edge computing · Task deployment · Reinforcement learning · terminal mobility

1 Introduction

In the edge computing environment, the limited scope of edge node services can lead to deteriorated communication conditions and reduced data transmission efficiency during terminal mobility, resulting in decreased service quality or even service interruptions. Therefore, designing dynamic deployment methods for Deep Neural Networks (DNN) inference tasks based on the terminal's location, network communication conditions, and the availability of resources in the accessible edge nodes becomes crucial to ensure service continuity.

H. Jin et al. (Eds.): IAIC 2023, CCIS 2060, pp. 133–143, 2024.
https://doi.org/10.1007/978-981-97-1332-5_11

Existing task deployment methods [1–4] primarily focus on balancing and optimizing indicators such as resource overhead, end-to-end delay, and load balancing, aiming to minimize the probability of service interruptions. However, the above-mentioned methods often assume that edge nodes have pre-cached the required functional components for services and possess the capability to execute various types of tasks. However, in scenarios with multiple types of coexisting DNN inference tasks, it is challenging to pre-cache all types of DNN service models due to the limited cache resources in edge nodes. Furthermore, during the dynamic deployment process of DNN inference tasks, when a mobile terminal initiates a task request to an edge node for the first time, the edge node needs to obtain the DNN service model from the cloud server. Due to the possibly large size of DNN models, the model loading delay cannot be ignored. Existing task deployment methods have not fully considered the limited availability of cache resources and the impact of DNN model loading time, making it difficult to guarantee the completion quality of DNN inference tasks.

Moreover, the mobility of terminals and the randomness of DNN inference task arrivals cause the supply-demand relationship of resources to continuously change [5–10]. Centralized complex solving algorithms have the disadvantages of being unable to make quick decisions and unable to adapt to dynamic environments, making them unsuitable for directly addressing the dynamic deployment problem of DNN inference tasks. Deep Reinforcement Learning (DRL), as an AI algorithm with self-learning and online learning capabilities, can effectively perceive the dynamics of task requests and networks and can solve the dynamic deployment problem of DNN inference tasks with lower complexity, thereby effectively ensuring service continuity.

To address the above issues, this section proposes a DPPO-based dynamic deployment method for DNN inference tasks, achieving the integration and on-demand allocation of edge multidimensional resources. The specific details are as follows:

(1) Construct an edge-terminal collaborative architecture for dynamic deployment of DNN inference tasks. Based on the terminal's location, communication conditions, and the availability of resources in the accessible edge nodes, design a dynamic deployment process for DNN tasks, including DNN model caching, inference computation offloading, and communication and computation resource allocation. Design deployment cost models and end-to-end delay models that include computation energy consumption, transmission energy consumption, and cache overhead, and construct the DNN inference task deployment problem that aims to optimize deployment costs and end-to-end delay.

(2) Considering the complex decision space and dynamic network environment, further convert the proposed multi-objective optimization problem into a Markov Decision Process (MDP) and construct the state space, action space, and reward function. Combining distributed thinking, propose a DPPO-based dynamic deployment algorithm for DNN inference tasks to achieve fast solving.

(3) Validate the proposed method from multiple perspectives such as the delay performance and deployment cost of DNN inference tasks. Experimental results demonstrate that the proposed method can provide reasonable deployment decisions to solve the optimization problem mentioned above, effectively reducing end-to-end delay and deployment costs of tasks.

2 System Model and Problem Description

2.1 Network Model

The edge-terminal collaborative architecture for dynamic deployment of DNN inference tasks is shown in Fig. 1. It consists of a cloud server, N mobile terminals represented by the set $\mathcal{N} = \{1, 2, ..., N\}$, and M edge nodes represented by the set $\mathcal{M} = \{1, 2, ..., M\}$. The cloud server has abundant computing and storage resources. Therefore, in this section, it is assumed that the cloud server is responsible for the training process of the DNN model and stores the trained DNN service models. The scenario considered in this section involves mobile terminals with mobility and dynamically arriving DNN inference tasks. The system's runtime is divided into several time slots, represented by the set $t \in \mathcal{T} = \{0, 1, ..., T - 1\}$, and the length of each time slot is denoted by τ. The dynamic deployment decisions for DNN inference tasks are executed at the beginning of each time slot. To meet the requirements of different scenarios, the decision time and the length of time slots can be dynamically adjusted. Due to the limited coverage of edge nodes, the mobile terminals can access different edge nodes in different time slots. Let $\mathcal{G}(t) = \{\mathcal{N}, \mathcal{M}_n(t)\}$ represent the network topology state of the edge in time slot t, and $\mathcal{M}_n(t) = \{(m, n) | m \in \mathcal{M}\}$ represent the set of edge nodes that the mobile terminal n can access in time slot t. F_n represents the computing resource of the mobile terminal, and F_m, S_m, and B_m represent the computing, caching, and wireless bandwidth resources of the edge node, respectively.

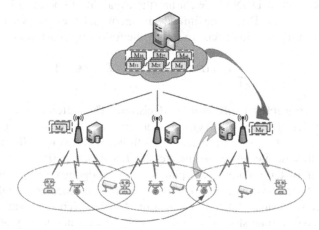

Fig. 1. The dynamic deployment architecture for edge-terminal collaborative DNN inference tasks.

2.2 Dynamic Deployment Decision Model

Let $\mathcal{K} = \{1, 2, ..., K\}$ represent the set of DNN inference tasks. Let (d_k, c_k, s_k) represent a triplet that represents the input data size, computing resource requirement, and cache resource requirement of a specific DNN inference task in time slot i, mobile terminal and edge node. Let $\mathcal{U}(t) = \{u_{n,k}(t) | n \in \mathcal{N}, k \in \mathcal{K}\}$ represent the arrival status of DNN inference task requests from mobile terminal j in time slot i, where $u_{n,k}(t) \in \{0, 1\}$. When $u_{n,k}(t) = 1$ represents the k class of DNN inference task requests generated by mobile terminal n in time slot t.

There is a location coupling relationship between inference computation and DNN models, hence the cache resources in edge nodes become a necessary factor for deploying DNN inference tasks. Specifically, the edge node is only capable of performing the corresponding inference computation when the required DNN model has been pre-cached in the edge node. Therefore, when a terminal moves within the coverage range of an edge node and the edge node does not possess the required DNN service model for that terminal, a DNN model cache request needs to be sent to the cloud server. The cloud server then delivers the required DNN model and loads it into the edge node. It should be noted that in this chapter, the process of delivering and loading the DNN model is collectively referred to as DNN model caching.

In summary, the dynamic deployment process of DNN inference tasks in mobile scenarios mainly includes DNN model caching decision, inference computation offloading decision, wireless bandwidth allocation, and computing resource allocation. The descriptions are as follows:

- **DNN model caching decision:** Caching DNN models from the cloud server to the edge node. Let $X(t) = \{x_{m,k}(t) | m \in \mathcal{M}, k \in \mathcal{K}\}$ represent the DNN model caching decision of the edge node m in time slot t, where $x_{m,k}(t) \in \{0, 1\}$. When the edge node caches the k class of DNN model, it has the capability to perform DNN inference tasks of the same type. Due to the limited amount of cache resources in the edge node, the model caching decision needs to satisfy the following constraints.

$$\sum_{k \in \mathcal{K}} x_{m,k}(t) s_k \leq S_m, \ \forall t \in \mathcal{T}, \ m \in \mathcal{M}$$

- **Inference computation offloading decision:** The inference computation of the mobile terminal is offloaded to a specific edge node. Let $Y(t) = \{y_{m,n}(t) | m \in \mathcal{M}_n(t), n \in \mathcal{N}\}$ represent the offloading decision of the inference computation of the mobile terminal in time slot t, where $y_{m,n}(t) \in \{0, 1\}$ represents the offloading of the inference computation to the edge node m in time slot t. Let $y_n(t) \in \{0, 1\}$ represent whether the mobile terminal n executes the inference computation locally in time slot t. The constraint for the inference computation of the mobile terminal in time slot t is that it can only be executed locally or offloaded to one edge node. This constraint is described as follows.

$$\begin{cases} \sum_{m \in \mathcal{M}_n(t)} y_{m,n}(t) + y_n(t) = 1, \ \forall t \in \mathcal{T}, \ n \in \mathcal{N} \\ \sum_{m \in \mathcal{M}_n(t)} y_{m,n}(t) \leq 1, \ \forall t \in \mathcal{T}, \ n \in \mathcal{N} \end{cases}$$

As mentioned before, there is a location coupling relationship between inference computation and DNN models. Therefore, the following constraints need to be satisfied:

$$\sum_{k=1}^{K} u_{n,k}(t)y_{m,n}(t) \leq x_{m,n}(t), \ \forall t \in \mathcal{T}, \ n \in \mathcal{N}, \ m \in \mathcal{M}_n(t)$$

- **Bandwidth allocation decision:** Allocate wireless bandwidth resources from edge node to terminal. Let $W(t) = \{w_{m,n}(t)|m \in \mathcal{M}_n(t), n \in \mathcal{N}\}$ represent the wireless bandwidth resources allocated from the edge node m to the mobile terminal n in time slot t. When $y_{m,n}(t) = 1$, the mobile terminal n offloads the inference computation to the edge node m, the transmission process of task data consumes bandwidth resources. Due to the limited total amount of wireless bandwidth resources in the edge node, the bandwidth allocation decision needs to satisfy the following constraints.

$$\sum_{n\in\mathcal{N}} y_{m,n}(t)w_{m,n}(t) \leq B_m, \ \forall t \in \mathcal{T}, \ m \in \mathcal{M}_n(t)$$

- **Computing resource allocation decision:** Allocating computing resources from edge node to mobile terminal. Let $z(t) = \{z_{m,n}(t)|m \in \mathcal{M}, n \in \mathcal{N}\}$ represent the allocation of computing resources from the edge node m to the mobile terminal n in time slot t. Due to the limited total amount of computing resources in the edge node, the computing resource allocation decision needs to satisfy the following constraints.

$$\sum_{n\in\mathcal{N}} y_{m,n}(t)z_{m,n}(t) \leq F_m, \ \forall t \in \mathcal{T}, \ m \in \mathcal{M}_n(t)$$

2.3 Delay Model

The caching of DNN models, offloading of inference computations, bandwidth, and computing resource allocation decisions directly impact the end-to-end delay of DNN inference tasks. In this section, we will model the computation delay, task data, and DNN model transmission delay for DNN inference tasks on mobile terminals and edge nodes. The specific description is as follows.

(1) Computing delay

When $y_n(t) = 1$, the mobile terminal n performs local inference computations, the computation delay is related to the computing resources of the mobile terminal and can be represented as:

$$T_n^{com}(t) = \frac{y_n(t) \sum_{k\in\mathcal{K}} u_{n,k}(t)c_k}{F_n}, \ \forall t \in \mathcal{T}, \ n \in \mathcal{N}$$

When $y_{m,n}(t) = 1$, the mobile terminal n offloads the inference computations to edge node m, the computation delay is related to the computing resources allocated at the edge node and can be represented as:

$$T_{n,m}^{com}(t) = \frac{y_{m,n}(t) \sum_{k\in\mathcal{K}} u_{n,k}(t)c_k}{z_{m,n}}, \ \forall t \in \mathcal{T}, \ n \in \mathcal{N}$$

The end-to-end delay of DNN inference tasks among all mobile terminals in time slot t is represented as:

$$T(t) = \sum_{n \in \mathcal{N}} \left\{ T_n^{com}(t) + T_{n,m}^{com}(t) + \max\left\{ T_n^{tra}(t) + T_{n,md}^{tra}(t) \right\} \right\}, \forall t \in \mathcal{T}$$

$T_n^{tra}(t)$ and $T_{n,md}^{tra}(t)$ respectively represent the data transmission delay between mobile terminals and edge nodes, as well as the model transmission time between cloud servers and edge nodes.

(2) Deployment Cost

In the context of terminal mobility, the deployment cost of DNN inference tasks mainly includes the computing power consumption, transmission energy consumption, and cache resource cost of both mobile terminals and edge nodes. The specific description is as follows.

$$E_n(t) = E_n^{com}(t) + E_n^{tra}(t), \forall t \in \mathcal{T}, n \in \mathcal{N}$$

$$E_m(t) = \tau P_m(t), \forall t \in \mathcal{T}, m \in \mathcal{M}$$

(3) Caching Cost

The total deployment cost of DNN inference tasks generated by all mobile terminals in time slot t is represented as:

$$Q(t) = \sum_{n \in \mathcal{N}} E_n(t) + \sum_{m \in \mathcal{M}} E_m(t) + \sum_{n \in \mathcal{N}} V_n(t)$$

2.4   The Optimization Problem Model

Based on the above, the optimization objective of the DNN inference task deployment problem in the context of terminal mobility includes end-to-end delay and deployment cost, represented as:

$$J(x, y, w, z) = \frac{1}{T} \sum_{t \in \mathcal{T}} (\omega T(t) + (1 - \omega)Q(t))$$

The problem of deploying DNN inference tasks that are optimal in terms of deployment cost and end-to-end delay can be described as follows: Based on the current edge network topology state and DNN inference task requests, the goal is to minimize the weighted sum of end-to-end delay and deployment cost by solving the decisions of DNN model caching, inference computation offloading, bandwidth allocation, and computing resource allocation, subject to various constraints. The aforementioned optimization problem can be represented as:

$$\min_{x,y,w,z} J(x, y, w, z)$$

$C_1 : x_{m,k}(t), y_{m,n}(t), y_n(t) \in \{0, 1\}, \forall t \in \mathcal{T}, n \in \mathcal{N}, m \in \mathcal{M}$

$C_2 : \sum_{k \in \mathcal{K}} x_{m,k}(t)s_k \leq S_m, \forall t \in \mathcal{T}, m \in \mathcal{M}$

$C_3 : \sum_{m \in \mathcal{M}_n(t)} y_{m,n}(t) + y_n(t) = 1, t \in \mathcal{T}, n \in \mathcal{N}$

$C_4 : \sum_{m \in \mathcal{M}_n(t)} y_{m,n}(t) \leq 1, \forall t \in \mathcal{T}, n \in \mathcal{N}$

$C_5 : \sum_{k \in \mathcal{K}} u_{n,k}(t)y_{m,n}(t) \leq x_{m,n}(t), \forall t \in \mathcal{T}, n \in \mathcal{N}, m \in \mathcal{M}_n(t)$

$C_6 : \sum_{n \in \mathcal{N}} y_{m,n}(t)w_{m,n}(t) \leq B_m, \forall t \in \mathcal{T}, m \in \mathcal{M}_n(t)$

$C_7 : \sum_{n \in \mathcal{N}} y_{m,n}(t)z_{m,n}(t) \leq F_m, \forall t \in \mathcal{T}, m \in \mathcal{M}_n(t)$

Constraint C_1 represents the value range for DNN model caching decisions and inference task offloading decisions. Constraint C_2 represents the limited total cache resource of edge nodes. Constraint C_3 and C_4 indicate that the inference computation of the mobile terminal can only be executed locally or offloaded to one edge node. Constraint C_5 states that the mobile terminal can offload the corresponding inference computation task to an edge node only if the DNN model parameters are already cached in that edge node. Constraint C_6 and C_7 respectively denote the limited bandwidth and computing resource of edge nodes.

3 Algorithm Design

In the mobile terminal mobility scenario, DNN model caching decisions $x_{m,k}(t)$, inference computation offloading decisions $y_{m,n}(t)$ and $y_n(t)$, bandwidth allocation decisions $w_{m,n}(t)$, and computing resource allocation decisions $z_{m,n}(t)$ need to change with the variation of the edge network topology. The DRL mechanism can learn strategies to maximize rewards through continuous interaction with the environment. The DRL-based dynamic deployment framework for DNN inference tasks includes three components: the environment, the agent, and the reward. The environment includes the edge network topology state and DNN inference task requests. The agent interacts with the environment, starting from a state, selecting actions based on the policy, and receiving rewards. Actions, rewards, and environment states form batch samples used to update the Actor-Critic (AC) network.

When dealing with the dynamic deployment problem of DNN inference tasks, traditional AC algorithms may suffer from the challenge of determining the learning rate, leading to slow convergence. Therefore, based on the idea of Proximal Policy Optimization (PPO), which is a distributed and proximal policy optimization algorithm, this study designs a dynamic deployment algorithm for DNN inference tasks based on Distributed PPO (DPPO). PPO, as a reinforcement learning algorithm based on the AC framework, uses regularization to limit the update magnitude of policies, addressing the issue of determining the traditional policy gradient update step size. To further improve convergence speed, the DPPO algorithm combines the distributed approach by utilizing multiple threads to collect data in the environment, with multiple threads sharing a global PPO network. This effectively solves the challenge of determining the learning rate and enhances convergence efficiency.

The dynamic deployment problem of DNN inference tasks is modeled as an MDP, which includes four components: the state space st, the action space ac, the reward function rd, and the state transition probability P. In the context of reinforcement learning, the task corresponds to the quadruple $MDP = (st, ac, \pi, rd)$, where the policy π is obtained through continuous exploration in the environment. Based on this policy, the optimal action can be determined to be executed in state $st(t)$. The quality of a policy is determined by the long-term rewards obtained by executing it, and the objective of reinforcement learning is to find a policy that maximizes the accumulated rewards over time (Table 1).

Table 1. The DPPO-based algorithm for dynamic deployment of DNN inference tasks.

Algorithm 1 The DPPO-based algorithm for dynamic deployment of DNN inference tasks
Input: Edge node information, terminal information, DNN inference task request ;
Output: DNN model caching decision, task offloading decision, bandwidth and computing resource allocation decision ;
1:    Random initialization of system state and global network parameters
2: **for** $epsiode \in (1,2,...,EP)$;
3:      **for** $sub\_epsiode \in \{1,2,...,EP_s\}$
4:          Each thread executes actions based on the global PPO policy $\pi(st(t))$ for DNN model caching $x_{m,k}(t)$, inference calculation offloading $y_{m,n}(t)$ and $y_n(t)$, bandwidth resource allocation $w_{m,n}(t)$, and computing resource allocation $z_{m,n}(t)$.
5:          Each thread obtains rewards and the next state, and saves the current state, action, and reward as a sample.
6:      **end for**
7:      Each thread synchronizes and uploads the collected data to the global network.
8:      Interact with the environment to obtain node load status information.
9:      Update Critic network parameters based on backpropagation.
10:      **if** $sub\_epsiode\%circle == 0$
11:          Update Actor2 using the parameters from Actor1.
12: **end for**

4 Simulation and Performance Analysis

To evaluate the performance of the proposed method in this section, two task dynamic deployment algorithms were selected for comparison: Q learning-based task deployment algorithm and an Asynchronous Advantage Actor-Critic (A3C) based task deployment algorithm. The mechanisms of the two algorithms are briefly introduced as follows:

Q learning-based task deployment algorithm: This algorithm is a value iteration-based reinforcement learning algorithm. It stores Q values in a Q-table that maps states to actions and selects actions that can achieve the maximum reward based on the Q-values. This algorithm does not consider DNN model caching decisions.

A3C-based task deployment algorithm: This algorithm utilizes a multi-threaded computing pattern, placing AC networks in multiple threads to interact with the environment. It obtains action policies through the Actor network and evaluates actions based on the Critic network. After a period of learning, it uses previous learning experiences to guide subsequent learning and interaction. This algorithm also does not consider DNN model caching decisions.

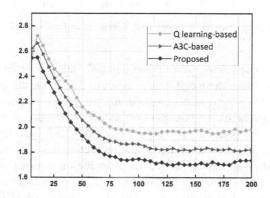

Fig. 2. Comparison of Average Deployment Costs at Different Time Slots

In this section, two metrics, average task delay and average deployment cost, are selected for comparative analysis of the three algorithms. The specific results and descriptions are as follows. Figure 2 shows the comparison of average deployment costs of the three algorithms at different time slots when the weight value is set to 0.8. From the graph, it can be observed that as the time slots increase, all three reinforcement learning algorithms can learn better strategies for DNN model caching, inference offloading, communication, and computing resource allocation, leading to a gradual reduction in deployment costs. The proposed algorithm consistently maintains the lowest average deployment cost. For example, at the 175th time slot, the average deployment cost of the proposed algorithm is 15.6% and 7.5% lower than the Q learning-based algorithm and the A3C-based algorithm, respectively. The reason for this finding can be attributed to the fact that the A3C-based algorithm does not consider the resource consumption of DNN model caching, thus having limited optimization for the task deployment cost. The Q learning-based algorithm selects actions with the highest reward based on the Q values in the Q-table, which can lead to suboptimal solutions in high-dimensional action spaces. This is why it achieves the lowest task deployment cost. The proposed algorithm takes into account the impact of DNN model caching on the task deployment cost and optimizes the DNN model caching, inference offloading, communication, and computing resource allocation decisions in a joint manner to achieve the lowest task deployment cost.

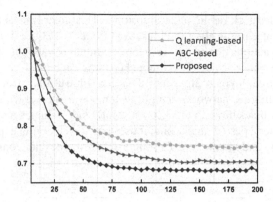

Fig. 3. Comparison of Average Task Latencies at Different Time Slots

Figure 3 depicts the comparison of average task delay at different time slots for the three algorithms when the weight value is set to 0.8. From the graph, it is evident that compared to the two comparative algorithms, the proposed algorithm consistently achieves the lowest average task delay performance. For example, at the 200th time slot, the task delay of the proposed algorithm is 9.1% and 3.1% lower than the Q learning-based algorithm and the A3C-based algorithm, respectively. The reason for this finding is that the comparative algorithms do not consider the impact of DNN model loading delay on task delay performance, thus unable to further optimize the average task delay. In contrast, the proposed algorithm considers the influence of DNN model caching decisions on average task delay and provides multiple options for DNN service models in the inference offloading decision to optimize the average task delay.

5   Conclusion

With the widespread adoption of edge computing, Deep Neural Networks (DNN) inference tasks are gradually deployed on edge computing nodes. The inference and decision-making process of intelligent services is moved to the edge side, reusing edge resources to provide ubiquitous services. However, during the service process, due to constrained edge resources or factors such as terminal mobility, DNN inference tasks may experience long delays or service interruptions, affecting the timeliness and continuity of the services. To address the problem of deteriorated communication conditions and reduced data transmission efficiency during terminal mobility, which leads to decreased service quality or even interruptions, a dynamic deployment method for DNN inference tasks based on distributed proximal policy optimization (DPPO) is proposed. Building upon an edge-terminal collaborative architecture for dynamic deployment of DNN inference tasks, this method takes into account the terminal's location, communication conditions, and the availability of resources in accessible edge nodes. The process involves DNN model caching, inference computation offloading, as well as communication and computation resource allocation. The experimental results demonstrate that the proposed method can adapt to the dynamic environment of the edge and achieve the integration and on-demand allocation of edge multidimensional resources, effectively ensuring service continuity.

References

1. Shi, J., Du, J., Wang, J., et al.: Priority-aware task offloading in vehicular fog computing based on deep reinforcement learning. IEEE Trans. Veh. Technol. **69**(12), 16067–16081 (2020)
2. Li, Q., Wang, S., Zhou, A., et al.: QoS driven task offloading with statistical guarantee in mobile edge computing. IEEE Trans. Mob. Comput. **21**(1), 278–290 (2022)
3. Liu, J., Ren, J., Zhang, Y., et al.: Efficient dependent task offloading for multiple applications in MEC-cloud system. IEEE Trans. Mob. Comput., 1 (2021)
4. Chen, R., Wang, X.: Maximization of value of service for mobile collaborative computing through situation aware task offloading. IEEE Trans. Mob. Comput., 1 (2021)
5. Tan, L., Kuang, Z., Zhao, L., et al.: Energy-efficient joint task offloading and resource allocation in OFDMA-based collaborative edge computing. IEEE Trans. Wirel. Commun. **21**(3), 1960–1972 (2022)
6. Nath, S., Wu, J.: Dynamic computation offloading and resource allocation for multi-user mobile edge computing. In: GLOBECOM 2020 - 2020 IEEE Global Communications Conference, pp. 1–6 (2020)
7. Chang, Z., Liu, L., Guo, X., et al.: Dynamic resource allocation and computation offloading for IoT fog computing system. IEEE Trans. Ind. Inf. **17**(5), 3348–3357 (2021)
8. Song, Z., Liu, Y., Sun, X.: Joint task offloading and resource allocation for NOMA-enabled multi-access mobile edge computing. IEEE Trans. Commun. **69**(3), 1548–1564 (2021)
9. Zheng, J., Cai, Y., Wu, Y., et al.: Dynamic computation offloading for mobile cloud computing: a stochastic game-theoretic approach. IEEE Trans. Mob. Comput. **18**(4), 771–786 (2019)
10. Zhang, K., Leng, S., He, Y., et al.: Cooperative content caching in 5G networks with mobile edge computing. IEEE Wirel. Commun. **25**(3), 80–87 (2018)

Graph-Cut Based DNN Inference Task Partitioning and Deployment Method

Xiaohou Shi, Yaqi Song, and Meiling Dai[✉]

China Telecom Research Institute Beijing, Beijing, China
daiml1@chinatelecom.cn

Abstract. As users demand higher inference accuracy, the number of network layers and neurons in Deep Neural Network (DNN) models continues to grow, resulting in increasingly demanding requirements for computational power, storage, and other resources for DNN inference tasks. On the edge side, partitioning resource-intensive DNN inference tasks into multiple dependent subtasks and deploying them to different nodes has become a crucial approach to ensuring task computation efficiency. To address the problem of fine-grained partitioning of DNN inference tasks with directed acyclic graph (DAG) topology, a graph-cut-based method for DNN inference task partitioning and deployment is proposed. Firstly, a distributed edge-terminal collaborative architecture is constructed to model the partitioning and deployment of DNN inference tasks with DAG topology. Then, the problem of optimal partitioning and deployment of DNN inference tasks with minimal latency and energy consumption is formulated. Finally, graph-cut-based algorithms for DNN inference task partitioning and computation resource allocation are designed. Experimental results demonstrate that the proposed method optimally utilizes the limited and distributed resources at the edge, effectively ensuring service timeliness.

Keywords: Edge computing · Task deployment · Graph-cut · Task Partitioning

1 Introduction

The rapid development of deep learning technology has facilitated the widespread adoption of various intelligent applications. These applications, driven by Deep Neural Networks (DNNs), often require substantial computing and storage resources to meet latency and inference accuracy requirements. However, due to the limited resources of devices in terms of energy, computing power, and storage, relying solely on devices to perform DNN inference processes is unrealistic. DNN models typically consist of different types of neural network layers, and data between each layer is transmitted in the form of matrices. In some cases, the data volume of intermediate layers is significantly smaller than the original input data volume. Leveraging this characteristic, DNN inference tasks can be partitioned into multiple subtasks and leveraged with the resources of both devices and edge servers in a distributed deployment. This approach facilitates efficient utilization of communication and computing resources in the edge computing environment [1–6].

H. Jin et al. (Eds.): IAIC 2023, CCIS 2060, pp. 144–154, 2024.
https://doi.org/10.1007/978-981-97-1332-5_12

It has been found that DNN inference tasks are no longer limited to chain structures, but also widely adopt Directed Acyclic Graph (DAG) structures. For instance, the GoogleNet and ResNet networks, which won the ImageNet challenge, both employ DAG structures. Subtasks of such reasoning tasks exhibit complex data dependencies and calculation sequences, making existing task partition and deployment methods designed for chain structures less suitable for these DNN reasoning tasks. Moreover, there is a conflicting relationship between optimizing energy consumption and task delay, exacerbating the challenge due to the high energy consumption and low latency requirements of DNN inference tasks. Current approaches for DNN inference task partition and deployment primarily optimize either delay or energy consumption separately, necessitating enhanced models and algorithm designs that comprehensively consider both aspects [7–10]. As a result, there is a conflict between optimizing energy consumption and minimizing task delay, as DNN inference tasks require both low latency and energy efficiency. Current approaches tend to optimize one aspect at the expense of the other.

To address these challenges, this paper focuses on the partition and deployment methods for DAG-structured DNN inference tasks and designs an iterative optimization algorithm incorporating the minimum cut algorithm and computing resource allocation algorithm to optimize task delay and system energy consumption. The specific research objectives are outlined as follows:

(1) Analysis of data dependencies and calculation sequences between subtasks in DAG-structured DNN inference tasks, and the development of a decision model for partitioning and deployment. This includes proposing calculation and transmission delay models for DNN reasoning tasks, along with calculation and communication energy consumption models. Based on these models, a multi-objective optimization problem considering task delay and energy consumption is formulated.

(2) Considering the problem as a mixed-integer nonlinear problem, it is decomposed into two sub-problems—DNN inference task partitioning and computing resource allocation. An iterative optimization algorithm is designed. Firstly, an auxiliary graph is constructed to transform the task partition problem into a minimum cut problem. A single DNN inference task partition algorithm based on Yuri Boykov is proposed. Secondly, computing resource allocation is adjusted based on the partitioning decision.

(3) The proposed method is validated using various types of DNN inference tasks from diverse perspectives, including task delay and energy consumption. Experimental results demonstrate that the proposed method can make informed partitioning and resource allocation decisions for different types of DNN inference tasks, effectively balancing task delay and energy consumption.

Overall, this research aims to provide advanced techniques in the domain of DAG-structured DNN inference task partition and deployment, offering optimization strategies that balance task delay and energy consumption.

2 System Model and Problem Description

In this section, a DAG-structured DNN inference task partition and deployment decision model are first proposed. On this basis, this paper constructs the computing, communication and caching models of DNN inference tasks, and proposes a multi-objective optimization problem model for task delay and system energy consumption.

2.1 Partition Decision Model

Figure 1(a) shows the layer structure of a typical Inception V4. Figure 1(b) shows the DNN inference task of the DAG structure corresponding to the model. Taking the red dotted line in the figure as an example, the DNN inference task is partitioned into two parts, the former part is deployed locally on the device, and the latter part is deployed on other devices or edge servers. It can be seen from the figure that there are two relationships between subtasks: 1) Dependency relationship, the latter subtask needs to wait for the result of the former subtask. As shown in the figure, subtask v_2 cannot perform calculations until the output data of subtask v_1 is obtained; 2) Parallel relationship, subtasks v_{10} and v_{11} can be executed in different CPU cores at the same time to improve computational efficiency.

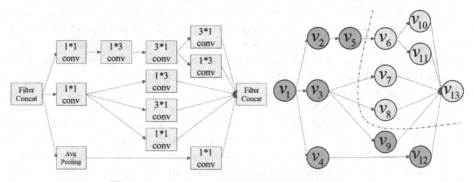

Fig. 1. Example of partition of DNN inference tasks

It is defined that $G = (V, Y)$ represents the DNN inference task of the DAG structure, the vertex set $V = \{v_1, v_2, ..., v_k\}$ in G represents the subtask set of the DNN inference task, and $v_k = (c_k, d_k, s_k)$ represents the k th subtask of the DNN inference task, where c_k, d_k and s_k represent the computing resource requirements, input data volume, and cache resource requirements of the subtask v_k respectively. The set of edges $Y = \{y_{1,2}, y_{1,3}, ..., y_{i,j}\}$ represents the dependency between subtasks, and $y_{i,j}$ represents the calculation result of the subtask v_i is the input data of the subtask v_j, that is, v_j can only be calculated after the calculation of v_i is completed. The input data of a subtask may come from multiple subtasks. Taking the subtask v_{13} in Fig. 3 as an example, after obtaining the calculation results of the subtasks v_7, v_8, v_9, v_{10}, v_{11} and v_{12}, v_{13} can perform inference calculate.

2.2 Deployment Decision Model

Figure 2 shows the DNN inference task deployment architecture of distributed edge-device collaboration, including multiple devices and an edge server. For example, devices such as wearable devices, monitoring devices, and smartphones can collect user data in their respective application scenarios, and have certain computing and storage resources to perform some DNN inference tasks locally. D2D communication is supported between devices. The edge server connected to the base station configures resources such as communication, storage, and computing. Therefore, this section considers deployment decisions for two DNN inference tasks: multi-end collaboration and side-end collaboration, which are described in detail below.

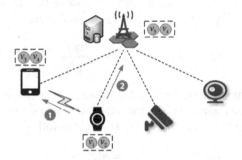

Fig. 2. DNN inference task deployment architecture

(1) Multi-device collaboration: Partition the DNN inference task into two parts, and deploy them to the local and other devices with idle resources respectively. D2D communication technology is used between devices to realize data transmission.

(2) Device-to-device collaboration: Partition the DNN inference task into two parts and deploy them to the local and edge servers respectively. The cellular communication technology is used to realize data transmission between the device and the edge node.

Define the variable $\alpha_k \in \{0, 1, ..., N\}$ to indicate the deployment location of the subtask v_k, $\alpha_k = 0$ to indicate that v_k is deployed on the edge server, $\alpha_k = n, n \in \{1, 2, ..., N\}$ to indicate that v_k is deployed on the device, and N to indicate the number of device.

2.3 Computational Model

According to the partition and deployment decision model of DNN inference tasks, the task computing delay and energy consumption models are constructed. In the field of deep learning, FLOPS (floating-point operations per second) are usually used to characterize the computing resources of devices and edge servers. Definition f_n represents the computing resources allocated by the device n to the current DNN inference task, which satisfies $f_n \leq f_n^{peak}$. The computation time of the subtask v_k at the device n is expressed as follows.

$$\tau_k^n = \frac{c_k}{f_n}$$

Therefore, the total computation latency of the DNN inference task in the device n is expressed as follows:

$$T_n^{com} = \sum_{k \in K} I\{\alpha_k = n\}\tau_k^n, \, n > 0$$

Among them, $I\{\cdot\}$ represents the indicator function, if and only if the conditions in $\{\cdot\}$ are satisfied, $I\{\cdot\} = 1$, otherwise it takes the value 0. The computational energy consumption of subtask v_k in device n is expressed as follows.

$$e_k^n = \kappa c_k (f_n)^2 = \kappa \frac{(c_k)^3}{(\tau_k^n)^2}$$

where κ represents the effective switched capacitance associated with the chip architecture. Combining the above formula, the total computing energy consumption of the DNN inference task in the device n can be expressed as follows.

$$E_n = \sum_{k \in K} I\{\alpha_k = n\}e_k^n, \, n > 0$$

Definition f^e represents the computing resources allocated by the edge server e to the current DNN inference task, satisfying $f^e \leq f^{peak}$. The computation time of the subtask v_k in the edge server e is expressed as follows.

$$\tau_k^e = \frac{c_k}{f^e}$$

Therefore, the total computation delay of DNN inference task in edge server e is expressed as follows.

$$T_e^{com} = \sum_{k \in K} I\{\alpha_k = 0\}\tau_k^n$$

The computational energy consumption of the subtask v_k in the edge server e is expressed as follows.

$$e_k^e = \kappa c_k (f_e)^2 = \kappa \frac{(c_k)^3}{(\tau_k^e)^2}$$

The total computational energy consumption of the DNN inference task in the edge server e is expressed as follows.

$$E_e = \sum_{k \in K} I\{\alpha_k = 0\}e_k^e$$

2.4 Communication Model

Next, this paper constructs the transmission delay and transmission energy consumption models of DNN inference tasks. D2D communication technology is supported between devices, and task data can be transmitted from the D2D sender i to the D2D receiver j. Assuming that the D2D communication limit distance of the device is dis_{max}^{d2d}, D2D communicates with the downlink of the multiplexed cell through the orthogonal signal allocated by the base station. The geographical distance between the device i and j is $dis_{i,j}$, the path loss is inversely proportional to the power of $\varepsilon(\varepsilon > 2)$ of the distance, the link will experience independent and identically distributed Rayleigh fading, and the fading coefficient obeys an exponential distribution with a mean of 1, then the power of the receiving device j is $P_j = P_i h dis_{i,j}^{-\varepsilon}$, where P_i is the transmit power of the device i. Therefore, it is defined that the signal-to-noise ratio of the interference signal received by the device j from the cell base station is expressed as:

$$SINR = \frac{P_j}{I_{BD} + \sigma^2}$$

Among them, I_{BD} is the interference of the cell base station to the receiving end, and σ^2 is the link additive white noise. In the D2D mode, the transmission rate between the devices i and j is denoted as $R_{i,j}^{d2d} = B_{i,j}^{d2d} \log_2(1 + SINR)$, where $B_{i,j}^{d2d}$ represents the wireless bandwidth obtained by the D2D pair. Therefore, the transmission delay of the data of the DNN inference task between the devices i and j is expressed as follows.

$$T^{d2d} = \sum_{k \in K} \sum_{u \in pred(k)} I\{\alpha_u = i\} I\{\alpha_k = j\} \frac{d_{u,k}}{R_{i,j}^{d2d}}, \ dis_{i,j} \le dis_{max}^{d2d}, i > 0, j > 0$$

Among them, $pred(k)$ represents the set of previous subtasks adjacent to the subtask v_k, and $d_{u,k}$ represents the size of the data transferred from the subtask v_u to v_k. The data transmission rate from the device i to the edge server e is expressed as follows.

$$R_{i,e} = B_{i,e}^e \log_2(1 + SINR)$$

Among them, $SINR = P_e/(I_{DB} + \sigma^2)$, I_{DB} is the impact of D2D communication on the base station. $P_e = P_i h dis_{i,e}^{-\varepsilon}$ is the received power of the edge server, where $dis_{i,e}$ is the distance from the edge server e to the device i, and $B_{i,e}^e$ represents the wireless bandwidth allocated by the device i. If the maximum access distance of the base station is dis_{max}^e, the data transmission delay of the DNN inference task is expressed as follows:

$$T^{d2e} = \sum_{k \in K} \sum_{u \in pred(k)} I\{\alpha_u = i\} I\{\alpha_k = 0\} \frac{d_{u,k}}{R_{i,e}^e}, \ dis_{i,e} \le dis_{max}^e, i > 0$$

2.5 Caching Model

Define S_n and S^e to represent the cache resources of device n and edge server e, respectively. If you decide to deploy the subtask v_k to a device or edge server, you first need to cache the model parameters corresponding to the subtask to the device or edge server. Therefore, the cache resource occupancy of the device or edge server is denoted as $\sum_{k \in K} I\{\alpha_k = n\} s_k, n > 0$ and $\sum_{k \in K} I\{\alpha_k = 0\} s_k$.

2.6 Optimization Problem Model

To sum up, under the distributed device-to-device collaborative architecture, the optimization goals of DNN inference task partition and deployment include energy consumption and task delay, which are expressed as follows.

$$F = \omega_t(T_n^{com} + T_e^{com} + T^{tra}) + \omega_e(E_n + E_e + E_{tra})$$

Among them, ω_t and ω_e represent the weights of task delay and energy consumption, respectively, and satisfy $\omega_t + \omega_e = 1$. The energy consumption and task delay optimization problem model constructed in this paper is expressed as follows.

$$\min_{\alpha, f} F$$
$$C_1 : \alpha_k \in \{0, 1, ..., N\}$$
$$C_2 : f_n \le f_n^{peak}$$
$$C_3 : f^e \le f^{peak}$$
$$C_4 : \sum_{k \in K} I\{\alpha_k = n\}s_k \le S_n, n > 0$$
$$C_5 : \sum_{k \in K} I\{\alpha_k = 0\}s_k \le S^e$$

where the constraint condition C_1 represents the value range of the DNN inference task partition and deployment decision, the constraint conditions C_2 and C_3 represent the upper limit of the computing resources allocated to the task by the device and the edge server, and the conditions C_4 and C_5 represent the device and edge server cache capacity There is an upper limit. In summary, the optimization problem model can be described as: by solving the DNN inference task partition and computing resource allocation method, the weighted sum of the delay and related energy consumption of the task is minimized.

3 Algorithm Design

The multi-objective optimization problem is a mixed integer nonlinear problem, including the solution of integer variables and continuous variables. Therefore, this section proposes an iterative optimization algorithm including Yuri Boykov-based DNN inference task partition algorithm and computational resource allocation algorithm, described as follows.

(1) When optimizing the DNN inference task partition, a deployment decision is randomly selected, and the computing resource allocation decision f_n or f^e of the device or edge server is randomly set.
(2) Build an auxiliary graph and use the maximum flow minimum cut algorithm to solve the partition decision α_k of the DNN inference task. According to the partition decision, the computing resource allocation decision f_n or f^e is solved.
(3) Iterate the above process continuously until the minimum value of the objective function is obtained, and finally obtain the DNN inference task partition and computing resource allocation decision. The specific algorithm flow is shown in Table 1.

Table 1. Iterative Optimization Algorithm Pseudocode

Iterative Optimization Algorithm
Input: DNN inference tasks, device and edge server information
Outputs: task partitioning and deployment decisions: α, computing resource allocation decisions: f_n and f^e
1. for $n \in \{0,1,2,...,N\}$
2. Random selection of deployment decisions, fixed computing resource allocation decisions: f_n and f^e
3. Combined with $G=(V,Y)$, construct the auxiliary graph G', and set the weights of the auxiliary graph
4. Using Yuri Boykov max-flow min-cut algorithm to solve DNN inference task partition decision α_k
5. According to the partition decision, solve the computing resource allocation decision f_n and f^e
6. According to the objective function, update the partitioning and computing resource allocation decisions
7. end for
8. Complete the deployment of the current DNN inference task according to the partition and resource allocation decisions

3.1 Auxiliary Graph Construction

From the perspective of graph theory, the partition problem of DNN inference task $G = (V, Y)$ can be described as: Find a set of edges $Y_c \in Y$, so that the graph G becomes two disconnected parts after being divided according to Y_c, and the former part is deployed in the initiating request The device is local, and the latter part is deployed on other devices or edge servers. Therefore, this section solves the problem of DNN inference task partition through the maximum flow minimum cut algorithm, including auxiliary graph construction and the design of the DNN inference task partition algorithm based on Yuri Boykov. The specific description is as follows.

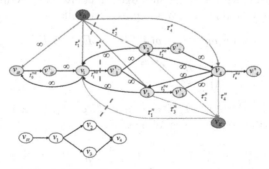

Fig. 3. Auxiliary diagram G' construction example

As shown in Fig. 3, in order to simplify the description, define $G = (V, Y)$ to represent the DNN inference task with four subtasks, set v_{st} to represent the virtual entry subtask of the DNN inference task, and its computing resource and cache resource requirements are set to zero. On the basis of $G = (V, Y)$, add source point v_D and sink point v_E. v_D represents the device that generates the DNN inference task, and v_E represents the device or edge server deployed by the latter subtask in the DNN inference task. In Fig. 3-2, first assume that v_E represents an edge server. For any vertex in $G = (V, Y)$, set the corresponding vertex mmm in the auxiliary graph v_k, V' and Y' represent the set of all vertices and edges in the auxiliary graph G', respectively.

3.2 A Single DNN Inference Task Partition Algorithm Based on Yuri Boykov

The traditional max-flow min-cut algorithm based on augmented path is to search the path from the source point to the sink point in the graph, and traverse all the paths through breadth-first search (or breadth-first search) to obtain the maximum flow path. Since it needs to traverse all vertices in the graph, its computational efficiency is relatively low. Taking the Push-Relabel algorithm as an example, its time complexity is $O(|Y||V|^2)$ [XX]. Therefore, this section divides a single DNN inference task based on Boykov's max-flow min-cut algorithm. The algorithm constructs two search trees starting from the source point and the sink point, and reuses the existing tree structure to improve the algorithm. Purpose of efficiency.

Boykov's algorithm first constructs two non-intersection trees with source v_D and sink v_E. In the tree V_D, the edge set points from the parent node to the child node, and the edge in the tree V_E points from the child node to the parent node. The nodes of tree V_D and tree V_E exist in two states: active and passive. The main dynamic node represents the outer boundary node of the tree, and the search tree can still be expanded from this node, while the passive node is located inside the tree. The algorithm mainly includes three steps: growth stage, augmentation stage and adoption stage. After obtaining the partition decision of a single DNN inference task according to the Yuri Boykov maximum flow minimum cut algorithm, the deployment decision of the subtask is also determined accordingly. On this basis, this section allocates the computing resources in the device and edge server to optimize the task delay. And energy consumption.

At this time, the transmission energy consumption and transmission time of the DNN inference task between the device and the edge server are also determined. Therefore, a multi-objective optimization problem is expressed as follows.

$$\min_{\alpha, f}\{\omega_t T_n^{com} + \omega_t T_e^{com} + \omega_e E_n + \omega_e E_e\}$$
$$C_1 : 0 \leq f_n \leq f_n^{peak}$$
$$C_2 : 0 \leq f^e \leq f^{peak}$$

First, we can get $T_n^{com} \propto \tau_k^n$, $T_e^{com} \propto \tau_k^e$ and $\tau_k^n \propto 1/f_n$, $\tau_k^e \propto 1/f^e$, therefore, T_n^{com} and T_e^{com} can be further expressed as follows:

$$T_n^{com} + T_e^{com} = \eta_n \frac{1}{f_n} + \eta_e \frac{1}{f^e}, \eta_n > 0, \eta_e > 0$$

$E_n \propto (f_n)^2$, $E_e \propto (f^e)^2$ can be obtained, and according to formula (3–12) can be obtained:

$$E_n + E_e = \theta_n(f_n)^2 + \theta_e(f^e)^2, \theta_n > 0, \theta_e > 0$$

To sum up, the optimization objective in Eq. (3–16) can be equivalent to:

$$F = \eta_n \frac{1}{f_n} + \eta_e \frac{1}{f^e} + \theta_n (f_n)^2 + \theta_e (f^e)^2$$

The Hessian matrix of Eq. (3–19) is expressed as follows:

$$H = \begin{bmatrix} 2\eta_n \frac{1}{(f_n)^3} + 2\theta_n & 0 \\ 0 & 2\eta_e \frac{1}{(f^e)^3} + 2\theta_e \end{bmatrix}$$

Since $f_n > 0$ and $f^e > 0$, the Hessian matrix H of the objective function is a positive definite matrix, so the function in Eq. (3–16) has a minimum value, and solving this function obtains the optimal resource allocation decision corresponding to the current cut set.

4 Simulation and Performance Analysis

In this section, the specific parameters and environment settings of the simulation experiment are given first. Secondly, three algorithms compared with the algorithm proposed in this paper are introduced. Finally, the experimental results are analyzed in detail and the conclusions are described, which verifies the advantages of the algorithm proposed in this paper.

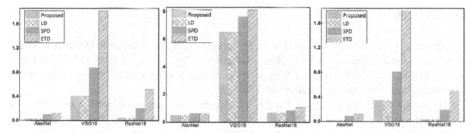

Fig. 4. Comparison of objective functions, energy consumption and delay of the four algorithms ($\omega_t = 0.01$, $\omega_e = 0.99$)

Figure 4 shows the comparison of the objective function value, energy consumption and delay of the four algorithms when deploying three DNN inference tasks when the weights are set to $\omega_t = 0.01$ and $\omega_e = 0.99$. As can be seen from the figure, the algorithm and LD proposed in this chapter achieve the minimum objective function value, energy consumption and task delay. Analyze the main reasons for the above situation: Under the current weight setting, energy consumption is the main optimization target. Since LD deploys the DNN inference task locally in the device, there is no wireless transmission energy consumption, so the optimal deployment scheme under the current weight setting is obtained. SPD divides the DNN inference task and completes the deployment. This process will generate transmission delay and transmission energy consumption, so the objective function value of this method is large. ETD deploys all the DNN inference tasks to the edge server, and the original data of the task needs to be transmitted through the wireless link, so the objective function value of this method is the largest.

5   Conclusion

This paper proposes a single DNN inference task partition and deployment method based on Yuri Boykov. First, the data dependencies and computation order between subtasks are analyzed, and a decision model for the partition and deployment of DAG-structured DNN inference tasks is constructed. Considering the constraints of cache resources, the calculation, transmission delay model and energy consumption model of DNN inference tasks are constructed. On this basis, a multi-objective optimization problem model for task delay and energy consumption is proposed. Considering that this problem is a mixed integer nonlinear problem, the optimization problem is decomposed into two sub-problems of DNN inference task partition and computing resource allocation. The iterative optimization algorithm is designed. First, the auxiliary graph is constructed and the task partition problem is transformed into a minimum cut problem. A DNN inference task partition algorithm based on Yuri Boykov is proposed. Second, the computing resource allocation is adjusted according to the partitioning decision. Combined with various types of DNN inference tasks, the proposed method is verified from the perspectives of task delay and system energy consumption. The experimental results show that the proposed method can make reasonable partition and resource allocation decisions in the face of different types of DNN inference tasks, and effectively balance the task delay and energy consumption.

References

1. Krizhevsky, A., Sutskever, I., Hinton, G.E.: ImageNet classification with deep convolutional neural networks. In: Advances in Neural Information Processing Systems, vol. 25 (2012)
2. Simonyan, K., Zisserman, A.: Very deep convolutional networks for large-scale image recognition. arXiv preprint arXiv:14091556 (2014)
3. He, K., Zhang, X., Ren, S., et al.: Deep residual learning for image recognition. In: 2016 IEEE Conference on Computer Vision and Pattern Recognition (CVPR), pp. 770–778 (2016)
4. Howard, A.G., Zhu, M., Chen, B., et al.: MobileNets: efficient convolutional neural networks for mobile vision applications. arXiv preprint arXiv:170404861 (2017)
5. Mao, H., Yao, S., Tang, T., et al.: Towards real-time object detection on embedded systems. IEEE Trans. Emerg. Top. Comput. **6**(3), 417–431 (2018)
6. Redmon, J., Divvala, S., Girshick, R., et al.: You only look once: unified, real-time object detection. In: 2016 IEEE Conference on Computer Vision and Pattern Recognition (CVPR), pp. 779–788 (2016)
7. Liu, W., et al.: SSD: single shot MultiBox detector. In: Leibe, B., Matas, J., Sebe, N., Welling, M. (eds.) ECCV 2016. LNCS, vol. 9905, pp. 21–37. Springer, Cham (2016). https://doi.org/10.1007/978-3-319-46448-0_2
8. Zhang, Z., Wang, C.: SaPus: self-adaptive parameter update strategy for DNN training on multi-GPU clusters. IEEE Trans. Parallel Distrib. Syst. **33**(7), 1569–1580 (2022)
9. Rehr, R., Gerkmann, T.: SNR-based features and diverse training data for robust DNN-based speech enhancement. IEEE/ACM Trans. Audio Speech Lang. Process. **29**, 1937–1949 (2021)
10. Shin, D., Kim, G., Jo, J., et al.: Low complexity gradient computation techniques to accelerate deep neural network training. IEEE Trans. Neural Netw. Learn. Syst., 1–15 (2021)

Task Offloading with Dual-Mode Switching in Multi-access Edge Computing

Xiaoliang Zhang[✉], Jiaqi Duan, Mei Yan, and Shunming Lyu

State Grid Information and Telecommunication Branch, Beijing, China
sddavidzhang@163.com

Abstract. Task offloading is an important mechanism in edge computing that can reduce the execution latency and dropping rate of computation tasks. However, the existing deep learning-based task offloading methods generate offloading decisions with high randomness and poor quality in the early stage before the model converges. The bid-based task offloading methods, in turn, have difficulty in utilizing task execution history information to guide offloading decisions. To overcome the shortcomings of the existing task offloading methods, we design a dual-mode switching method for task offloading in the multi-access edge computing environments. The method dynamically switches between the deep reinforcement learning-based decision mode and the dynamic bidding-based decision mode. The method utilizes a global bidding mechanism to fine tune the raw offloading decisions made by the two modes to reduce doom-to-fail task offloading decisions. The experimental results on the simulator show that the proposed dual-mode switching task offloading method is able to achieve a low task dropping rate across the entire execution process. The average execution latency of the tasks gradually decreases and converges as time grows.

Keywords: Multi-Access Edge Computing · Task Offloading · Mode Switching · Deep Reinforcement Learning · Dynamic Bidding

1 Introduction

1.1 Background

With the rapid development of information technology, complex Internet applications pose serious performance challenges for both mobile devices and cloud computing centers. The mobile devices may not be able to process all tasks locally with low latency due to limited computational resources. To facilitate efficient task processing, the mobile edge computing (MEC) [1] is introduced, where mobile devices or edge servers can offload their computation tasks to nearby edge servers or cloud computing centers for processing. The task offloading in MEC [2] can significantly reduce task processing latency and the rate of dropped tasks for latency sensitive tasks. Based on its advantages, the task offloading has been applied in many scenarios in MEC such as drone inspection [3], smart power grids [4], and smart cites [5] and so on.

Along with the increasingly developed network connection and growing computing demand, the multi-access edge computing [6] with multiple edge nodes and multiple cloud centers (as shown in Fig. 1) is getting more and more attention. It makes efficient collaboration between the cloud and the edge becomes complex.

The offloading strategy between the cloud and the edge is a core problem to solve in task offloading. The strategy must make two kinds of decisions. The first is whether the edge server should offload its tasks to a cloud center. The second is to which cloud center the edge server should offload its tasks if the edge server decides to perform offloading. In practice, the processing capacity of all edge servers and cloud centers is limited. If a large number of edge servers offload their tasks to the same cloud center, the load of that cloud center may become very high. The offloaded tasks may experience large processing delays or even be dropped beyond the deadline, which is often unacceptable in practice. Therefore, designing an efficient offloading strategy that can fully utilize the resources at both the edge and the cloud in a multi-access edge computing environment is the main objective of this paper.

1.2 Related Work

The existing research works on task offloading strategies in the edge computing environment can be classified as heuristic methods, learning-based methods, and dynamic offloading mechanisms.

Heuristic Methods. The heuristic methods use predefined heuristic rules to generate the offloading decision. Li et al. [7] propose an adaptation-based edge server selection method, where the adaptation calculation is related to the remaining time of the task and the server's computational capacity. However, the adaptation calculation is simple and it is difficult to generate efficient computational task offloading schemes for complex computational resource usage states. Neto et al. [10] proposed an estimation-based approach in which each device makes offloading decisions based on estimated processing and transmission capabilities, but the approach does not take into account the underlying queuing system, which assumes that the processing and transmission of each task should always be completed within a single time slot (or before the next task arrives), which is difficult to guarantee in practice. Lyu et al. [11] proposed a Lyapunov-based algorithm to ensure the stability of the task queue, however, the method requires the environment of task divisibility, i.e., the task can be arbitrarily partitioned into multiple parts, which may not be realistic due to the fact that there are often dependencies among the tasks.

Learning-Based Methods. The learning-based methods use deep-learning models or reinforcement-learning-based models to make decisions based on the task execution history. Tang et al. [8] propose a reinforcement learning-based task offloading model in edge computing, which utilizes deep-Q network to assist in computational task offloading decision making, but the model generates poor quality decision schemes and high task discarding rate when it does not converge in the pre-training period. Zhao et al. [13] propose a distributed offloading algorithm based on deep reinforcement learning to solve the wireless channel contention between mobile devices, but the algorithm on each mobile device needs the quality of service information of the other mobile devices, and the method does not take into account the fact that in reality there is not a wireless connection between all mobile devices. The learning-based methods.

Dynamic Offloading. The dynamic offloading methods adjusts their offloading decisions according to the actual execution states of the edge servers and the cloud centers. Lu et al. [9] propose a dynamic offloading model for computing tasks based on an auction strategy, using the greedy method as the auction strategy, but this strategy completely ignores the computing power of the edge and terminal devices, and fails to fully utilize the resources on the edge side. Lee et al. [12] proposed an algorithm based on online optimization techniques to minimize the maximum latency of a task, which considers latency-tolerant tasks, whereas the objective of this paper will focus on dealing with deadline-based latency-sensitive tasks, which is somewhat contradictory because deadlines affect the load level of the edge nodes, which in turn affects the latency of the offloading task.

Some approaches [14–16] focus on designing the reward mechanism for participants in task offloading. The cloud centers will not unconditionally provide computing resources to edge servers unless they can be sufficiently rewarded. A reward mechanism encourage them to participate in offloading tasks.

It is difficult for the existing methods to generate high-quality offloading decisions during the entire offloading process. The quality of the heuristic methods is not as good as the learning-based methods since they cannot adaptively adjust their policies according to the actual execution of the offloading tasks. The learning-based methods suffer from poor offloading decisions before the learning model converges. Though the dynamic offloading methods can fine tune the static offloading decisions, they cannot generate high-quality offloading decisions on a macro level.

1.3   Contributions

To overcome the drawbacks of the existing methods and generate high quality task offloading decisions during the whole execution process, we propose a dual-mode switching task offloading method for the multi-access edge computing. The method contains two decision modes: the reinforcement-learning-based decision mode and the dynamic bid decision mode. The method uses a probabilistic switch to control the switching between modes. It fine tunes the raw offloading decision with a global task bidding queue to balance workloads among cloud computing centers with free resources. We evaluate the performance of the proposed method and compare it with the existing deep-learning-based method and the bidding-based method in a simulator. The experimental results show that the proposed method is able to achieve a low task dropping rate across the entire execution process, while the existing deep-learning method suffers from the high dropping rate at the early stage and the bidding-based method fails to decrease the dropping rate. The performance of the proposed method is consistent in homogeneous and heterogeneous environments.

2 System Model

In this work, we consider the task offloading between a set of cloud centers $N = \{c_1, c_2, ..., c_N\}$ and edge servers $M = \{e_1, e_2, ..., e_M\}$ (i.e., multi-access edge computing). The offloading decision happens at the beginning of all the time slots $T = \{t_1, t_2, ...\}$ and the duration of each time slot is Δ seconds.

2.1 Multi-access Edge System

In this work, we design our strategy based on the classic two-tier system model in the multi-access mobile edge computing as shown in Fig. 1. An edge server receives computation tasks from multiple mobile devices connected to it. The tasks received by the edge server can either be processed locally or offloaded to a cloud center remotely. We assume that at the beginning of each time slot, the edge server receives a new task with a certain probability. When a new task arrives, the local scheduler of the edge server first needs to decide whether to process the task locally or offload it to a cloud center. If the edge server decides to process the task locally, its scheduler places the task in the computing queue for processing. Otherwise, the scheduler further needs to decide which cloud center to offload the task to. The scheduler then places the task into the transmission queue for offloading. The tasks in the transmission queue are sent to the corresponding cloud center over a network link one by one. For the computing (or transmission) queue, we assume that if the task is completed in one time slot, the next task in the queue will be processed (or transmitted) at the beginning of the next time slot.

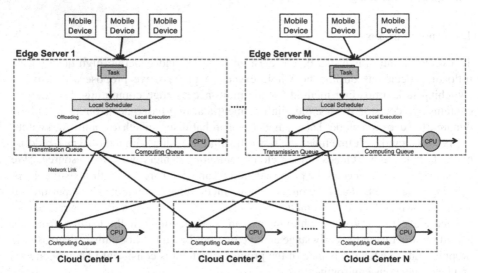

Fig. 1. The two-tier system model of a multi-access edge system.

2.2 Task Model and Task Execution

The tasks in the system model are assumed to be indivisible. A task is either executed or dropped. We assume that a task can either be executed locally at an edge server or offloaded to a cloud center. The task will not be executed both at the edge server or the cloud center.

At the beginning of each time slot, a new task arrives at an edge server with a certain probability τ. We assume that each edge server can only accept a task at the beginning of each time slot. If a new task arrives at the beginning of the time slot $t \in T$ for the edge server $m \in M$, we use $p_m(t)$ to denote the unique index of the task. $b_m(t)$ is the number of bits of the task. $d_m(t)$ is the deadline of the task, i.e. the maximum number of time slots that the task can wait before it finishes. Once the task $p_m(t)$ reaches its cutoff time (the end of the time slot $t + d_m(t) - 1$), if the task is not finished, it is dropped from the system.

For the task $p_m(t)$ that arrives at the edge server m at the time slot t, the local scheduler of the edge server determines the execution position of the task $x_m(t) \in \{0, 1\}$, where $x_m(t)$ is 1 if the scheduler chooses to execute the task locally and $x_m(t)$ is 0 if the scheduler chooses to offload the task. If $x_m(t)$ is 1, the task is appended to the computing queue of the edge server $x_m(t)$. If $x_m(t)$ is 0, the scheduler further needs to decide to which cloud center the task should be offloaded. We use $y_{m,n}(t) \in \{0, 1\}$ to represent offloading the task $p_m(t)$ to the cloud center n ($y_{m,n}(t) = 1$) or not ($y_{m,n}(t) = 0$). If $y_{m,n}(t)$ is 1, the task is appended to the transmission queue of the edge server. It is required that each task with $x_m(t) = 0$ can only be offloaded to one cloud center,

$$\sum_{n \in N} y_{m,n}(t) = 1, \forall m \subset M, t \in T$$

In each edge server, the computing and transmission queues work in a FIFO way. If a task finishes execution (or transmission) in a time slot, the next task in the queue will start in the next time slot. Each task may be executed or transmitted for several time slots. We use $W_m^{comp}(t)$ to denote the number of time slots that the task $p_m(t)$ waits for in the computation queue. Assume f_m is the computation power of the edge server m and ρ is the number of CPU cycles required to process each unit of data, the delay of $p_m(t)$ is defined as $D_m^{comp}(t) = W_m^{comp}(t) + \lceil \frac{b_m(t)}{\Delta f_m/\rho} \rceil$. Similarly, we define $W_m^{tran}(t)$ and $D_m^{tran}(t)$ for the tasks in the transmission queue, where $D_m^{tran}(t) = W_m^{tran}(t) + \lceil \sum_{n \in N} \frac{y_{m,n}(t)b_m(t)}{B_{m,n}\Delta} \rceil$ and $B_{m,n}$ is the bandwidth of the network link between the edge server m and the cloud server n.

In the cloud center, the cloud center organizes a computation queue for each edge server. The tasks received from the same edge server are appended to the corresponding computation queue. The cloud center processes all the queues concurrently, but the tasks in the same queue are processed in a FIFO way. We use $W_{m,n}^{comp}(t)$ to represent the waiting time of the task $p_m(t)$ in the cloud center n in the computation queue. Assume the computation power of the cloud center n is f_n, ρ is the number of CPU cycles required to process each unit of data, and k is the number of concurrent tasks that the cloud center n can execute, the delay of the task $p_m(t)$ at the cloud center n is defined as $D_{m,n}^{comp}(t) = W_{m,n}^{comp}(t) + \lceil \frac{b_m(t)\rho k}{\Delta f_m} \rceil$.

Example. The duration of each time slot Δ is 0.1 s. The edge server 2 receives a task $p_2(t)$ with $b_2(t)$ is 4 Mb. If the task is offloaded to the cloud center 1 to execute after waiting for 2 time slots. The bandwidth between the cloud center and the edge server is 20 Mbps. Then, the transmission delay $D_2^{tran}(t) = 2 + \lceil \frac{4}{20*0.1} \rceil = 4$ time slots. Assume the computation power of the cloud center 1 f_1 is 35 GHz, $\rho = 0.3$ in the cloud center, and k $= 4$. If there is no task before $p_2(t)$ in the computation queue, the delay of the task on the cloud server is $D_{2,1}^{comp}(t) = 0 + \lceil \frac{4*0.3*4}{0.1*35} \rceil = 2$. Therefore, the total delay of the task $p_2(t)$ is $s_{total} = D_2^{tran}(t) + D_{2,1}^{comp}(t) = 6$ time slots. If $s_{total} \leq d_2(t)$, the task finishes with the total delay of 6 time slots. Otherwise, the task $p_2(t)$ is dropped.

3 Dual-Mode Task Offloading

In order to improve the quality of the offloading decisions before the decision learning model converges, we propose the *dual-mode offloading mechanism* in the local scheduler of the edge server. The local scheduler includes two decision modes for task scheduling: the reinforcement-learning-based mode and the dynamic bidding mode. The local scheduler uses the learning-based mode to generate a rough decision and fine tune the decision further will the dynamic bidding mode. In this section, we introduce the two modes and then elaborate on how to switch between the two modes.

3.1 Reinforcement-Learning-Based Decision Mode

In the learning-based decision mode, we deploy follow the Q-learning framework to train a reinforcement learning model (like the deep Q-network-based model [8]) on each edge server to make the task offloading decisions. The target of the reinforcement model is to minimize the average total delay s_{total} of all tasks.

Figure 2 shows the architecture of the reinforcement learning model for task offloading. The model consists three parts: the input layer, a deep neural network, and a output layer. The input layer receives feature vectors from several components in the system, including the task feature vectors, system state feature vectors, and the task execution history vectors. Those feature vectors describe the system state and the history of the executed tasks. The input feature vectors and combined and fed into the deep neural network. The deep neural network learns a mapping from the state feature vectors to Q-values of different offloading actions. The Q-values predicts the long-term expected costs (the total delay of the tasks) of the offloading decision.

The input layer considers three kinds of features. The task feature includes the number of bits of the task $b_m(t)$, the deadline of the task $d_m(t)$, and the type of the task. The system state features include the queue-related features and the cloud center state features. The queue-related features contain the time slots that the task has been waiting for $W_m^{comp}(t)$, $W_m^{tran}(t)$, and the number of time slots that each task may further wait before being processed. The cloud center state features include the lengths of the

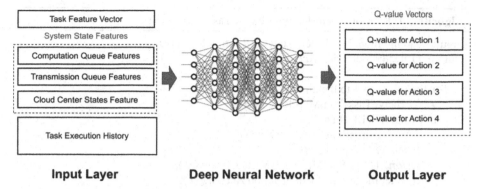

Input Layer	**Deep Neural Network**	**Output Layer**

Fig. 2. The architecture of the reinforcement learning model for task offloading.

computation queues of the cloud center. The task execution history is a matrix that stores the workload history of each cloud center in the past few time slots. The matrix can be used by the deep recurrent neural networks (like LSTM) to extract features for the execution history.

The deep neural network part can be any neural network models that support Q-learning framework. In this work, we adopt the Q-network-based model [8] which consists of LSTM layer to extract history features and two fully connected layers and advantage and value layer to transform the hidden representation vectors.

The output layer predicts a Q-value for each task offloading action. The offloading action for the task $p_m(t)$ is denoted as $Action_m(t) = (x_m(t), y_{m,1}(t), ..., y_{m,n}(t))$. The Q-value predicts the expected total task delay of the offloading decision. We regard the total delay of a dropped task as the penalty delay which is larger than the average delay of an executed task. The optimization target of the reinforcement learning is to first minimize the task dropping rate and then minimize the task total delay.

Algorithm 1 shows the workflow of the reinforcement learning mode working on each edge server. For each time slot, if the server receives a valid task, it uses the reinforcement learning model L to generate the local best offloading $Action_m(t)$. At the end of the time slot, the edge server collects the task execution events happening during this time slot, including task completion or task dropping. The events will be appended to the history matrix H of the edge server. The learning model L is also updated according to the new events. After enough time slots, the learning model L will finally converge.

Algorithm 1: Reinforcement-learning-based Task Offloading

Input: Learning model L, System states S, History Matrix H;
Output: Offloading decision $Action_m(t)$.

1: **for** time slot $t := 1$ to the number of slots **do**
2:   $T :=$ ReceiveTaskFromMobileDevices();
3:   **if** T is not NULL **then**
5:     $IV :=$ ConstructInputVectors(T, t, S);
6:     $ActionSet := L$.predict(IV);
7:     $Action_m(t) :=$ FindMinimalQValue($ActionSet$);
8:     Conduct the task offloading action $Action_m(t)$;
9:   Conduct the local computation and transmission;
10:   $e :=$ CollectTaskExecutionEvents(t); // a task finishes or is dropped.
11:   H.update(e); // Update the history matrix
12:   L.update(e, H); // Update the learning model

3.2 Dynamic Bid Decision Mode

In our system, we setup a global task bidding queue Q to handle the remaining tasks that still have not determined the offloading target as shown in Fig. 3. In the global task bidding queue, the tasks will be sorted in the descending order of their sizes $b_m(t)$. The tasks in the queue will be bid by the cloud centers. For a task t, each cloud center will calculate the total delay $s_{total,c}$ of the task if it is offloaded to the cloud center c. If $s_{total,c} < d_m(t)$ which means that the task can be finished before the deadline, the computation task t can be offloaded to the cloud center c, and the total delay $s_{total,c}$ will be used as the cost to "bid" of the cloud center. Otherwise $s_{total,c} \geq d_m(t)$, the task can not be offloaded to the cloud center c for execution and its cost is regarded as -1.

For every task T in the global task bidding queue Q, the system finds out the sets of cloud centers to which the task T can be offloaded to, $S = \{n1, n2, ...\}$. If the set S is not empty, we select the cloud center c with the highest total delay (i.e., the highest bid in the auction) as the winner of the auction. We offload the task T to the winner cloud center by adding the task T to the transmission queue of the edge server. If the set S is empty, the auction of the task fails. We return the task back to the edge server from which the task comes from, and add the task to the computation queue of the edge server.

The dynamic bidding mechanism offloads computing tasks in the global auction queue to cloud centers with weaker computing capabilities as preferentially as possible to improve the resource utilization of these cloud centers, while retaining the computation resources of cloud centers with strong computing capabilities to cope with the subsequent sudden large computing tasks that may occur.

3.3 Dual-Mode Switch

In order to avoid the reinforcement learning-based mode from producing poor quality offloading decisions before the model converges, we propose a dual-mode switch mechanism for task offloading decision. Algorithm 2 shows its workflow.

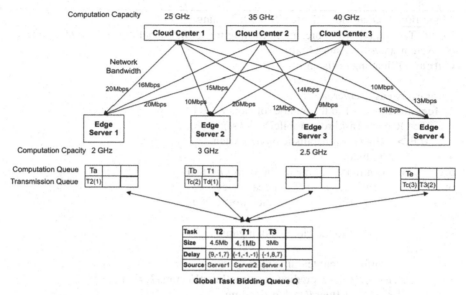

Fig. 3. The global task bidding queue in the dynamic bid decision mode.

In the dual-mode switch mechanism, we use a probabilistic switch to control the switching between modes. At the beginning of each time slot t, we generate a random number θ from a uniform probability distribution. We compare θ with $\varepsilon = \min(\alpha t, \beta)$, where t is the time slot ID, α is a user-given preference parameter and β is the maximum switch threshold. If θ is greater than ε, the system chooses the pure bidding mode. Otherwise, the system chooses the reinforcement learning mode. As the time slot increases, the value of ε grows linearly until it reaches maximized β. In the early stage, the value of ε is small and the reinforcement learning-based model tends to generate poor quality decisions. At this point, the system favors the pure bidding mode with greater probability. As time goes by, ε gradually becomes larger, and the reinforcement learning mode gradually dominates the decision.

The offloading decisions made by the two modes may still be rough. They may choose to run the tasks in an edge server or a cloud center that is clearly incapable of completing the task on time. At this time, we use the global bidding queue to help balance the workloads among cloud centers. In the pure bidding mode, the scheduler estimates the total task delay of running the task locally (line 5 of Algorithm 2). If the task cannot finish in time, the scheduler adds the task to the global bidding queue to try offloading via bidding. In the reinforcement learning mode, if the scheduler estimates that the target cloud center cannot complete the task in time, the scheduler will also add the task to the global bidding queue (line 17 of Algorithm 2). The global bidding will try to offload the tasks to capable cloud centers as much as possible.

Algorithm 2: Dual-Mode Switch Task Offloading

Input: Task $p_m(t)$, Reinforcement learning model L, Global dynamic bidding queue Q, System states S, History Matrix H;

Output: Offloading decision;

// For every edge server m

1: **for** time slot $t := 1$ to the number of slots **do**
2:     $T := \text{ReceiveTaskFromMobileDevices}()$;
3:     **if** $\theta > \varepsilon$ **then** // Probabilistic mode switch
4:         // Pure bidding mode
5:         **if** T can finish in time on the server m **then**
6:             $Action_m(t) := 0$; // Local execution
7:             Add T to the computation queue of m;
8:         **else**
9:             Add T to the global bidding queue Q;
10:    **else**
11:        // Reinforcement learning mode
12:        $Action_m(t) := L.\text{predict}(\text{ConstructInputVectors}(T, t, S))$;
13:        **if** $x_m(t) = 1$ **then** // task offloading
14:            **if** $s_{total,c} < d_m(t)$ **then** // the target cloud center is able to process
15:                Add T to the transmission queue of m;
16:            **else**
17:                Add T to the global bidding queue Q;
18:        **else** // no offloading
19:            Add T to the computation queue of m;
20:    $\text{ConductGlobalBidding}(Q)$; // Fine tune offloading decision among servers
21:    Conduct the local computation and transmission of m;
22:    $e := \text{CollectTaskExecutionEvents}(t)$;
23:    $H.\text{update}(e)$; // Update the history matrix
24:    $L.\text{update}(e, H)$; // Update the reinforcement learning model

---

4 Experiments

We conduct experiments in a multi-access MEC simulator to evaluate the performance of the proposed dual-mode switch task offloading mechanism.

4.1 Experimental Setup

All the experiments were conducted in a Windows 11 (64 bit edition) server equipped with the Intel i9-13980HX CPU, 16 GB memory, 1 TB HDD, and NVIDIA GeForce RTX 4080 GPU. We implemented all the algorithms in Python 3.6.2 with Tensorflow 1.4.0 and numpy 1.15.0.

We implemented the reinforcement learning-based mode with the neural network model proposed in [8]. The learning rate was 0.01 and the discount factor in the reinforcement learning was 0.9. The history matrix H stored 500 records. The model was updated in batch for every 10 time slots.

We simulated the experiments with three kinds of environment as shown in Table 1. The preference parameter α was 0.0025 and the switch threshold β was 0.99. The duration of a time slot Δ is 0.1s. The deadline of every task is 10 time slots. The task arrived at each edge server with the probability of 0.3.

Table 1. MEC environments used in the experimental evaluation.

Environment Type	Homogeneous	Heterogeneous	Hybrid
# Edge Servers	50	50	50
# Cloud Centers	10	10	10
Computing Capability of Edge Servers	2.5 GHz	[2 GHz, 3 GHz]	25@2 GHz, 25@3 GHz
Computing Capability of Cloud Centers	41.8 GHz	[10 GHz, 55 GHz]	5@30 GHz, 5@50 GHz
Network Bandwidth	14 Mbps	[8 Mbps, 20 Mbps]	14 Mbps
Task Size	[2 Mb, 5 Mb]	[2 Mb, 5 Mb]	[2 Mb, 5 Mb]
Cycles per Bit ρ	0.297 GHz	0.297 GHz	0.297 GHz

4.2 Comparison with the Existing Methods

We compare our dual-mode method with the deep reinforcement learning-based method [8] and the bidding-based method [9] in the heterogeneous environment. We ran the simulation for 60,000 time slots. Figure 4 shows the task dropping rate and Fig. 5 shows the average task delay of the three methods.

The deep reinforcement learning-based method suffers from high task dropping rate in the early stage. The deep learning model does not converge and most of the offloading decisions are generated with randomness. The task dropping rate of the bidding-based method is stable, but it is much higher than the other two methods. The bidding-based method cannot take advantage of the task execution information to guide the decision. Our method combines the advantages of the two modes together. It achieves a very low task drop rate even at the early stage with the help of the fine tuning brought by the global bidding queue. Our method finally converges to the same dropping rate as the deep reinforcement learning-based method. The deep-learning model helps reduce the average task delay. As the model converges, the average task delay of the deep reinforcement learning-based method and our method become much lower than the pure bidding-based method.

4.3 Effects of Edge Environments

In order to evaluate the adaptability of our method to different edge computing environment configurations, Fig. 6 shows the task dropping rate and the average delay during the

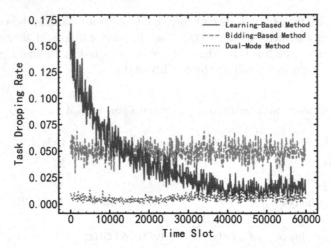

Fig. 4. The task dropping rate of the three task offloading methods.

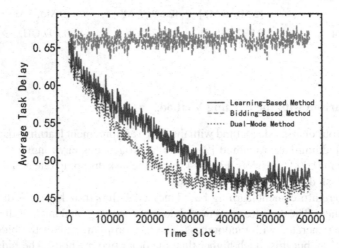

Fig. 5. The average task delay of the three task offloading methods.

simulations of 60000 time slots under three different environments. The experimental results show that our method performs consistently in both homogeneous and heterogeneous environments. The task dropping rate becomes stable even from the early stage of the execution.

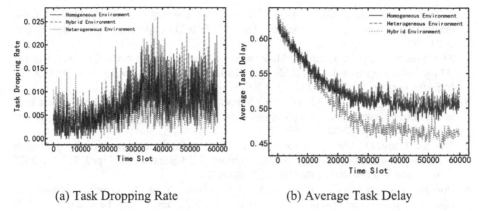

(a) Task Dropping Rate   (b) Average Task Delay

Fig. 6. The performance of the proposed method in homogeneous and heterogeneous environments.

5 Conclusion

Task offloading is an important mechanism in edge computing that can reduce task execution latency and dropping rate. However, the existing deep learning-based task offloading methods generate offloading decisions with high randomness and poor quality in the early stage before the model converges. The bidding-based task offloading methods, in turn, have difficulty in utilizing task execution history information to guide offloading decisions. To overcome the shortcomings of the existing task offloading methods, we design a dual-mode switching method for task offloading in the multi-access edge computing environments. The method dynamically switches between the deep reinforcement learning-based decision mode and the dynamic bidding-based decision mode. The method also utilizes a global bidding mechanism to fine tune the raw offloading decisions made by the two modes to avoid doom to fail offloading decisions. The experimental results on the simulator show that the proposed dual-mode switching task offloading method is able to achieve a low task dropping rate across the entire execution process, and the average task latency gradually decreases and converges as time grows. The performance of the proposed method is also consistent in both homogeneous and heterogeneous environments. In the future, we plan to investigate the dynamic mode switching based on the prediction rate of the deep-learning-based model.

Acknowledgement. This work is supported by the State Grid Science and Technology Project (Grant No. 5700-202318292A-1-1-ZN).

References

1. Mao, Y., You, C., Zhang, J., Huang, K., Letaief, K.B.: A Survey on mobile edge computing: the communication perspective. IEEE Commun. Surv. Tutor. **19**, 2322–2358 (2017). https://doi.org/10.1109/COMST.2017.2745201

2. Islam, A., Debnath, A., Ghose, M., Chakraborty, S.: A survey on task offloading in multi-access edge computing. J. Syst. Archit. **118**, 102225 (2021). https://doi.org/10.1016/j.sysarc. 2021.102225

3. McEnroe, P., Wang, S., Liyanage, M.: A survey on the convergence of edge computing and AI for UAVs: opportunities and challenges. IEEE Internet Things J. **9**, 15435–15459 (2022). https://doi.org/10.1109/JIOT.2022.3176400

4. Feng, C., Wang, Y., Chen, Q., Ding, Y., Strbac, G., Kang, C.: Smart grid encounters edge computing: opportunities and applications. Adv. Appl. Energy **1**, 100006 (2021). https://doi. org/10.1016/j.adapen.2020.100006

5. Xu, X., Huang, Q., Yin, X., Abbasi, M., Khosravi, M.R., Qi, L.: Intelligent offloading for collaborative smart city services in edge computing. IEEE Internet Things J. **7**, 7919–7927 (2020). https://doi.org/10.1109/JIOT.2020.3000871

6. Porambage, P., Okwuibe, J., Liyanage, M., Ylianttila, M., Taleb, T.: Survey on multi-access edge computing for internet of things realization. IEEE Commun. Surv. Tutor. **20**, 2961–2991 (2018). https://doi.org/10.1109/COMST.2018.2849509

7. Li, Z., Shi, L., Shi, Y., Wei, Z., Lu, Y.: Task offloading strategy to maximize task completion rate in heterogeneous edge computing environment. Comput. Netw. **210**, 108937 (2022). https://doi.org/10.1016/j.comnet.2022.108937

8. Tang, M., Wong, V.W.S.: Deep reinforcement learning for task offloading in mobile edge computing systems. IEEE Trans. Mobile Comput. **21**, 1985–1997 (2022). https://doi.org/10. 1109/TMC.2020.3036871

9. Lu, W., Wu, W., Xu, J., Zhao, P., Yang, D., Xu, L.: Auction design for cross-edge task offloading in heterogeneous mobile edge clouds. Comput. Commun. **181**, 90–101 (2022). https://doi.org/10.1016/j.comcom.2021.09.035

10. Neto, J.L.D., Yu, S.-Y., Macedo, D.F., Nogueira, J.M.S., Langar, R., Secci, S.: ULOOF: a user level online offloading framework for mobile edge computing. IEEE Trans. Mob. Comput. **17**, 2660–2674 (2018). https://doi.org/10.1109/TMC.2018.2815015

11. Lyu, X., et al.: Distributed online optimization of fog computing for selfish devices with out-of-date information. IEEE Trans. Wirel. Commun. **17**, 7704–7717 (2018). https://doi.org/10. 1109/TWC.2018.2869764

12. Lee, G., Saad, W., Bennis, M.: An online optimization framework for distributed fog network formation with minimal latency. IEEE Trans. Wirel. Commun. **18**, 2244–2258 (2019). https:// doi.org/10.1109/TWC.2019.2901850

13. Zhao, N., Liang, Y.-C., Niyato, D., Pei, Y., Wu, M., Jiang, Y.: Deep reinforcement learning for user association and resource allocation in heterogeneous cellular networks. IEEE Trans. Wirel. Commun. **18**, 5141–5152 (2019). https://doi.org/10.1109/TWC.2019.2933417

14. Liu, Y., Xu, C., Zhan, Y., Liu, Z., Guan, J., Zhang, H.: Incentive mechanism for computation offloading using edge computing: a Stackelberg game approach. Comput. Netw. **129**, 399–409 (2017). https://doi.org/10.1016/j.comnet.2017.03.015

15. Li, G., Cai, J.: An online incentive mechanism for collaborative task offloading in mobile edge computing. IEEE Trans. Wirel. Commun. **19**, 624–636 (2020). https://doi.org/10.1109/ TWC.2019.2947046

16. Zhang, D., et al.: Near-optimal and truthful online auction for computation offloading in green edge-computing systems. IEEE Trans. Mob. Comput. **19**, 880–893 (2020). https://doi.org/10. 1109/TMC.2019.2901474

Channel Allocation Scheme Based on NSGA-II for Frequency-Division-Multiplexing UHF RFID System

Jie Meng[1], Yuan Li[2], Yulu Zhang[2], Shuai Ma[2], Gui Li[3], Jian Li[1],
and Guangjun Wen[1(✉)]

[1] School of Information and Communication Engineering/Yibin Institute, University of
Electronic Science and Technology of China, Chengdu 611731, China
`wgj@uestc.edu.cn`

[2] Department of IoT Technology Research, China Mobile Research Institute, Beijing 100053,
China

[3] The 10th Research Institute of China Electronics Technology Group Corporation,
Chengdu 611731, China

Abstract. With the proposal and development of the internet of things (IoT), the
ultra-high frequency (UHF) passive radio frequency identification (RFID) system
based on backscatter communication technology has become one of the core technologies of the IoT due to its advantages of low power consumption and low cost.
However, the insufficiency of the system throughput has become the primary problem restricting the implementation of ubiquitous connectivity. Frequency division
multiplexing (FDM) UHF passive RFID is an effective method to improve the system throughput, and the quality of RFID terminal sub-carrier channel allocation
is the key. This paper adopts the multi-objective Pareto optimization rule to establish the channel allocation model of FDM UHF passive RFID system based on
the harmonic interference power and spectrum utilization, and uses the improved
NSGA-II optimization algorithm to iteratively obtain the Pareto solution set of
the system model. The simulation results show that in the setting channel capacity
interval, a set of channel allocation configuration are provided respectively, and the
channel allocation method proposed in this paper improves the system spectrum
utilization by 42.95% and 82.81% under the two different distribution conditions
of tags respectively, and achieves the maximum spectrum utilization of 62%.

Keywords: ultra-high frequency radio frequency identification · frequency
division multiplexing · Pareto optimization · NSGA-II

1 Introduction

The ultra-high frequency (UHF) passive Radio Frequency Identification (RFID) technology is an automatic identification technology that obtains data from electronic tags
through Radio Frequency (RF) signals with operating frequency of 860 MHz to 960
MHz. The reader is transmitting the carrier, and the passive tag captures the external

© The Author(s), under exclusive license to Springer Nature Singapore Pte Ltd. 2024
H. Jin et al. (Eds.): IAIC 2023, CCIS 2060, pp. 169–183, 2024.
https://doi.org/10.1007/978-981-97-1332-5_14

RF signal as the energy driver to modulate and reflect the RF signal to communicate with the reader [1]. For the advantages of low power consumption and low cost, UHF passive RFID system can meet the demands of wide coverage and long standby time in green IoT, UHF passive RFID is considered as one of the most effective solutions to realize ubiquitous connectivity, and widely used in intelligent management fields such as manufacturing, logistics, asset management, and public safety.

With the scale and density of tags increasing significantly, the demand of large-capacity communication and the limited number of accessed terminals have become the main contradiction of UHF passive RFID technology [2]. To improve the system through-put rate, many works expect to provide more dynamic access opportunities for tags. Literature [3] proposes FlipTracer, which establishes a graph model called single flip graph according to the highly dynamic and unpredictable characteristics of actual reflected signals, captures the conversion mode of collision signals, and addresses collision signals by tracking the single flip graph model to achieve large aggregation throughput. However, this method has high requirements on computing speed and tag time synchronization. Many works modify and upgrade hardware to improve system concurrency. Literature [4] proposed DigiScatter, which integrated inverse discrete Fourier transform (IDFT) into tag design to realize all digital baseband processing. The concurrency and data rate are adjusted by configuring the IDFT size, and the spectrum is dynamically allocated to generate efficient spectrum resource utilization. However, it undoubtedly increases the calculation overhead on the tag, and has certain exclusion on existing commercial tags, increasing the maintenance and update cost of the system. More works propose that multiple access technology is the key to improving spectrum utilization efficiency and total network throughput [5]. Most existing works on traditional orthogonal multiple access technologies focus on the multi-tag anti-collision algorithm based on time division multiplexing technology [6]. On the other hand, works based on Frequency Division Multiplexing technology focuses on the design of multi-reader access protocol, anti-collision algorithm, and anti-interference algorithm, which are basically implemented in Orthogonal frequency division multiplexing (OFDM) systems [7–9]. However, OFDM RFID is limited by the established resource scheduling granularity. Some works propose to adopt multi-carrier multiple access technology, and carried out research on sub-carrier allocation scheme based on multiple iterations [10], fair mechanism [11], normalized channel gain [12], game theory [13–15], intelligent optimization [16–18], and other mechanisms, which has been verified in broadband wireless communication systems. However, the backscattered signal of UHF passive RFID system is complex. For example, the reflection channel has high propagation loss, frequency sensitivity, and is affected by noise, as well as significant interference from RF signal sources, and multiple sub-carriers produce infinite harmonic interference. Relevant works from Keio University study these topics [19, 20], but only consider a single element, i.e., the channel allocation based on tag distance or bandwidth. These works fail to make full use of frequency band resources, and are not suitable for the application of scenarios with a large number of tags in UHF passive RFID systems.

To address the issue of the low spectrum utilization in the existing UHF passive RFID system, this paper designs a flexible and efficient channel allocation scheme for FDM UHF passive RFID system. By analyzing the conflict relationship between the system spectrum utilization rate and bit error rate (BER), a multi-object optimization problem (MOP) based on the requirements of practical application scenarios is formed according to the Pareto optimization rule. We establish the channel allocation model of FDM UHF passive RFID system, and further optimize the parameters of channel allocation model of UHF passive RFID system by using the improved NSGA-II algorithm. The spectrum utilization of RFID system is improved while maintaining the same BER as the existing schemes. The simulation results show that the proposed scheme can produce a channel allocation configuration with a spectrum utilization rate of up to 62% in the RFID systems containing 10 to 50 channels. Compared with the existing schemes, the spectrum utilization rates of the system under the two different distribution conditions are increased by 42.95% and 82.81%, respectively. This paper provides theoretical support for the application scenarios of large-scale-tag access and communication.

2   UHF Passive RFID System Channel Allocation Model

The FDM UHF passive RFID system is mainly composed of tags, reader, upper-layer application and its database, as shown in Fig. 1. Tags are used to store information about unique marking objects. Readers are connected to upper-layer applications to control communication processes and read or rewrite data inside tags. The upper-layer application provides control commands to readers and perform specific functions based on feedback data.

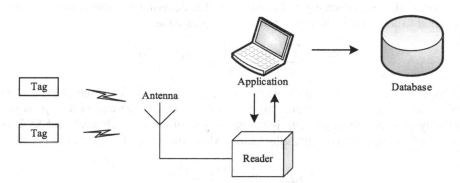

Fig. 1. RFID system structure diagram.

The working process of the FDM UHF passive RFID system in accordance with the ISO/IEC 18000 UHF RFID protocol [21] is as follows: the reader first radiates a continuous RF signal of a fixed frequency through the antenna, and the tag within the working range receives the continuous RF signal and is activated; The reader broadcasts the "select" command or select a specific tag according to whether the tag information is known. After receiving the command, the tag analyzes the initial channel allocation information such as the number of Miller code sub-carriers (M parameter value) and reflection rate BLF, or the dedicated channel allocation information. The reader broadcasts the "inventory" command after the specified time slot, and the tag reflects the command execution result to the reader in the independent sub-channel after receiving the command. The reader performs demodulation and decoding of the received reflected signals, and identifies the tag with a confirmation mark to establish a complete communication process. If a tag collision occurs, repeating the preceding steps. Performing other operations only after the reader completes the inventory of all tags in the coverage range. The reader receives tag information of each sub-channel at the same time and sends it to the upper-layer application for sub-sequent judgment.

If multiple tags in independent sub-channels reflect RF signals at the same time, the higher harmonics of the fundamental wave of one sub-channel will spread to other sub-channels, as shown in Fig. 2.

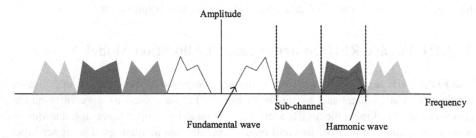

Fig. 2. Harmonic interference diagram.

For the affected sub-channels, harmonic interference is mainly reflected in the change of signal to noise ratio. Shannon's formula is expressed as

$$cap = \sum_i cap_i = \sum_i b_i log_2(1 + SNR_i) \tag{1}$$

where $i = 1, 2, \ldots, num$ represents the counting subscript of each sub-channel (corresponding to a tag), num represents the total number of sub-channels, cap_i is the capacity of each sub-channel, b_i is the bandwidth of each sub-channel, and SNR_i is the signal-to-noise ratio of each sub-channel. Meanwhile, the BER can be calculated as

$$BER = \frac{\sum_i erfc(|\sqrt{SNR_i}|)}{num} \tag{2}$$

The harmonics fall into the affected sub-channel as noise, while the signal-to-noise ratio of the affected sub-channel decreases, the capacity decreases and the BER increases. Therefore, the power intensity of the harmonic interference signal in the system has become an important factor affecting the key index of the RFID system besides the quality of the original channel.

Figure 3 shows the influence of different channel allocation configuration on the spectrum utilization and harmonic interference power of the system and their conflict relationship. If each sub-channel occupies relatively narrow resources or has a large distribution interval in the available frequency band, it can avoid the influence of partial carriers on other sub-channels, but there is a great waste in the utilization of system spectrum resources. As shown in Fig. 3(a), to suppress the harmonic interference in the system frequency band to the maximum extent, the sub-channels occupied by the 3rd and 5th harmonics generated by the fundamental wave with the center frequency f_1 are not allocated, so the available resources of the frequency band are significantly reduced. On the contrary, if each sub-channel is continuously distributed in the available frequency band, the system frequency band resources can be utilized to the maximum extent, but the mutual interference between sub-carriers is also intensified. As shown in Fig. 3(b), in order to make full use of the frequency band resources, continuous division is conducted, but this causes the 3rd and 5th harmonics generated by the fundamental wave with the center frequency of f_1 to overlap with the fundamental wave with the center frequency of f_3 and f_5, worsening the signal-to-noise ratio of the latter sub-channels, resulting in a sharp rise in the total interference power intensity in the frequency band and an increase in the average BER of the system.

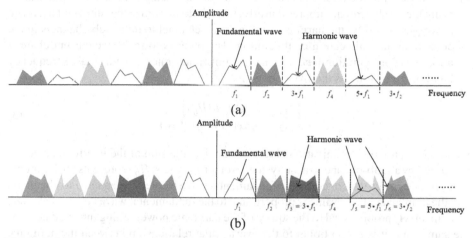

Fig. 3. Sub-channel conflict diagram of different channel allocation configurations (a) Channel allocation configurations with priority of harmonic interference power index; (b) Channel d allocation configurations with priority of spectrum utilization index.

Based on the above analysis, this paper proposes to establish a channel allocation model of FDM UHF passive RFID system based on harmonic interference power and spectrum utilization, to improve the system spectrum utilization without increasing the BER. Since the traditional multi-objective optimization problem is decomposed into multiple single-objective problems and solved by a mature single-objective optimization algorithm, it is impossible to maintain the BER of the existing system while improving the system spectrum utilization. Therefore, this paper adopts the Pareto optimization rule to solve the multi-objective optimization problem. It takes harmonic interference power and spectrum utilization as two optimization objectives in the multi-objective optimization problem, and the available frequency band resources of the system and the established tag distribution in the usage scenario as constraints, solves the solution set of the channel allocation configuration, according to the actual application's tolerance of each parameter and the demand for performance, adjust the weight of the optimization target, so as to select the optimal configuration. The specific model is established as follows:

1) Model input: available frequency band resource BW, number of sub-channels num, distribution mode of tag d, sub-channel center frequency search step Δf, sub-channel width search step Δb;
2) Model variables: band width occupied by each sub-channel $b_i = bh_i - bl_i$ (bh_i and bl_i represent the upper and lower boundaries of the sub-channel respectively), and its center frequency f_i;
3) Model output: channel allocation solution set, each solution is a division of the available frequency band resources of the system composed of the central frequency of each sub-channel and its corresponding bandwidth $[b_i, f_i]$;
4) Optimization objectives: one is to make the current channel allocation configuration bring the best harmonic suppression effect, that is, to minimize the sum of interference power generated by the harmonics of each sub-channel to other sub-channels in the current channel allocation configuration. The other is to make the current channel allocation configuration have the highest utilization rate of the available frequency band of the system, as the (3) is shown:

$$\begin{cases} G_1 = min\{sum(H_{ij})\} \\ G_2 = max\{sum(b_i)/BW\} \end{cases} \tag{3}$$

where G_1 represents optimization objective one, i.e., the sum of the interference power generated by all sub-channels in the system to other channels; H_{ij} represents the harmonic interference power generated by the fundamental wave located in channel i to channel j, and the amplitude of this value is proportional to the fundamental wave power of channel i and inversely proportional to the square of the harmonic power falling into channel j. At the same time, it is proportional to the overlapping relationship between the harmonic signal spectrum and the j channel. G_2 represents the optimization objective two, i.e., the utilization rate of the available frequency band of the system, which is specifically represented as the ratio of the sum of the occupied band width b_i by each sub-channel and total frequency band width BW in system. Obviously, the larger b_i is, the more fully the frequency band is utilized.

A set of sub-channel center frequency f_i and corresponding band width b_i are randomly initialized according to the number of system tags when use this model. Secondly, if the distribution of each channel meets the boundary control conditions, the current allocation configuration is obtained and the sub-sequent steps are continued; otherwise, the allocation scheme is re-generated. Thirdly, according to the information of reader transmit power and gain, tag reflection coefficient and distribution, path loss model, and channel overlap relationship, etc., the harmonic interference power generated by each channel in the available frequency band under the current allocation configuration is calculated. Finally, the two target values obtained from the current allocation configuration are calculated. The pseudo-code is shown in Table 1.

Table 1. System channel allocation model pseudo-code

System channel allocation model pseudo -code	
Input	Total channel bandwidth, number of sub-channels, distribution of sub-channels, search step size var;
Output	The current allocation configuration and its interference power and spectrum utilization;
Step1	create_x(var, num);

Step2	IF(check_x($1:num$))
	    Step3;
	ELSE
	    Step1;

Step3	$k = 3$;
	FOR($i = 1:num$)
	    $f_{harmonic} = k \times f_i$;
	    $b_{harmonic} = [bl, bh] = [f_{harmonic} - \frac{b_l}{2}, f_{harmonic} + \frac{b_l}{2}]$;
	    WHILE($[bl, bh] \cap BW \neq \emptyset$)
	        FOR($j = i + 1:num$)
	            $H_{ij} = \frac{[bl,bh] \cap [bl_j, bh_j]}{[bl,bh]} \times \frac{|P_i|}{k^2}$;
	        END
	        $k = k + 2$;
	    END
	END

Step4	sum(H_{ij});
Step5	sum(b_l) /BW;

3   Channel Allocation Optimization Method Based on NSGA-II

In this paper, improved NSGA-II optimization algorithm is proposed to iteratively optimize the model parameters for FDM UHF passive RFID system and search for Pareto optimal frontier channel allocation configuration. Consider the conventional multi-objective optimization problem:

$$min|f(x)| = min|f_1(x), f_2(x), \dots, f_i(x), \dots f_m(x)|$$

$$s.t. \begin{cases} g_j(x) \leq 0, j = 1, \dots p \\ h_k(x) \leq 0, k = 1, \dots q \end{cases} \quad (4)$$

Where $f(x)$ represents the optimization target, m represents the number of targets to be optimized, $x = (x_1, x_2, \dots, x_n)$ is an n-dimensional decision variable, $g_j(x)$ represents inequality constraints, and $h_k(x)$ represents equality constraints. The space R^n satisfying both equality and inequality constraints is called the decision space, and R^m is called the target space. If x_1 and x_2 are feasible solutions to the problem, and there is $f_i(x_1) < f_i(x_2)$ for any $f_i(x)$, then x_1 dominates x_2, when no solution can be dominated by other solutions in the same set and both are feasible, That is the Pareto optimal solution set, and its boundary is called Pareto optimal frontier [22]. Improved NSGA-II algorithm [23] is a multi-objective genetic algorithm based on fast non-inferior ranking. By introducing non-dominant ranking and congestion calculation, improved NSGA-II performs selection, crossover, variation, and screening to maintain a population of potential solutions, supports information formation and exchange in these directions, and searches by surface, to approach Pareto optimal solution set step by step.

In the problem presented in this paper, the harmonic interference power summations and the system spectrum utilization cannot be optimized simultaneously, and the computational complexity of the brute force search channel allocation configuration increases rapidly with the increase of the number of system tags (channels). Therefore, this paper proposes to apply improved NSGA-II algorithm in the channel allocation of FDM UHF passive RFID system, solve the Pareto optimal solution set of balancing two optimization objectives, evaluate the performance of all solutions in the solution set, and implement the optimal configuration according to the application requirements. Figure 4 shows the optimization algorithm process.

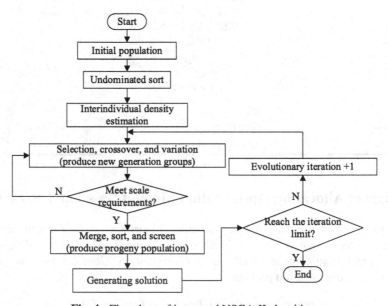

Fig. 4. Flowchart of improved NSGA-II algorithm.

This paper sets the population size to 50, the crossover ratio to 0.8, the mutation probability to 0.05, and the progeny population is composed of the new generation, whose size is 80% of the initial population, and the parent generation, whose size is 20% of the initial population.

Firstly, the initial population is established according to the channel allocation model, and the individual sequencing parameters are calculated one by one. Secondly, the domination set and the domination times of the individuals are initialized, the individuals are ranked according to the domination relations of the optimization objectives among the individuals. The fewer the number of dominations, the higher the rank. The individuals who are not dominated by any individual have a rank of 1. The dominations of individuals with a rank of 1 are successively traversed, the dominated times of each individual in the set is decreased and ranked downward. Thirdly, the inter-individual density is estimated, i.e., the degree of crowding among individuals in the same rank is calculated. The same objective function value of individuals at the same level is arranged in ascending order, and the crowding degree between the first and last individuals is set as infinite. The crowding degree between other individuals is calculated according to the ratio of the difference of the objective function value and the distance between the head and the tail. Thus, the parameters of the initial population were solved.

After that, generate new generation for iterative optimization. Firstly, performing the selection operation based on binary tournament, and two individuals are randomly selected from the parent generation to compare rank ranking, select individuals with higher rank, and select individuals with greater crowding at the same level. Secondly, performing a single point crossover based on random intersections with the results of the selection to generate two new individuals. After meeting the demand of size, the uniform mutation operation with probability of 0.05 is performed on it. Thirdly, calculating the series parameter values of the mutated individuals, and merging two populations, the progeny population are screened out in the order of decreasing rank and crowding degree. The highest-ranking group in the progeny generation are selected as the solution. Examples of data are given in the fourth section.

4   Simulation Result

4.1   Simulation Setup

In this section, MATLAB platform is used to simulate the tag distribution scene to verify the effectiveness of the channel allocation scheme in this paper.

According to the ISO/IEC 18000-6C protocol, this paper adopts sub-carrier modulation, and sets the information transmission rate from the tag to the reader as 5 kbps to 320 kbps, and the sub-carrier frequency range as 40 kHz to 640 kHz. The other parameters are shown in the Table 2.

4.2   Equidistant Distribution Simulation

Considering a FDM UHF passive RFID system containing 10 to 50 sub-channels, with the reader as the center of the circle, and each tag is equidistant from the reader 10 m. With 10 channels as the observation step, the Pareto solution set of channel allocation configuration is generated by 100 iterations under five channel numbers.

The two optimization target parameters (interference power and 1-utilization rate) of each solution as the horizontal and vertical coordinates are used to show the performance of the solution set, as shown in Fig. 5, and the blue marked solution in Fig. 5(a) is taken as an example, whose specific values are shown in Table 3.

Table 2. System parameter setting

parameter	value
Range of channel center frequency	40 kHz–640 kHz
Unit channel bandwidth	10 kHz
Range of the unit bandwidth used by each channel	1–4
Total available channel width of the system	1 MHz
RFID system work area	A circle with a reader as its center and a radius of 10m
Reader transmits power	1 W
System label environment distribution	Equidistant/random
Transmit/reflect gain	1
Tag reflection coefficient	0.1
Single path loss model	$1/\left(4\pi d^2\right)$

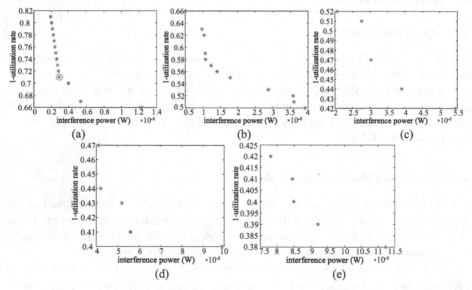

Fig. 5. System channel allocation scheme solution set performance diagram for (a) $num = 10$; (b) $num = 20$; (c) $num = 30$; (d) $num = 40$; (e) $num = 50$;

As can be seen from the sub-graphs in Fig. 5, for the solutions generated with the same number of channels, multiple solutions do not form a dominant relationship. Therefore, the optimal channel allocation configuration can be accurately positioned while meeting the system's requirements for a certain performance index. By comparing the sub-graphs in Fig. 5, with the increase of the sub-channel number in the system, the spectrum

Table 3. Example of the system channel allocation scheme (*num* = 10)

parameter	Value (KHz)
Each channel center frequency	[140, 230, 270, 295, 350, 380, 500, 570, 610, 640]
Each channel bandwidth	[30, 30, 40, 10, 30, 30, 30, 30, 20, 40]
Upper boundary of each channel	[155, 245, 290, 300, 365, 395, 515, 585, 620, 660]
Lower boundary of each channel	[120, 215, 250, 290, 335, 365, 485, 555, 600, 520]

utilization is fuller. When the system sub-channel number is 50, the spectrum utilization in the system can reach 62%. At the same time, the harmonic interference power only has slightly increase, which verifies the effectiveness of the proposed scheme.

To further analyze the performance of the channel allocation scheme proposed in this paper, the channel average spectrum utilization, transmission rate and BER achieved by the solution set in Fig. 5 are calculated, as shown in Fig. 6(a). At the same time, take the channel allocation scheme in literature [24] as the experimental control group, the system parameter settings are consistent with those in Table 2, and the simulation results are shown in Fig. 6(b).

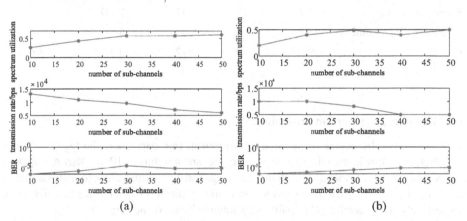

(a)   (b)

Fig. 6. Performance comparison of equidistant distributed channel allocation configurations by (a) proposed scheme in this paper; (b) scheme proposed by literature [24].

From the mean values of each element in the solution set, the quality of the system channel allocation scheme proposed by this paper is generally better than that of literature [24]. The specific numerical pairs are shown in Table 4 and Table 5. In terms of channel average spectrum utilization in Table 4, compared with scheme of literature [24], the proposed scheme has an improvement of 8.9% to 42.95%, and the improvement is more obvious under the condition of a large number of system sub-channels. In terms of the average channel transmission rate in Table 5, since the average BER of the two methods under any number of sub-channels is lower than 10^{-5}, it can be regarded as the harmonic interference is very small compared with the original noise of the system, when the channel allocation configuration of the two schemes is carried out under this noise intensity, the average transmission rate index of each sub-channel is mainly affected by the effective channel width occupied by the tag to transmits information, so the index has the same increase trend as the average spectrum utilization ratio of the system.

Table 4. Comparison of average spectrum utilization

Scheme	num				
	10	20	30	40	50
Proposed scheme	26.34%	43.66%	57.44%	57.18%	60%
Literature [24] scheme	19.94%	40.09%	48.94%	40%	50%
Contrast	+32.3%	+8.9%	+17.36%	+42.95%	+20%

Table 5. Comparison of average channel transmission rate (Kbps)

Scheme	num				
	10	20	30	40	50
Proposed scheme	13.17	10.92	9.57	7.148	6
Literature [24] scheme	9.97	10.02	8.16	5	5
Contrast	+32.1%	+8.98%	+17.28%	+42.96%	+20%

4.3 Random Distribution Simulation

To be closer to the actual application environment, this paper sets the tags randomly distributed in a circle with the reader as the center and a radius of 10 m, other parameters are same as those in Table 2. The channel average spectrum utilization, transmission rate and BER achieved by the proposed scheme and literature [24] scheme are calculated, respectively. The specific value pairs are shown in Table 6 and Table 7.

Table 6. Comparison of average spectrum utilization

Scheme	num				
	10	20	30	40	50
Proposed scheme	26.82%	45.3%	55.74%	57.8%	61.33%
Literature [24] scheme	24.86%	49.8%	30.49%	40%	50%
Contrast	+7.88%	−9.03%	+82.81%	+44.5%	+22.66%

Table 7. Comparison of average channel transmission rate (Kbps)

Scheme	num				
	10	20	30	40	50
Proposed scheme	13.29	9.37	8.32	6.22	5.63
Literature [24] scheme	11.09	10.85	4.5	4.35	4.3
Contrast	+19.84%	−13.64%	+84.89%	+42.99%	+30.93%

In the case that system tags are randomly distributed within the coverage range of readers, the proposed scheme still has a certain degree of advance compared with the literature [24] scheme. In terms of channel average spectrum utilization in Table 6, the proposed scheme can increase by 82.81% at the highest. When the number of channels is 20, the potential reason for the slight decrease may be that the number of iterations adopted in this experiment is insufficient, resulting in local optimization. In general, the proposed schemes can effectively generate channel allocation configuration with high spectrum utilization, and the improvement effect of spectrum utilization under the condition of a large number of sub-channels is generally higher than that under the condition of a small number of sub-channels. In terms of the average channel transmission rate in Table 7, although the interference situation of the system has worsened, it is still mainly affected by the effective channel width occupied by the tag to transmit information, which has the same trend of increase as the average spectrum utilization rate of each channel in the system. In addition, the average BER of the two schemes under any number of channels remains in the order of 10^{-1}, this is since under this distribution of tags in the experiment, a large number of tags are very close to readers, and the total interference intensity brought to the system also increases. However, compared with literature [24] scheme, the proposed scheme still has some advantages.

In summary, compared with the existing works, the channel allocation configuration generated by the proposed scheme significantly improves the spectrum utilization under the condition of guaranteeing the BER.

5   Conclusion

In view of the increasing access scale of tags and the low spectrum utilization in FDM UHF passive RFID system, this paper established a channel allocation model, and used improved NSGA-II optimization algorithm to iteratively optimize the system model parameters and search for the Pareto optimal frontier. In RFID systems containing 10 to 50 channels, the model was able to flexibly set the focus target according to the application needs, and efficiently formed a set of channel allocation configurations corresponding to multiple targets. Simulation result showed that compared with existing research schemes, the channel allocation configurations generated by the proposed scheme improved the average spectrum utilization rate by 8.9% to 42.95% for the equidistant distribution, and improved that of 7.88% to 82.81% for the equidistant distribution under the premise of ensuring the communication accuracy. Moreover, the highest spectrum utilization rate of the channel allocation configurations generated by the proposed scheme achieved 62%. It is necessary to say that, at present, the analysis of application requirements and the setting of the weights of each objective are carried out by human judgment, the performance of this method in terms of flexibility and autonomy still needs to be improved if the number of objectives is increased or the requirements change dynamically. Therefore, we hope to enhance the system's learning ability and autonomous judgment through more learning-based ideas, so that the system can truly adapt to a variety of applications automatically, which is also a priority for our future work.

Acknowledgements. This work was supported in part by National Key R&D Program (2018AAA0103203), in part by Guangdong Provincial Research and Development Plan in Key Areas (2019B010141001), in part by the joint project of China Mobile Research Institute& X-NET (Project Number: 2022H002),in part by National Natural Science Foundation of China (U22B2004, 61971113, 61901095, 62371106), in part by Sichuan Provincial Science and Technology Planning Program of China (2022YFG0230, 2023YFG0040), in part by the Fundamental Enhancement Program Technology Area Fund (2021-JCJQ-JJ-0667), in part by the Joint Fund of ZF and Ministry of Education (8091B022126), in part by the Fund Project of Intelligent Terminal Key Laboratory of Sichuan Province(SCITLAB-1015), and in part by Innovation Ability Construction Project for Sichuan Provincial Engineering Research Center of Communication Technology for Intelligent IoT (2303-510109-04-03-318020).

References

1. Huang, Y.L.: Internet of Things Radio Frequency Identification (RFID) Core Technology Course. Posts and Telecommunications Press, Beijing (2016)
2. Gong, W., Chen, S., Liu, J.C.: Towards higher throughput rate adaptation for backscatter networks. In: The IEEE 25th International Conference on Network Protocols (ICNP), Toronto, pp. 1–10. IEEE (2017)
3. Jin, M., He, Y., Meng, X., Zheng, Y.L., Fang, D.Y., Chen, X.J.: FlipTracer: practical parallel decoding for backscatter communication. In: The 23rd Annual International Conference on Mobile Computing and Networking (MobiCom 2017), pp. 275–287. Association for Computing Machinery, New York (2017)
4. Zhu, F.Y., Feng, Y.D., Li, Q.R., Tian, X.H., Wang, X.B.: DigiScatter: efficiently prototyping large-scale OFDMA backscatter networks. In: The 18th International Conference on Mobile Systems, Applications, and Services (MobiSys 2020), pp. 42–53. Association for Computing Machinery, New York (2020)
5. Rajoria, N., Kamei, H., Mitsugi, J., et al.: Multi-carrier backscatter communication system for concurrent wireless and batteryless sensing. In: International Conference on Wireless Communications and Signal Processing, Nanjing. IEEE (2017)
6. Wang Y.: Research on Efficient and Robust Anti-Collision Algorithm of Radio Frequency Identification Tags. Southwest Jiao tong University (2020)
7. Razaie, H., et al.: A fair reader collision avoidance protocol for RFID dense reader environments. Wirel. Netw. (2018)
8. Liu, T.T., Yang, C.Y., Yang, L.L.: A low-complexity sub-carrier-power allocation scheme for frequency-division multiple-access systems. IEEE Trans. Wirel. Commun. **9**(5), 1564–1570 (2010)

9. Sileh, I.K.: Adaptive Relaying Protocol Multiple-Input Multiple-Output Orthogonal Frequency Division Multiplexing Systems. University of Southern Queensland (2014)
10. Li, F.F.: Research on Dynamic Sub-Carrier Allocation in OFDMA System. Beijing University of Posts and Telecommunications (2007)
11. Jiang, W.L., Zhang, Z.Z., Sha, X.J., Sun, L.A.: Low complexity hybrid power distribution combined with sub-carrier allocation algorithm for OFDMA. J. Syst. Eng. Electron. **22**(6), 879–884 (2011)
12. Hou, Z., Wu, D., Cai, Y.M.: Sub-carrier and power allocation in uplink multi-cell OFDMA systems based on game theory. In: 12th International Conference on Communication Technology, pp. 1113–1116. IEEE (2010)
13. Zhang, Z.Y., Shi, J., Chen, H.H., et al.: A cooperation strategy based on Nash bargaining solution in cooperative relay networks. IEEE Trans. Veh. Technol. **57**(4), 2570–2577 (2008)
14. Zhang, G.P., Zhang, H.L., Zhao, L.Q., et al.: Fair resource sharing for cooperative relay networks using Nash bargaining solutions. IEEE Commun. Lett. **13**(6), 381–383 (2009)
15. Sanguanpuak, T., Rajatheva, R.: Power bargaining for amplify and forward relay channel. In: The 4th International Conference on Communications and Networking in China (ChinaCom 2009), pp.1−5 (2009)
16. Ahmadi, H., Chew, Y.H.: Adaptive sub-carrier-and-bit allocation in multiclass multiuser OFDM systems using genetic algorithm. In: IEEE International Symposium on Personal (2010)
17. Sharma, N., Anupama, K.R.: A novel genetic algorithm for adaptive resource allocation in MIMO-OFDM systems with proportional rate constraint. Wirel. Pers. Commun. **61**(1), 113–128 (2011)
18. Sharma, N., Tarcar, A.K., Thomas, V.A., et al.: On the use of particle swarm optimization for adaptive resource allocation in orthogonal frequency division multiple access systems with proportional rate constraints. Inf. Sci. **182**(1), 115–124 (2012)
19. Igarashi, Y.., Sato, Y.., Kawakita, Y., et al.: A feasibility study on simultaneous data collection from multiple sensor RF tags with multiple sub-carriers. In: IEEE International Conference on RFID. IEEE (2014)
20. Rajoria, N., Igarashi, Y., Mitsugi, J., et al.: Concurrent backscatter streaming from battery-less and wireless sensor tags with multiple sub-carrier multiple access. IEICE Trans. Commun. **E100.B**(12) (2017)
21. ISO/IEC JTC 1/SC 31 Automatic identification and data capture techniques. Information technology — Radio frequency identification for item management — Part 6: Parameters for air interface communications at 860 MHz to 960 MHz: ISO/IEC 18000-6: 2013, 27 September 2023. https://www.iso.org/standard/46149.html
22. He, L., Liu, Q.H.: Multi-Objective Optimization Theory and Continuous Method. Science Press, Beijing (2015)
23. Zheng, W., Doerr, B.: Better approximation guarantees for the NSGA-II by using the current crowding distance. GECCO, 611–619 (2022)
24. Rajoria, N., Kamei, H., Mitsugi, J., et al.: performance evaluation of variable bandwidth channel allocation scheme in multiple sub-carrier multiple access. IEICE Trans. Commun. **101**(2), (2017)

Research on Physical Layer of Passive IoT Communication Protocol Based on Cellular Fusion

Yi He[1], Yulu Zhang[2], Yuan Li[2], Shuai Ma[2], Gui Li[3], Junyang Liu[1], Haiwen Yi[1], Yue Liu[1], Guangjun Wen[1], Xu Zhang[1], and Jian Li[1(✉)]

[1] School of Information and Communication Engineering/ Yibin Institute, University of Electronic Science and Technology of China, Chengdu 611731, China
lj001@uestc.edu.cn
[2] Department of IoT Technology Research, China Mobile Research Institute, Beijing 100053, China
[3] The 10th Research Institute of China Electronics Technology Group Corporation, Chengdu 611731, China

Abstract. As transportation, logistics and other industries put forward the demand for low-cost item tracking, it is urgent to use the advantages of cellular networks to achieve seamless coverage of the connection of things, and realize the universal internet of everything in a true sense. In view of the shortcomings of traditional radio frequency identification (RFID) technology, such as limited communication distance, large interference, inability to continuous networking, high deployment and human operation and maintenance costs, and no support for positioning, this paper proposes a cellular fusion passive air interface protocol stack physical layer communication scheme. Based on the 18000-6C protocol, 256 quadrature amplitude modulation (QAM) is introduced in the downlink and an improved Pulse interval encoding (PIE) encoding is proposed. Tail biting convolutional codes (TBCC) and Polar code are used in the uplink to improve communication distance and optimize inventory efficiency. The signal level and error rate curve of the communication link are simulated. The simulation results show that both the base station and the label can send and receive data correctly. In the downlink, the bit error rate (BER) decreases with the increase of modulation depth. In the uplink, binary phase shift keying (BPSK) modulation has better error performance than amplitude shift keying (ASK) modulation, and Polar code has better error performance than TBCC.

Keywords: cellular communication · passive Internet of Things (IoT) · QAM · Polar code · TBCC

1 Introduction

The Internet of Things is one of the core driving forces for the development of future wireless communication systems [1]. In order to reduce the power consumption of IoT nodes, technologies such as Bluetooth low energy (BLE), long range radio (LoRa), narrow band internet of things (NB-IoT) and the 5th generation wireless systems (5G) reduced capacity (RedCap) have flourished in recent years. According to statistics, China's ultra-high frequency (UHF) RFID sales are expected to reach 11.5 billion in 2024, which means that the passive IoT represented by UHF RFID has huge application potential and research value.

Traditional passive RFID tags carry a small amount of data, mostly using low-order modulation such as amplitude keying, phase shift keying and other simple coding methods such as pulse coding. Although these methods are simple to implement, which limits the application range of the system. In order to improve the information transmission rate of tags, research can be conducted on BPSK, QAM for the uplink and downlink signals of tags, as well as multi-carrier modulation based on orthogonal frequency-division multiplexing (OFDM) [2]. References [3] proposed RFID tags that support BPSK and 16-QAM modulation, respectively. Similarly, references [4, 5] implemented a backscatter modulation scheme based on OFDM carriers. In addition, some more efficient coding methods, such as an encoding method based on bit energy consumption differences [6], are also expected to be applied to passive IoT to optimize modulation-coding depth and reduce signal reflection losses.

This paper mainly studies the scheme and simulates of the physical layer communication link of the passive communication protocol for cellular fusion. The downlink base station uses 256 QAM to construct ASK signals with different modulation depths, which are encoded by PIE and transmitted through antennas. The tag end receives data through operations such as envelope detection; The uplink tag side uses FM0, Miller, Polar or TBCC encoding and ASK or BPSK modulation to process data. The simulation results show that under the same conditions, the BER of BPSK modulation is lower than that of ASK modulation, and the BER of Polar encoding is lower than that of TBCC.

2   Key Technologies

2.1   Cellular Passive Communication Architecture

The composition of the cellular direct connection architecture is shown in Fig. 1, where the cellular base station communicates directly with the tag and sends radio frequency (RF) carrier signals and command signals to the tag.

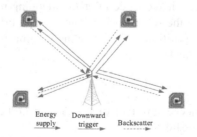

Fig. 1. Cellular direct connection architecture.

Among them, the RF carrier signal is used to provide energy to the tag, and the command signal carries the control information of the tag; The tags transmit information to the base station through backscattering. In the cellular direct connection architecture, passive tags are limited by the charging sensitivity of downlink energy supply signals, resulting in a contraction of downlink coverage capacity; The semi passive tag utilizes environmental energy collection and uplink enhancement technology to ensure that the demodulation sensitivity of the tag data is not limited to the wireless charging sensitivity, thereby achieving a balance between the downlink and uplink coverage capabilities of the tag.

2.2 Coding and Decoding

Polar

The reliability estimation of polarized channels can effectively identify good channels and complete channel sorting, which is the key to constructing high-performance polarization codes. In this paper, polarization weights are used to estimate the reliability of additive white gaussian noise (AWGN) polarized channels. The calculation is as follows:

$$W_i = \sum_{j=0}^{n-1} B_j * 2^{j/4} \tag{1}$$

where, W_i is the polarization weight used on the i-th input bit; i represents the sequence number of the input bit (for Polar codes with a length of $N = 2^n$, $i = 0, 1, 2, \ldots, N - 2, N-1$; n represents the number of bits required to represent the integer N); B_j represents the j-th bit of bit 0 or 1 that converts integer i into binary form.

After estimating the reliability of the sub-channels, we can select the first K bits of reliability ranking channel to place information bits, and place frozen bits on the remaining channel sequence position, and these two sets form the input u_1^N of the vector channel. This process is the mixing of information bits and frozen bits. The encoding of Polar codes also follows the encoding principle of general binary linear block codes. The encoding process of a code word with a code length of $N = 2^n$ can be represented by the following equation

$$x_1^N = u_1^N G_N \tag{2}$$

x_1^N is to input the encoded bit sequence after the mapping relationship, the mapping relationship is to complete the selection of channels, and information bits are placed in good channel sequence positions. This mapping relationship can be represented by generating a matrix, as follows:

$$G_N = F^{\otimes n} \tag{3}$$

$F^{\otimes n}$ is the n th Kronecker product of the Arikan kernel, and $F^{\otimes n}$ satisfies the following recursive relationship

$$F^{\otimes n} = F \otimes F^{\otimes(n-1)} \tag{4}$$

Based on the coding principle of Polar codes, after channel splitting, independent sub-channels become interdependent sub-channels, meaning that sub-channels with sequence numbers first will have an impact on the subsequent channels. Therefore, Arikan proposed a serial cancellation (SC) decoding algorithm [7], which assumes that the current result is correct when each bit is translated, and then performs the decoding of the next bit until all bits are decoded.

SC decoding uses a recursive approach to sequentially decode all bits, and the log-likelihood ration (LLR) value is calculated as Eq. (5):

$$L_N^{(i)}\left(y_1^N, \hat{u}_1^{i-1}\right) = \ln\left(\frac{W_N^{(i)}\left(y_1^N, \hat{u}_1^{i-1}|0\right)}{W_N^{(i)}\left(y_1^N, \hat{u}_1^{i-1}|1\right)}\right) \tag{5}$$

By definition, the value of LLR is the ratio of the probability of the sender sending '0' to the probability of sending '1' when the receiver obtains the value of $\left(y_1^N, \hat{u}_1^{i-1}\right)$. Obviously, if the LLR value is greater than 0, it means that the probability of originally sending "0" is slightly higher. Therefore, it is determined that the signal that this receiving end should receive is '0', and vice versa, it is '1'. The decision rule is shown in Eq. (6).

$$\hat{u}_i = \begin{cases} 0, \text{ if } L_N^{(i)}\left(y_1^N, \hat{u}_1^{i-1}\right) \geq 0 \\ 1, \text{ if } L_N^{(i)}\left(y_1^N, \hat{u}_1^{i-1}\right) < 0 \end{cases} \tag{6}$$

In the information sent by the sender, there is a portion of frozen bits without information content. Assuming that these frozen bits are all set to 0 during transmission, when decoding this bit, even if the LLR is less than 0, it is hard determined to be '0'. Define functions f and g respectively as Eqs. (7) and (8), and the calculation of LLR can be completed through these two equations:

$$f(a, b) = \ln\left(\frac{1 + e^{a+b}}{e^a + e^b}\right) \tag{7}$$

$$g(a, b, u_s) = (-1)^{u_s} a + b \tag{8}$$

Among them, $a, b \in \mathbb{R}$, $u_s \in \{0, 1\}$. The LLR recursive operation formula is shown in Eq. (9):

$$\begin{cases} L_N^{(2i-1)}\left(y_1^N, \hat{u}_1^{2i-2}\right) = f\left(L_{N/2}^{(i)}\left(y_1^{N/2}, \hat{u}_{1,o}^{2i-2} \oplus \hat{u}_{1,e}^{2i-2}\right), L_{N/2}^{(i)}\left(y_{N/2+1}^N, \hat{u}_{1,e}^{2i-2}\right)\right) \\ L_N^{(2i)}\left(y_1^N, \hat{u}_1^{2i-1}\right) = g\left(L_{N/2}^{(i)}\left(y_1^{N/2}, \hat{u}_{1,o}^{2i-2} \oplus \hat{u}_{1,e}^{2i-2}\right), L_{N/2}^{(i)}\left(y_{N/2+1}^N, \hat{u}_{1,e}^{2i-2}\right), \hat{u}_{2i-1}\right) \end{cases} \tag{9}$$

Among them, $\oplus$ represents modular two addition operation.

The termination condition for recursion is $N = 1$, at which point $L_1^{(1)}(y_j) = \ln\left(\frac{W(y_j|0)}{W(y_j|1)}\right)$. At this point, it still belongs to the definition formula because the transition probability is unknown and does not have operability in practice. When introducing the Gaussian approximation method, the logarithmic likelihood ratio of the received symbol y is defined as Eq. (10):

$$L(y) = \ln\frac{p(y|0)}{p(y|1)} = \frac{2y}{\sigma^2} \tag{10}$$

Among them, y is the received symbol and σ^2 is the noise variance. Therefore, the initial value of the logarithmic likelihood ratio at the decoding end can be easily obtained.

When channel polarization is insufficient, SC decoding can lead to serious error propagation. In response to this situation, a successive cancellation list (SCL) decoding algorithm has been proposed, which can effectively solve the problems of the SC algorithm [8]. According to the decoding principle, the logical relationship between each bit can be clearly displayed through a full binary tree. The decoding rules for a polar code tree with a code length of 4 are shown in Fig. 2. Taking a node v in the tree (except for the leaf node) as an example, if it represents u_1^i, its left and right nodes represent $\left(u_1^i, u_1^{i+1} = 0\right)$ and $\left(u_1^i, u_1^{i+1} = 1\right)$ respectively. The SC decoding code tree is only related to the code length N, and the decoding process is from the root node to the leaf node, selecting the most suitable path according to the LLR value to sequentially decode each bit. The red line in the figure represents a complete decoding path. It can be seen that when making decisions on each layer of the tree, the LLR value of each node is used, and the largest path is always selected for backward decoding. The final result is $u_1^N = [0, 0, 1, 1]$.

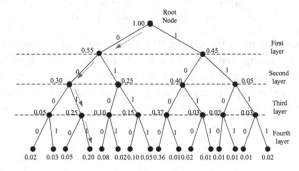

Fig. 2. Binary tree decoding of Polar code with code length 4.

Considering the code tree in Fig. 2, the SC algorithm only retains the path with the highest transfer probability when decoding each bit, and directly makes decisions on the bits. SCL decoding improves on its foundation by allowing multiple decoding paths to be retained and the next bit to be decoded when making a decision on a certain bit, without making a hard decision on it. The overall process is similar to the SC principle, extending from the root node, with each child node extending two branches to search down until reaching the leaf node. If the reserved path is L, when the total number of search paths is not greater than L, all possible decoding results are retained. However, when the total number is greater than L, pruning is performed according to certain rules, retaining only L paths. L is called the SCL decoding width.

When the L value is 1, it means that each layer of the decoding tree only retains a unique path, which is the SC algorithm; When $L \geq 2^N$, it is equivalent to preserving all paths of the decoding tree, which is maximum likelihood decoding. From this, it can be seen that SCL decoding actually includes these two decoding methods. When $2^N > L \geq 2$ is selected, it is crucial to select the $2L$ paths that need to be preserved among the L paths. Obviously, relying solely on the logarithmic likelihood ratio of each bit is not feasible, and it is necessary to select more suitable indicators for evaluating path costs. In the SCL decoding algorithm, this metric is represented as the Path Metric (PM).

For any path $l \in \{1, 2, \ldots, L\}$ and transmission bit $i \in \{1, 2, \ldots, N\}$ in SCL decoding, the PM value is defined as Eq. (11):

$$PM_l^{(i)} = \sum_{j-1}^{i} \ln\left(1 + \exp\left(-(1 - 2\hat{u}_j[l]) \cdot L_N^{(j)}\right)\right) \tag{11}$$

Among them, $L_N^{(j)}[l] = \ln\left(\dfrac{W_N^{(i)}\left(y_1^N, \hat{u}_1^{i-1}[l]|0\right)}{W_N^{(i)}\left(y_1^N, \hat{u}_1^{i-1}[l]|1\right)}\right)$. If the probability of 0 and 1 bits sent by the sender is the same, then for any two different paths $l_1, l_2 \in \{1, 2, \ldots, L\}$, if and only if $PM_{l_1}^{(i)} > PM_{l_2}^{(i)}$ holds, there is the following Eq. (12) relationship:

$$W_N^{(i)}\left(y_1^N, \hat{u}_1^{i-1}[l_1]\hat{u}_1^i[l_1]\right) < W_N^{(i)}\left(y_1^N, \hat{u}_1^{i-1}[l_2]\hat{u}_1^i[l_2]\right) \tag{12}$$

From the above equation, it can be inferred that the path metric PM has an inverse relationship with the path transition probability. For any decoding path, the higher the transition probability, the higher the likelihood that the path is the correct path, and the smaller the PM value. Based on this, the most suitable L path can be selected from the $2L$ paths for subsequent decoding. To simplify the operation, Eq. (11) can be approximated as Eq. (13):

$$PM_l^{(i)} \approx \begin{cases} PM_l^{(i-1)}, \hat{u}_i[l] = \delta\left(L_N^{(i)}[l]\right) \\ PM_l^{(i-1)} + \left|L_N^{(i)}[l]\right|, \hat{u}_i[l] \neq \delta\left(L_N^{(i)}[l]\right) \end{cases} \tag{13}$$

Among them, $\delta(x) = \frac{1}{2}(1 - sign(x))$, $PM_l^{(0)} = 0$. If frozen bits are taken into account during decoding, the above equation can be written as Eq. (14):

$$PM_l^{(i)} \approx \begin{cases} PM_l^{(i-1)}, \hat{u}_i[l] = \delta\left(L_N^{(i)}[l]\right) \\ PM_l^{(i-1)} + \left|L_N^{(i)}[l]\right|, \hat{u}_i[l] \neq \delta\left(L_N^{(i)}[l]\right) \\ +\infty, u_i \text{ is a frozen bit but decoding error} \end{cases} \tag{14}$$

The decision on information bits is shown in Eq. (15):

$$\hat{u}_i = \delta\left(L_N^{(i)}\left(y_1^N, \hat{u}_1^{i-1}\right)\right) \tag{15}$$

Cyclic redundancy check is a common channel error detection technique with strong operability and significant effectiveness, which has been widely applied in practical systems. It adds a cyclic redundancy check (CRC) sequence at the end of the sender's message through a specific calculation method, and the receiver also uses this method to verify whether the recovered data is correct. In the channel encoding of PBCH in 5G systems, a cascade of CRC and polar codes, called CRC aided (CA) polar codes, is used. This simple concatenation greatly improves the performance of polar codes.

The CRC aided successive cancellation list (CA-SCL) decoding algorithm has been proposed for this cascading approach. Firstly, SCL decoding is performed on the received data [9]. Among the reserved multiple paths, the first path that passes CRC verification is selected as the output decoding result, rather than selecting the path with a smaller PM value. The CA polar encoding and decoding process is shown in Fig. 3. Assuming the original information length is k and the added CRC sequence length is m, the complete K polar code information bits are obtained, where $K = k + m$. After polar encoding, a polar code of length N is obtained.

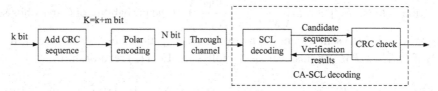

Fig. 3. CA-SCL encoding and decoding flowchart.

TBCC

The structural diagram of the TBCC encoder is shown in Fig. 4, where the number of registers $m = 6$, the bit rate $R = 1/3$, the input information bit stream of the encoder is u, and the output is $\left(c^{(0)}, c^{(1)}, c^{(2)}\right) \backslash* \text{MERGEFORMAT}$ [10].

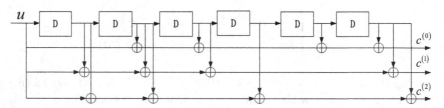

Fig. 4. TBCC encoder.

For the convolutional codes mentioned above that only have feedforward, there are usually two ending methods. One method is to use the tail bit for zeroing, which means that after the information bit encoding is completed, a zero string with a length equal to the length of the encoder register is continued to be transmitted to the encoder, causing the encoder's state to terminate at all zero. The tail bits generated during the zeroing period will be transmitted through the channel, resulting in a loss of bit rate due to these additional bits. Specifically, if an information block with a length of K is passed through a convolutional encoder with a number of registers m and a bit rate of k/n, the length of the output codeword is $(K + m)n$, and therefore the effective bit rate becomes $Kk/n(K + m)$.

Passive tags are transmitted using short codes, with a common EPC code length of 96 bits. Therefore, the initial value of the bitwise convolutional code encoder in the system is set to the last six bits of the input stream as shown in Fig. 5.This way, the end state of all encoded registers will be the same as the start state before encoding. This encoding method does not transmit additional bits, thereby improving encoding efficiency.

u_{K-1}	u_{K-2}	u_{K-3}	u_{K-4}	u_{K-5}	u_{K-6}

Fig. 5. The initial state value of the shift register.

Assuming the length of the sent information sequence is expressed as K, i represents the current number of iterations, and I is defined as the maximum number of iterations. $\Gamma_l^i(s)$ represents the cumulative measure of the surviving path when it reaches states at time l of the i th iteration. The specific steps of the TBCC decoding algorithm are as follows:

Step 1: Initialize. Initialize the metric value of all states to 0, so that the number of iterations is $i = 1$, which is $\Gamma_0^1(s) = 0, s \in \{0, \ldots, 2^m - 1\}$.

Step 2: Perform the first Viterbi decoding, sort all metrics in the ending state, and trace back from the largest state. If this path is a bit tail path, stop decoding and output this decoding sequence. If this path is not a tail biting path, choose the path with the second metric value to start backtracking until the tail biting path is found, stop decoding, and output this decoding sequence. If there is no tail biting path after tracing back from all states at the end, proceed to step 3.

Step 3: The initial measurement values of all states in the second Viterbi decoding are the final measurement values of all states at the end of the first decoding, i.e. $\Gamma_0^i(s) = \Gamma_K^{i-1}(s), s \in \{0, \ldots, 2^m - 1\}$. Perform the second Viterbi decoding and proceed to Step 4.

Step 4: Then start backtracking from the path with the highest metric value among all paths after the end of the second decoding, and output the first $K/2$ bits of the decoding from the K th bit to the $3K/2$ nd bit. Output the $K/2 + 1$ bit to the $K - 1$ bit as the next $K/2$ decoding bits.

2.3 Modulation and Demodulation

Due to the backscatter communication used by passive terminals, which achieves communication functions by changing the impedance matching state, the circuit is in a matched and unmatched state, so the downlink adopts ASK modulation. The base station utilizes PIE encoded DSB-ASK, SSB-ASK, or PR-ASK to modulate radio frequency carriers and communicate with one or more tags.

Traditional passive RFID tags carry relatively small amounts of data and often use load modulation. By adjusting the impedance of the tag antenna interface, the reflection coefficient of the antenna interface is changed to achieve low order modulation methods such as ASK and PSK; This extremely low complexity modulation method, on the one hand, enables the label to achieve signal demodulation by only supporting simple channel estimation and equalization operations, greatly simplifying the circuit and reducing computational power consumption; On the other hand, it also causes the problem of low throughput in passive systems, limiting the application range of passive tags. In order to improve the information transmission rate of tags, it is possible to consider upgrading the passive tag circuit to support more efficient modulation methods such as BPSK and QPSK under energy constraints.

3   Communication Link Design

3.1   Data Frame Design

The downlink data frame is shown in Fig. 6(a) and consists of four parts. Energy signal is mainly used to provide energy for labels; Delimiter signal is used to obtain downlink synchronization signals, determine the start of downlink data frames, determine the number of Bit repetitions, and determine the code rate of equal length precoding; PDSCH is used to determine the data between the frame header and frame footer; Postamble signal is used to determine the end of the downlink data frame.

The uplink data frame is shown in Fig. 6(b) and consists of four parts. Pilot tone signal is used to determine the arrival of uplink data frames; Preamble signal is used to determine the start of the uplink data frame. Obtain the uplink synchronization signal, determine the start of the uplink data frame, determine the number of bit repetitions, and determine the code rate for equal length precoding; PUSCH is used to determine the data between the frame header and frame footer, and according to the high-level protocol, it only determines the number of bit repeat encodings; Middle signal is used to maintain synchronization and determine the boundaries of Polar encoding; Postamble signal is used to determine the end of the uplink data frame.

Energy signal	Delimiter signal	PDSCH	Postamble signal

(a)

Pilot tone signal	Preamble signal	PUSCH	Postamble signal

(b)

Fig. 6.   Frame structure (a) downlink data frame design;(b) uplink data frame design.

3.2   Downlink Design

The simulation flowchart of the Discrete Fourier Transform Spread OFDM (DFT-s-OFDM) downlink is shown in Fig. 7. The base station sends command data frames, which, after PIE coding, undergo 12-point bit repetition, then 256 QAM mapping to construct ASK modulation signals, such as mapping 1 to 00111111 and 0 to 00001111.

After 12-point DFT calculation and frequency domain mapping, and after 16-point IFFT calculation and CP addition, the signal is up-converted to 900 MHz through a pulse shaping filter. After channel introduction of Gaussian white noise and sideband interference, bandpass filters remove out-of-band noise. The signal's envelope is then detected by envelope detection, further filtered by a lowpass filter, sampled and judged, and then decoded by PIE coding to retrieve the original data sent by the base station.

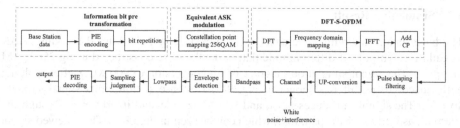

Fig. 7. Downlink simulation flowchart.

3.3 Uplink Design

The simulation flowchart for the spread spectrum of the uplink is shown in Fig. 8. The binary data stream generated by the tag, after FM0, Miller, Polar or TBCC coding, is multiplied by an m-sequence to achieve direct sequence spread spectrum effect, then modulated by BPSK or ASK. After channel introduction of Gaussian white noise and sideband interference, the receiver side uses bandpass filters to remove out-of-band noise. After coherent demodulation and low-pass filtering, de-spreading, sampling, and judgment are performed, followed by decoding to retrieve the original data sent by the tag.

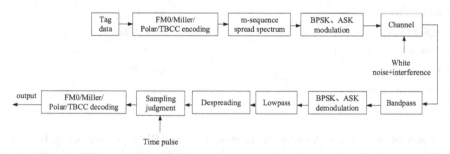

Fig. 8. Uplink simulation flowchart.

4 Simulation Result

4.1 Simulation Setup

The simulation parameter values are shown in Table 1. Considering that if the carrier frequency and symbol rate differ too much, array operations will increase. Therefore, in the simulation, parameters are scaled down: sampling frequency 30 MHz, downlink carrier frequency 9 MHz, uplink carrier frequency 9 MHz, downlink symbol period 0.72 ms, uplink symbol period 0.3 ms, base station transmission signal strength 46 dBm, tag transmission signal strength −36 dBm, downlink path loss 76 dB, uplink path loss 85 dB.

4.2 Simulation Result

The base station randomly generates data frames of length 80. After PIE coding, ASK modulation symbols of different modulation depths were constructed using 256 QAM mapping. The signal simulation diagram of up-conversion to 900 MHz after molding with raised cosine filter is shown in Fig. 9(a).The base station emits signal strength 46 dBm.The signal introduces noise and 930 MHz sideband interference through the channel, as well as 76 dB path loss, which can be seen in Fig. 9(b). The received signal at the tag end is filtered by bandpass filtering to remove out-of-band noise, the signal envelope is detected by envelope detection and the high-frequency component is filtered by lowpass filter, as shown in Fig. 9(c). As can be seen from Fig. 9(d), the tag performs PIE decoding on the sampled decision signal, and compared with the signal by the base station, the tag can correctly receive the data from the base station.

Table 1. Parameter settings of communication link simulation

Parameter	Value
Signal intensity of the base station	46 dBm
Downlink path loss	76 dB
Signal intensity of the tag	−36 dBm
Uplink path loss	85 dB
Sampling frequency	3000 MHz
Downlink symbol period	7.2 us
Carrier frequency	900 MHz
Uplink symbol period	3 us

The relationship between ASK modulation depth and BER is shown in Fig. 10(a). The modulation depth decreases from 87.5% to 66.7%, 36.4%, and the signal-to-interference ratio (SINR) is 0 dB and 10 dB, respectively. From the figure, it can be seen that as the modulation depth increases, the BER decreases, and as the SINR increases, the BER also shows a downward trend. The BER of Polar decoding is shown in Fig. 10(b), which shows that the BER performance of CA-SCL is better than SCL than that of SC, and BER of Polar and TBCC decoding is shown in Fig. 10(c), from which it can be seen that Polar's BER performance is better than TBCC.

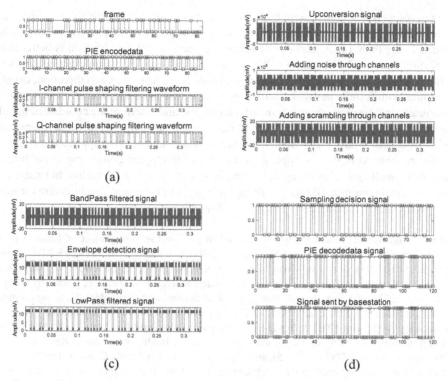

Fig. 9. Downlink waveform (a) Base station side signal processing; (b) Downlink channel processing; (c) Label end signal processing; (d) decoding results.

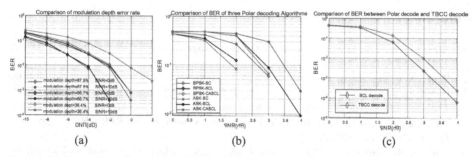

Fig. 10. BER graph (a) Comparison of modulation depth BER; (b) Comparison of BER of three Polar decoding Algorithms; (c) Comparison of BER between Polar decode and TBCC decode.

5   Conclusion

In this paper, a passive communication scheme and air interface protocol prototype integrating cellular communication and identification is designed based on the existing 18000-6C protocol and combined with the 5G communication protocol. It creatively proposes using 256 QAM modulation to construct ASK signals with different modulation depths and balanced PIE encoding. The communication waveforms and bit error rate curves were simulated, with both the base station and tags correctly demodulating each other's data. In the designed communication link, the BER performance of Polar code is better than TBCC and the BER performance of PSK modulation is better than ASK modulation. The next step will be to design and simulate the interaction protocol of the new protocol, as well as conduct hardware tests. This research direction, deeply integrated with cellular networks, will fully leverage the existing cellular infrastructure's advantages in coverage, receive/transmit sensitivity, and interference coordination, aiding the passive IoT in addressing current pain points, achieving leapfrog development, and is expected to have significant application value and extensive application space in the future. This work will be applied in the field of intelligent computing in the future. Currently, intelligent computing faces challenges such as big scenarios, big data, big problems, and ubiquitous demands. By optimizing communication architecture through new protocols, a high degree of ubiquity and intelligence can be achieved.

Acknowledgements. This work was supported in part by the joint project of China Mobile Research Institute& X-NET (Project Number: 2022H002), in part by National Key R&D Program (2018AAA0103203), in part by Guangdong Provincial Research and Development Plan in Key Areas (2019B010141001), in part by National Natural Science Foundation of China (61971113, 61901095, 62371106), in part by Sichuan Provincial Science and Technology Planning Program of China (2022YFG0230, 2023YFG0040), in part by the Fundamental Enhancement Program Technology Area Fund (2021-JCJQ-JJ-0667), in part by the Joint Fund of ZF and Ministry of Education (8091B022126), in part by the Fund Project of Intelligent Terminal Key Laboratory of Sichuan Province(SCITLAB-1015), and in part by Innovation Ability Construction Project for Sichuan Provincial Engineering Research Center of Communication Technology for Intelligent IoT (2303–510109-04–03-318020).

References

1. Atzori, L., Iera, A., Morabito, G.: The internet of things: a survey. Computer Network **54**(15), 2787–2805 (2010)
2. Wang, X.Y., Yigitler, H., Duan, R., Menta, E.Y., Jantti, R.: Coherent multiantenna receiver for BPSK-modulated ambient backscatter tags. IEEE Internet Things J. **9**(2), 1197–1211 (2021)
3. Thomas, S.J., Reynolds, M.S.: A 96 Mbit/sec, 15.5 pJ/bit 16-QAM modulator for UHF backscatter communication. In: IEEE International Conference on RFID (RFID), pp. 185–190. IEEE, Orlando (2012)
4. Correia, R., Carvalho, N.B.: Dual-band high order modulation ambient backscatter. In: IEEE/MTT-S International Microwave Symposium – IMS, pp.270–273. IEEE, Philadelphia (2018)
5. Tang, A., Kim, Y., Kim, Y., Virbila, G., Chang, M.F.: A 5.8 GHz 1.77 mW AFSK-OFDM CMOS backscatter transmitter for low power IoT applications. In: IEEE/MTT-S International Microwave Symposium – IMS, pp. 259–261. IEEE, Philadelphia (2018)
6. Zhang, Y.F., Li, E., Zhu, Y.H., Chi, K.K., Tian, X.Z.: Energy-efficient prefix code based backscatter communication for wirelessly powered networks. IEEE Wireless Commun. Lett. **8**(2), 348–351 (2019)
7. Arikan, E.: Channel polarization: a method for constructing capacity-achieving codes for symmetric binary-input memoryless channels. IEEE Trans. Inform. Theory **55**(7), 3051–3073 (2009)

8. Zhang, Z.Y., Zhang, L., Wang, X.B., Zhong, C.J., Poor, H.V.: A split-reduced successive cancellation list decoder for polar codes. IEEE J. Sel. Areas Commun. **34**(2), 292–302 (2016)

9. Wang, Q., Fu, P.P., Zhang, S.Z.: A comparison of concatenated polar codes with different interleaving and decoding schemes. In: International Conference on Computer and Communication Systems (ICCCS), pp. 570–574. IEEE, Shanghai (2020)

10. King, J., Ryan, W., Hulse, C., Wesel, R.D.: Efficient maximum-likelihood decoding for TBCC and CRC-TBCC codes via parallel list Viterbi. In: International Symposium on Topics in Coding (ISTC), pp.1–5. IEEE, Brest (2023)

Hybrid Control Strategy for Rumor Spread in Social Networks

Haiyan Zi[✉], Shixing Wu, and Kaixin Wu

Institute of Systems Security and Control, College of Computer Science and Technology, Xi'an University of Science and Technology, Xi'an 710054, China
21208223059@stu.xust.edu.cn

Abstract. To address the rampant spread of rumors in society, this study investigates control strategies for rumor propagation on social networks and proposes a hybrid control strategy rumor propagation model (IS2TRB) incorporating protective and blocking measures. Additionally, an optimal control model is developed to maximize the control of rumor propagation while minimizing control costs. Theoretical analyses regarding the existence and uniqueness of optimal control are provided. Numerical simulations validate that implementing optimal control incurs lower costs compared to hybrid control strategies, single protective strategies, and single blocking strategies. This approach effectively reduces the scale of rumor propagation, offering a robust strategy reference for combating the issue of rumor dissemination.

Keywords: Rumor propagation · Hybrid Control · Optimal Control

1 Introduction

Rumors can be defined as unverified or unsubstantiated information, stories, or discussions often disseminated through word of mouth and contemporary channels such as social media. Rumors have the potential to mislead people's judgments and decisions, even causing harm to both societal and individual interests [1, 2]. For instance, rumors related to the spread of viruses may lead to widespread misinformation, prompting unnecessary protective measures and resulting in the inefficient allocation of societal resources. The escalating issue of online violence fueled by rumors has further exacerbated the situation, sometimes creating an environment where falsehoods are indistinguishable from truth. With the rapid development of social media and the internet, rumors spread at an accelerated pace and broader scope, posing new challenges to societal governance and information dissemination [3].

In response to the challenges, various research studies and measures have been introduced, including the establishment of mechanisms for debunking rumors [4] and intensifying media regulation [5]. Continuous improvement and refinement of social media platforms aim to mitigate the spread of rumors on these platforms. However, the generation, propagation, and impact of rumors constitute a dynamic process involving extensive interactions and information dissemination among numerous individuals. This

H. Jin et al. (Eds.): IAIC 2023, CCIS 2060, pp. 198–210, 2024.
https://doi.org/10.1007/978-981-97-1332-5_16

process encompasses numerous nonlinear relationships and stochastic elements, necessitating the application of dynamic methods for description and analysis. Furthermore, conventional singular control measures often prove insufficient to effectively curb the dissemination of rumors. Therefore, the imperative arises to propose a comprehensive hybrid control strategy.

In recent years, strategies for controlling the spread of rumors have garnered widespread attention in academia and society as effective means of addressing the dissemination of misinformation. The primary objective of these strategies is to curb the propagation of rumors, thereby reducing their scope and impact to uphold social stability and public order. Within this domain, two main types of rumor control strategies have been identified [6, 7]: the interruption of crucial nodes in the rumor propagation network (blocking strategy) and the dissemination of truthful information to clarify rumors (protective strategy) [8, 9]. Introducing blocking measures into the model allows for targeted weakening of the influence of rumor spreaders, preventing the continued proliferation of rumors. Simultaneously, employing protective strategies serves to better shield genuine information from the interference of rumors [10]. However, the simultaneous adoption of both strategies may entail certain costs. Therefore, to minimize control costs, we introduce the theory of optimal control.

Optimal control theory employs mathematical modeling and optimization methods to find control strategies that optimize a given objective function under specified constraints. In the realm of research, optimal control has been successfully applied in various fields such as economics, engineering, ecology, and more [11]. Furthermore, optimal control methods are applicable to addressing the issue of rumor propagation. Through mathematical modeling, one can devise the most effective intervention strategies to control the spread of rumors, thereby mitigating their negative impact [12, 13].

Please note that the first paragraph of a section or subsection is not indented. The first paragraphs that follows a table, figure, equation etc. does not have an indent, either.

In this paper, we employ the optimal control approach to address the issue of rumor control, emphasizing a more practical perspective. Firstly, we consider the influence of the external environment on the population and propose a rumor propagation model incorporating a hybrid control strategy based on the mechanisms and influencing factors of rumor dissemination. Secondly, we formulate an objective cost function and conduct an in-depth study of the optimal control problem. Lastly, through numerical simulations, we compare the density of propagators under different strategies and validate that the optimal control cost is lower than that of the hybrid control strategy, single protective strategy, and single blocking strategy, thus confirming the effectiveness of the proposed model.

2 Model Establishment and Analysis

In order to further comprehend the patterns of rumor propagation and mitigate the impact of rumor dissemination, this paper proposes a hybrid control strategy to effectively reduce the spread of rumors. The hybrid control strategy involves targeted measures to control the scope or speed of rumor propagation. Specifically, it combines protective and blocking strategies. The specific rules for the hybrid control strategy are outlined as follows:

Protective Strategy: Implement an open and transparent information dissemination mechanism to enable the public to promptly understand the truth. Timely release positive information to mitigate the impact of rumor propagation.

Blocking Strategy: Respond swiftly and take emergency measures to promptly halt the spread of rumors. Restrict, block, and exclude rumor spreaders to prevent the dissemination of rumors at the source.

The state transformation diagram of IS2TRB model is shown in Fig. 1.

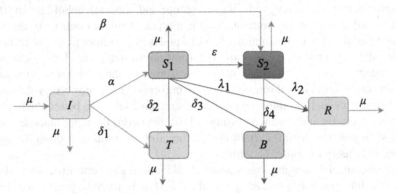

Fig. 1. Hybrid Control Strategy Model IS2TRB

Table 1. Meaning of each parameter

parameters	Meaning
α	The probability of I being infected by S_1
β	The probability of I being infected by S_1
ε	The probability of S_1 being infected by S_2
μ	The five populations exit at the same rate
λ_1	The immunity rate of S_1
λ_2	The immunity rate of S_2
δ_1	The truth propagation rate of I
δ_2	The truth propagation rate of S_1
δ_3	The blocking rate of S_1
δ_4	The blocking rate of S_1

In our social network model, we categorize the population into the following six states:

Ignorant (I): Representing those who are unaware of the rumor. Ordinary Spreaders (S_1): Representing users who lack an understanding of the veracity of the information and unintentionally spread false information due to curiosity or other motivations. Malicious

Spreaders (S_2): Representing users consciously disseminating false information, including misleading the public, inciting emotions, and damaging others' reputations. Immune (R): Representing users who are no longer interested in the rumor and cease to propagate information. Truth Spreaders (T): Implemented for protective strategy, actively disseminating truthful information to clarify rumors. Blockers (B): Implemented for blocking strategy, intervening with rumor spreaders to prevent the further dissemination of rumors. Among them, the significance of each parameter is shown in Table 1.

Based on the above analysis, the differential dynamical system of the IS2TBR model can be obtained, as shown in Eq. (1).

$$\begin{cases} \dfrac{dI(t)}{dt} = \mu - \alpha I(t)S_1(t) - \beta I(t)S_2(t) - \delta_1 I(t)T(t) - \mu I(t) \\[2mm] \dfrac{dS_1(t)}{dt} = \alpha I(t)S_1(t) - \varepsilon S_1(t) - \delta_2 S_1(t)T(t) - (\delta_3 + \lambda_1 + \mu)S_1(t) \\[2mm] \dfrac{dS_2(t)}{dt} = \beta I(t)S_2(t) + \varepsilon S_1(t) - (\delta_4 + \lambda_2 + \mu)S_2(t) \\[2mm] \dfrac{dT(t)}{dt} = \delta_1 I(t)T(t) + \delta_2 S_1(t)T(t) - \mu T(t) \\[2mm] \dfrac{dB(t)}{dt} = \delta_3 S_1(t) + \delta_4 S_2(t) - \mu B(t) \\[2mm] \dfrac{dR(t)}{dt} = \lambda_1 S_1(t) + \lambda_2 S_2(t) - \mu R(t) \end{cases} \tag{1}$$

Theorem 1: The solutions of system (1) are bounded.

Proof: Let $N(t)$ be the total number of individuals involved in rumor propagation at time t; then we have:

$$N(t) = I(t) + S_1(t) + S_2(t) + T(t) + B(t) + R(t) \tag{2}$$

From formula (1) and formula (2) we can get:

$$\frac{dN}{dt} = \frac{dI(t)}{dt} + \frac{dS_1(t)}{dt} + \frac{dS_2(t)}{dt} + \frac{dT(t)}{dt} + \frac{dB(t)}{dt} + \frac{dR(t)}{dt} = \mu - \mu N(t)$$

Then, integrating both sides of the above equation, we can obtain:

$$N(t) = (N_0 - 1)e^{-ut} + 1$$

When $t \to +\infty$, we have

$$\lim_{t \to +\infty} N(t) = 1$$

It can be observed that system (1) is bounded. Proposition proved. In summary, the positive invariant set of system (1) is:

$$\Omega = \left\{ I(t), S_1(t), S_2(t), T(t), B(t), R(t) \in R_+^6 : I(t) + S_1(t) + S_2(t) + T(t) + B(t) + R(t) \le 1 \right\}$$

Since the first five equations of system (1) do not involve $R(t)$, to reduce the dimension of system (1) and facilitate calculations, we can simplify system (1) to the following equivalent form:

$$
\begin{cases}
\dfrac{dI(t)}{dt} = \mu - \alpha I(t)S_1(t) - \beta I(t)S_2(t) - \delta_1 I(t)T(t) - \mu I(t) \\[2mm]
\dfrac{dS_1(t)}{dt} = \alpha I(t)S_1(t) - \varepsilon S_1(t) - \delta_2 S_1(t)T(t) - (\delta_3 + \lambda_1 + \mu)S_1(t) \\[2mm]
\dfrac{dS_2(t)}{dt} = \beta I(t)S_2(t) + \varepsilon S_1(t) - (\delta_4 + \lambda_2 + \mu)S_2(t) \\[2mm]
\dfrac{dT(t)}{dt} = \delta_1 I(t)T(t) + \delta_2 S_1(t)T(t) - \mu T(t) \\[2mm]
\dfrac{dB(t)}{dt} = \delta_3 S_1(t) + \delta_4 S_2(t) - \mu B(t)
\end{cases}
\tag{3}
$$

3 Analysis of Hybrid Control Strategy

3.1 Hybrid Control Strategy is Proposed

In the context of rumor propagation, optimal control analysis can be employed to explore how to minimize the scale of rumor dissemination. It aids in identifying economically efficient strategies to control rumors, thereby minimizing the number of propagators and corresponding strategic costs. Thus, control over rumors can be achieved through various strategies, such as truth dissemination and rumor blocking, each incurring different costs. The purpose of this study is to effectively control rumors while minimizing costs and losses. The influence rate δ_1 of truth spreaders and the suppression rate δ_4 of blockers are proposed as optimal control variables. System (3) can be reformulated as follows:

$$
\begin{cases}
\dfrac{dI(t)}{dt} = \mu - \alpha I(t)S_1(t) - \beta I(t)S_2(t) - \delta_1(t)I(t)T(t) - \mu I(t) \\[2mm]
\dfrac{dS_1(t)}{dt} = \alpha I(t)S_1(t) - \varepsilon S_1(t) - \delta_2 S_1(t)T(t) - (\delta_3 + \lambda_1 + \mu)S_1(t) \\[2mm]
\dfrac{dS_2(t)}{dt} = \beta I(t)S_2(t) + \varepsilon S_1(t) - (\delta_4(t) + \lambda_2 + \mu)S_2(t) \\[2mm]
\dfrac{dT(t)}{dt} = \delta_1(t)I(t)T(t) + \delta_2 S_1(t)T(t) - \mu T(t) \\[2mm]
\dfrac{dB(t)}{dt} = \delta_3 S_1(t) + \delta_4(t)S_2(t) - \mu B(t)
\end{cases}
\tag{4}
$$

Based on the above analysis, let $\mu = (\delta_1(t), \delta_4(t))$, and consider the following set as the admissible control set:

$$
\Gamma = \left\{ \mu | \underline{\delta}_1 \le \delta_1(t) \le \overline{\delta}_1, \underline{\delta}_4 \le \delta_4(t) \le \overline{\delta}_4, t \in [0, T] \right\}
\tag{5}
$$

where: $T > 0$, and assume the control variables are a set of Lebesgue integrable functions on the interval $[0, T]$, with $\underline{\delta}_1, \underline{\delta}_4, \overline{\delta}_1, \overline{\delta}_4$ being positive constants in the range of $[0,$

1], representing the upper bounds of the control variables.In this context, construct the objective function as follow:

$$Min J\left(\delta_1(t), \delta_4(t)\right) = \int_0^T \left[a(I(t) + S_1(t) + S_2(t)) + \frac{1}{2}b_1\delta_1^2(t) + \frac{1}{2}b_2\delta_4^2(t) \right] dt, \quad (6)$$

where $a, b_1, b_2 > 0$, a is the cost weight coefficient for the ignorant individuals, ordinary spreaders, and malicious spreaders. b_1 and b_2 are the weight coefficients for the control variables $\delta_1\ (t)$ and $\delta_4\ (t)$, respectively.

3.2 Existence of Optimal Control

According to the discussion on the existence of optimal control solutions based on Fleming and Rishel's theorem, Theorem 1 proves the existence of optimal control solutions. First, the Lagrangian function is defined as follows:

$$L(I, S_1, S_2, \delta_1, \delta_4) = a(I(t) + S_1(t) + S_2(t)) + \frac{1}{2}b_1\delta_1^2(t) + \frac{1}{2}b_2\delta_4^2(t) \quad (7)$$

Theorem 2: To minimize the number of propagators, the expression for the optimal control u^* that minimizes the objective functional is found on the constraint Γ as follows:

$$J\left(u^*\right) = \min_{u \in \Gamma} J\left(u\right)$$

Proof: To ensure the existence of an optimal control for system (4), the following five conditions must be satisfied:

(i) There exists $\mu \in \Gamma$ such that system (4) has a solution.
(ii) The control set Γ is a closed convex set.
(iii) The right-hand side of system (4) is constrained by linear functions.
(iv) The integrand of the objective function is convex on the control set Γ.
(v) There exists a positive constant $\rho > 1, c_1, c_2 > 0$ such that

$$L(I, S_1, S_2, \delta_1, \delta_4) \geq c_1\left(|\delta_1(t)|^\rho + |\delta_4(t)|^\rho\right) - c_2$$

According to Lukes' Theorem **2**, the boundary of the system (3) is in the control set, and the condition i is satisfied. By definition, the control set Γ is convex closed, and condition ii is satisfied. By definition, the equation of state is a linear function, and condition iii is satisfied. By $\frac{\partial^2 L}{\partial \delta_1^2} \geq 0, \frac{\partial^2 L}{\partial \delta_4^2} \geq 0$, the Lagrange function is convex, and condition iv is satisfied.

The following proof condition v:

$$L(I, S_1, S_2, \delta_1, \delta_4)$$

$$= a(I(t) + S_1(t) + S_2(t)) + \frac{1}{2}b_1\delta_1^2(t) + \frac{1}{2}b_2\delta_4^2(t)$$

$$\geq \frac{1}{2}b_1\delta_1^2(t) + \frac{1}{2}b_2\delta_4^2(t)$$

$$\geq \frac{\min b_i(i = 1, 2)}{2}(\delta_1^2(t) + \delta_4^2(t))$$

$$\geq \frac{b_{\min}}{2}|(\delta_1(t), \delta_1(t))|^2 - c_2$$

At $c_1 = \frac{\min b_i(i=1, 2)}{2}$, $\forall c_2 > 0$, $\rho = 2$, condition v is satisfied.

3.3 Construction of Optimal Control

According to Pontryagin's Maximum Principle, the Hamiltonian function constructed from system (3) and the objective function is as follows:

$$H = a(I(t) + S_1(t) + S_2(t)) + \frac{1}{2}b_1\delta_1^2(t) + \frac{1}{2}b_2\delta_4^2(t)$$

$$+\gamma_1(t)\frac{dI(t)}{dt} + \gamma_2(t)\frac{dS_1(t)}{dt} + \gamma_3(t)\frac{dS_2(t)}{dt} + \gamma_4(t)\frac{dT(t)}{dt} + \gamma_5(t)\frac{dB(t)}{dt} + \gamma_6(t)\frac{dR(t)}{dt}.$$

where $\gamma_i(t)(i = 1, 2, ...6)$ is the set of adjoint functions. Utilizing Pontryagin's Maximum Principle, we can obtain:

$$\begin{cases} \frac{d\gamma_1(t)}{dt} = -\frac{dH}{dI(t)} = -a + \gamma_1(t)(\alpha S_1(t) + \beta S_2(t) + \delta_1(t)T(t) + \mu) - \gamma_2(t)\alpha S_1(t) - \gamma_3(t)\beta S_2(t) \\ \qquad\qquad - \gamma_4(t)\delta_1 T(t) \\ \frac{d\gamma_2(t)}{dt} = -\frac{dH}{dS_1(t)} = -a + \gamma_1(t)\alpha I(t) + \gamma_2(t)(-\alpha I(t) + \varepsilon + \delta_2 T(t) + \delta_3 + \lambda_1 + \mu) - \gamma_3(t)\varepsilon \\ \qquad\qquad - \gamma_4(t)\delta_2 T(t) - \gamma_5(t)\delta_3 - \gamma_6(t)\lambda_1 \\ \frac{d\gamma_3(t)}{dt} = -\frac{dH}{dS_2(t)} = -a + \gamma_1(t)\beta I(t) + \gamma_3(t)(-\beta I(t) + \delta_4(t) + \lambda_2 + \mu) \\ \qquad\qquad - \gamma_5(t)\delta_4 - \gamma_6(t)\lambda_2 \\ \frac{d\gamma_4(t)}{dt} = -\frac{dH}{dT(t)} = \gamma_1(t)\delta_1(t)I(t) + \gamma_2(t)\delta_2 S_1(t) + \gamma_4(-\delta_1(t)I(t) - \delta_2 S_1(t) + \mu) \\ \frac{d\gamma_5(t)}{dt} = -\frac{dH}{dB(t)} = \gamma_5\mu \\ \frac{d\gamma_6(t)}{dt} = -\frac{dH}{dR(t)} = \gamma_6\mu \end{cases}$$

where the transversality condition is given by:

$$\gamma_1(T) = \gamma_2(T) = \gamma_3(T) = \gamma_4(T) = \gamma_5(T) = \gamma_6(T) = 0$$

The necessary condition for the control equation is:

$$\frac{\partial H}{\partial \delta_1} = 0, \frac{\partial H}{\partial \delta_4} = 0.$$

According to the Hamiltonian function and Pontryagin's Maximum Principle, the control equation is derived as:

$$\frac{\partial H}{\partial \delta_1(t)} = b_1 \delta_1^*(t) - \gamma_1 I^*(t) T^*(t) + \gamma_4 I^*(t) T^*(t) = 0$$

$$\frac{\partial H}{\partial \delta_4(t)} = b_2 \delta_4^*(t) - \gamma_3 S_2^*(t) + \gamma_5 S_2^*(t) = 0$$

Further, the optimal control variable can be obtained as:

$$\delta_1^*(t) = \frac{\gamma_1 I^*(t) T^*(t) - \gamma_4 I^*(t) T^*(t)}{b_1}$$

$$\delta_4^*(t) = \frac{\gamma_3 S_2^*(t) - \gamma_5 S_2^*(t)}{b_2}$$

According to the definition of the admissible control set $\Gamma = \left\{ \mu \mid \underline{\delta}_1 \leq \delta_1(t) \leq \overline{\delta}_1, \underline{\delta}_4 \leq \delta_4(t) \leq \overline{\delta}_4, t \in [0, T] \right\}$, the components of μ^* in the optimal control can be expressed as:

$$\delta_1^*(t) = Max\{Min\{ \frac{\gamma_1 I^*(t) T^*(t) - \gamma_4 I^*(t) T^*(t)}{b_1}, \overline{\delta}_1 \}, \underline{\delta}_1 \}$$

$$\delta_4^*(t) = Max\{Min\{ \frac{\gamma_3 S_2^*(t) - \gamma_5 S_2^*(t)}{b_2}, \overline{\delta}_4 \}, \underline{\delta}_4 \}$$

Thus, the expression for the optimal control system is as follows:

$$
\begin{cases}
\frac{dI(t)}{dt} = \mu - \alpha I(t) S_1(t) - \beta I(t) S_2(t) - \delta_1(t) I(t) T(t) - \mu I(t) \\
\frac{dS_1(t)}{dt} = \alpha I(t) S_1(t) - \varepsilon S_1(t) - \delta_2 S_1(t) T(t) - (\delta_3 + \lambda_1 + \mu) S_1(t) \\
\frac{dS_2(t)}{dt} = \beta I(t) S_2(t) + \varepsilon S_1(t) - (\delta_4(t) + \lambda_2 + \mu) S_2(t) \\
\frac{dT(t)}{dt} = \delta_1(t) I(t) T(t) + \delta_2 S_1(t) T(t) - \mu T(t) \\
\frac{dB(t)}{dt} = \delta_3 S_1(t) + \delta_4(t) S_2(t) - \mu B(t) \\
\frac{dR(t)}{dt} = \lambda_1 S_1(t) + \lambda_2 S_2(t) - \mu R(t) \\
\frac{d\gamma_1(t)}{dt} = -\frac{dH}{dI(t)} = -a + \gamma_1(t)(\alpha S_1(t) + \beta S_2(t) + \delta_1(t) T(t) + \mu) - \gamma_2(t)\alpha S_1(t) - \gamma_3(t)\beta S_2(t) - \gamma_4(t)\delta_1 T(t) \\
\frac{d\gamma_2(t)}{dt} = -\frac{dH}{dS_1(t)} = -a + \gamma_1(t)\alpha I(t) + \gamma_2(t)(-\alpha I(t) + \varepsilon + \delta_2 T(t) + \delta_3 + \lambda_1 + \mu) - \gamma_3(t)\varepsilon - \gamma_4(t)\delta_2 T(t) - \gamma_5(t)\delta_3 - \gamma_6(t)\lambda_1 \\
\frac{d\gamma_3(t)}{dt} = -\frac{dH}{dS_2(t)} = -a + \gamma_1(t)\beta I(t) + \gamma_3(t)(-\beta I(t) + \delta_4(t) + \lambda_2 + \mu) - \gamma_5(t)\delta_4 - \gamma_6(t)\lambda_2 \\
\frac{d\gamma_4(t)}{dt} = -\frac{dH}{dT(t)} = \gamma_1(t)\delta_1(t) I(t) + \gamma_2(t)\delta_2 S_1(t) + \gamma_4(-\delta_1(t) I(t) - \delta_2 S_1(t) + \mu) \\
\frac{d\gamma_5(t)}{dt} = -\frac{dH}{dB(t)} = \gamma_5 \mu \\
\frac{d\gamma_6(t)}{dt} = -\frac{dH}{dR(t)} = \gamma_6 \mu
\end{cases}
$$

4 Numerical Simulation and Analysis

4.1 Comparison of Different Control Strategies

Figure 2 illustrates the evolution of the density of ordinary spreaders under various control strategies, and Fig. 3 depicts the evolution of the density of malicious spreaders under different control strategies. From the two figures, it is evident that implementing the hybrid control strategy can minimize the density of rumor spreaders, validating the effectiveness of the model.

4.2 Optimal Control Analysis

The superiority of the hybrid control strategy lies in its ability to suppress rumor propagation by adjusting the values of the parameters δ_1, δ_2, δ_3 and δ_4. When $\delta_1 = \delta_2 = \delta_3 = \delta_4 = 1$, it indicates that the implementation intensity of the hybrid control strategy is maximized, achieving the best control effect but also accompanied by higher costs.

To address rumor propagation more flexibly in practice, we introduce optimal control to dynamically adjust the values of control parameters, achieving maximum control of rumor propagation while minimizing control costs. In the optimization process, we choose δ_1 and δ_4 as optimal control variables for the protection strategy and blocking strategy, respectively, for easier computation and analysis.

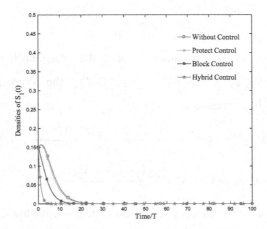

Fig. 2. Evolution of the Density of Normal Spreaders under Different Control Strategies

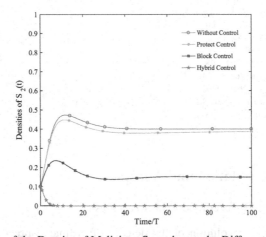

Fig. 3. Evolution of the Density of Malicious Spreaders under Different Control Strategies

The values of the optimal control variables $\delta_1^*(t)$ and $\delta_4^*(t)$ over time are chosen as $\alpha = 0.4, \beta = 0.8, \mu = 0.08, \varepsilon = 0.1, \delta_2 = 0.8, \delta_3 = 0.2, \lambda_1 = 0.1, \lambda_2 = 0.05, a = 0.8, b_1 =$

0.1, $b_2 = 0.1$. The boundary conditions are set as: $(I(0), S_1(0), S_2(0), T(0), R(0), B(0)) = (0.6, 0.15, 0.1, 0.15, 0, 0)$. The values of the optimal control variables $\delta_1^*(t)$ and $\delta_4^*(t)$ over time are shown in Fig. 4.

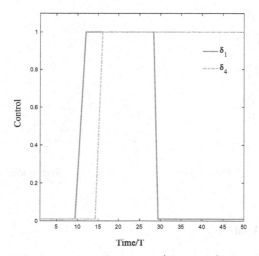

Fig. 4. Control variables $\delta_1^*(t)$ and $\delta_4^*(t)$

The goal of optimal control is to minimize the number of individuals in the population who are ignorant I, ordinary spreaders S_1, and malicious spreaders S_2, while maximizing the presence of truth spreaders T and blockers B, effectively controlling the spread of rumors. Figure 5 and Fig. 6 compare the optimal control strategy with the hybrid control strategy, confirming the effectiveness of implementing optimal control.

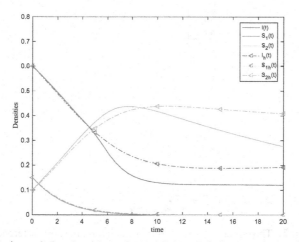

Fig. 5. The evolution of the densities of $I(t)$, $S_1(t)$, and $S_2(t)$ under optimal control and hybridcontrol.

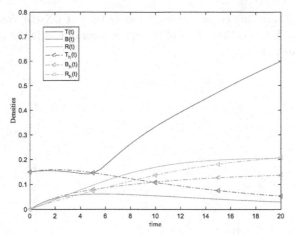

Fig. 6. The evolution of the densities of $T(t)$, $B(t)$, and $R(t)$ under optimal control and hybridcontrol.

4.3 Cost-Benefit Analysis

In this section, we conduct a cost-benefit analysis to assess the economic feasibility and benefits of the optimal control strategy for rumor spreading.

Figure 7 and Fig. 8 respectively depict the influence of different terminal times t_f and cost weighting factors a on the cost difference J_h, J_p, J_b compared to J. It is observed that the cost difference J is always greater than 0. The experiments demonstrate that the cost of the optimal control is always the lowest.

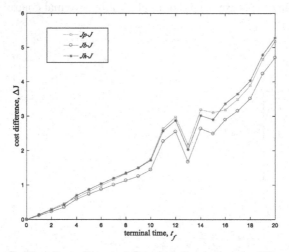

Fig. 7. The effect of terminal time variation on the cost difference

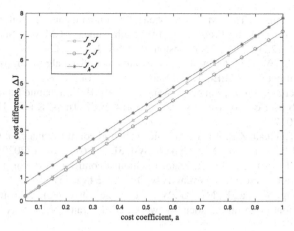

Fig. 8. The effect of cost coefficient variation on the cost difference

5 Conclusion

This paper investigates control strategies for rumor propagation on social networks and proposes a hybrid control strategy rumor propagation model (IS2TRB). Additionally, we establish an optimal control model aimed at minimizing control costs. Theoretical analyses regarding the uniqueness and existence of optimal control are conducted. Through numerical simulations, we validate that implementing optimal control results in lower costs compared to hybrid control strategies, single protective strategies, and single blocking strategies. The model effectively reduces the scale of rumor propagation, providing a robust strategy reference for controlling the issue of rumor spread.

References

1. He, Z., Cai, Z., Yu, J., Wang, X., Sun, Y., Li, Y.: Cost-efficient strategies for restraining rumor spreading in mobile social networks. IEEE Trans. Veh. Technol. **66**(3), 2789–2800 (2017)
2. Yang, L.-X., Li, P., Yang, X., Wu, Y., Tang, Y.Y.: On the competition of two conflicting messages. Nonlin. Dyn. **91**(3), 1853–1869 (2018)
3. Nash, R., Bouchard, M., Malm, A.: Investing in people: the role of social networks in the diffusion of a large-scale fraud. Soc. Netw. **35**(4), 686–698 (2013)
4. Zhang, Y., Xu, J.: A dynamic competition and predation model for rumor and rumor-refutation. IEEE Access **9**, 9117–9129 (2021). https://doi.org/10.1109/ACCESS.2020.3047934
5. Chen, H., et al.: The research on propagation modeling and governance strategies of online rumors based on behavior–attitude. Internet Res. **32**(2), 620–639 (2022). https://doi.org/10.1108/INTR-08-2020-0480
6. Wen, S., Jiang, J., Xiang, Y., Yu, S., Zhou, W., Jia, W.: Toshut them up or to clarify: restraining the spread of rumors in online social networks. IEEE Trans. Parallel Distrib. Syst. **25**(12), 3306–3316 (2014)
7. Sharma, K., Qian, F., Jiang, H., Ruchansky, N., Zhang, M., Liu, Y.: Combating fake news. ACM Trans. Intell. Syst. Technol. **10**(3), 1–42 (2019)

8. Nguyen, N.P., Yan, G., Thai, M.T., Eidenbenz, S.: Containment of misinformation spread in online social networks. In: Proceedings of the 4th Annual ACM Web Science Conference on—WebSci 2012, pp. 213–222, Evanston, IL, USA, June (2012)
9. Tong, G.A., Wu, W., Guo, L., et al.: An efficient randomized algorithm for rumor blocking in online social networks. IEEE Trans. Netw. Sci. Eng. **14** (2017)
10. Wang, B., Chen, G., Fu, L., Song, L., Wang, X.: DRIMUX: dynamic rumor influence minimization with user experience in social networks. IEEE Trans. Knowl. Data Eng. **29**(10), 2168–2181 (2017)
11. Ding, L., Hu, P., Guan, Z.H., Li, T.: An efficient hybrid control strategy for restraining rumor spreading. IEEE Trans. Syst. Man Cybern. Syst. **51**(11), 6779–6791 (2020)
12. Huo, L., Lin, T., Fan, C., Liu, C., Zhao, J.: Optimal control of a rumor propagation model with latent period in emergency event. Adv. Diff. Eq. **54** (2015)
13. Wang, X., Wang, X., Hao, F., Min, G., Wang, L.: Efficient coupling diffusion of positive and negative information in online social networks. IEEE Trans. Netw. Serv. Manage. **16**(13), 1226–1239 (2019)

Data Stream Aggregation Mechanism Under Local Differential Privacy

Jingjing Wang(✉) and Yongli Wang

Nanjing University of Science and Technology, Nanjing 210000, China
lemonljj@njust.edu.cn

Abstract. In the age of intelligence and big data, the collection and sharing of personal data is becoming more and more common and easy, and with it comes concerns about privacy leakage. How to collect reliable data while protecting user privacy has become an important topic. In order to ensure the accuracy of the results, the privacy budget allocated to each data point should be larger, so the number of data points to be allocated with privacy budget can be reduced. Based on this idea, this paper proposes a data stream aggregation mechanism based on local differential privacy, where the data provider maximizes user privacy by fitting the data stream, perturbing salient points, and ultimately reporting the noisy data, which protects the privacy of the individual data at the source of the data generation; the data collector, after obtaining the noisy data, restores the original data stream through data reconstruction, which provides a data analysis and statistical provides a feasible solution for data analysis and statistics.

In order to evaluate the performance and effectiveness of the mechanism, this paper conducts extensive experiments and comparisons on a physical activity monitoring dataset. The experimental results show that the mechanism can well balance privacy protection and data availability, and also achieves a smaller mean square error compared with existing methods, demonstrating better performance.

Keywords: Local Differential Privacy · Data Stream · Aggregation

1 Introduction

With the rapid development of information technology and the popularization of mobile devices, a variety of online services continue to collect personal data, among which there is no lack of personal sensitive and private information that users do not want to disclose, such as location data, health monitoring data (heart rate, blood pressure, etc.), and search history. Although privacy protection related laws have been introduced in recent years, most applications do not strictly comply with them [1]. Therefore, this collection process inevitably brings the risk of privacy leakage, and once the privacy leakage occurs, the consequences are often irreversible, and the data collector also faces the possibility of reputation damage or being sued by users [2]. To address such challenges, local differential privacy (LDP) has been proposed [3].

Applying local differential privacy protection to the aggregation of data streams aims to protect individual privacy [4] and simultaneously support the aggregated analysis of

H. Jin et al. (Eds.): IAIC 2023, CCIS 2060, pp. 211–225, 2024.
https://doi.org/10.1007/978-981-97-1332-5_17

data. It provides a solution for LDP protection for data stream aggregation by allowing multiple data providers to process their own data locally with differential privacy protection and integrate the protected data to obtain statistical results, ensuring that the real data does not leave the user's device.

However, there are limitations to this approach; these data are often rarely static. When data is increasing over time, it needs to be constantly processed [5]. When the amount of data is large, not only does it require a larger privacy budget, it often introduces excessive noise as well, and once the privacy budget is fully consumed, it means that the data is no longer protected by differential privacy [6], making data availability difficult to guarantee.

Li et al. [7] proposed a single-attribute numerical stream data protection mechanism based on LDP based on data protection for wearable devices. The mechanism identifies salient points from the collected data stream, adds noise only to these salient points and sends them to the data collector, who finally reconstructs the data for subsequent analysis and processing. The limitation of Li et al. is that it is only applicable to data streams that remain unchanged or fluctuate drastically within a short period of time. Because if the data stream is fluctuating for a long time, the number of salient points will be very large. At this time, the privacy budget saved is very limited if we want to guarantee the desired effect.

In order to solve this problem, we propose a novel method to fit the data stream using machine learning algorithms to generate a similar but different data stream before identifying salient points, and then based on the fitted data stream, we remove the redundant points in it before identifying salient points. Since the fitted data stream is smoother than the original data stream, together with the removal of redundant points, the number of salient points that need to be selected can be greatly reduced, saving privacy budget. However, at the same time, the use of fake data streams introduces additional errors, for which can be compensated by reducing the noise scale, especially when the privacy budget is small or the size of the data provider is small. In addition, we add noise using the perturbation mechanism proposed by Duchi JC et al. which preserves the maximum statistical utility for a given privacy level through the theory of great-minor minima.

- We propose a data stream aggregation mechanism that satisfies local differential privacy, by which each user reports a noisy data stream based on the fitted data stream. As a result, not only the privacy budget is reduced, but the data provider's private data is also protected.
- We propose a salient point perturbation method, which first removes some redundant data points that have less impact on the fitted data stream based on the derivatives of the data points, and takes the remaining points as salient points, which makes the final selected data points represent the whole data stream more accurately, thus reducing the statistical bias, and the collector can also obtain relatively accurate statistical results.
- We have conducted a large number of experiments, and the results show that our proposed method has better performance compared with the current methods.

Roadmap. We present the preliminary content in Sect. 2, introduce our method in Sect. 3, and evaluate it experimentally in Sect. 4. Section 5 summarizes the whole paper.

2 Preliminaries

2.1 Theoretical Foundation

Local Differential Privacy

Local Differential Privacy (LDP), as a variant of differential privacy [7], provides methods to realize privacy protection on individual devices or terminals, which is schematically shown in Fig. 1. Unlike the traditional differential privacy approach, the goal of LDP is to process data on individual devices to protect individual privacy and provide aggregated statistical results to a third party on this basis. Its formalization is defined as follows.

Definition $((\in, \delta) - LDP)$ [9]. A randomized algorithm M satisfies LDP if and only if, for any two records $x, x' \in D$ in its domain of definition D, and for any possible output Y, there are.

$$\Pr[M(x) \in Y] \le e^{\in} \Pr[M(x') \in Y] + \delta \tag{1}$$

where $\Pr[\cdot]$ denotes the likelihood of privacy being compromised and $\in$ is the privacy budget, which is used to measure the privacy protection strength. Since many LDP protocols are built on randomized responses, typically $\delta = 0$ [10], this can be simplified to $\in -LDP$. Algorithm M is applied independently to each user record and does not require a trusted data collector.

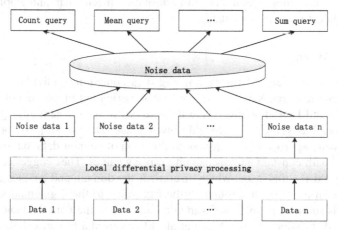

Fig. 1. Schematic diagram of local differential privacy

LDP has many application scenarios, such as data collection on mobile devices, location data analysis, and personalized recommendation systems. By performing data processing and privacy protection on individual devices, LDP provides a more flexible and decentralized approach to privacy protection, enabling individuals to better control their data privacy.

In addition, differential privacy has some important properties that facilitate the design of more sophisticated algorithms to solve more complex problems.

Theorem 1 (Sequential Composition). Several subroutines satisfying $\in_1, \in_2, \cdots, \in_k$ $-DP$ satisfy $\in -DP$ after combining them, where $\in= \sum_i$.

Theorem 2 (Parallel combination). With a set of privacy mechanisms $M = \{M_1, M_2, \cdots, M_k\}$, each M_i satisfies $\in_i -DP$ on disjoint records of the entire dataset, then M satisfies $\in -DP$, where $\in= \max\{\in_1, \in_2, \cdots, \in_k\}$ [11].

Duchi Perturbation Mechanism

Regarding data protection under differential privacy, Duchi JC et al. [12] look at the minimax optimization criterion (i.e., finding the estimation process that minimizes the maximum risk under a given privacy-preserving constraint), and propose a data perturbation mechanism that satisfies LDP to maximize the protection of individual privacy while maintaining high-quality estimation.

The main idea of the Duchi perturbation mechanism is that, given a single data value $x \in [-1, 1]$, each user provides two fixed values $\frac{e^\in+1}{e^\in-1} \frac{e^\in+1}{e^\in-1}$ or $-\frac{e^\in+1}{e^\in-1} \frac{e^\in+1}{e^\in-1}$ according to x, as in Eq. (2), and the probability of providing both is $\frac{e^\in-1}{2e^\in+2} \cdot x + \frac{1}{2} \frac{e^\in-1}{2e^\in+2} \cdot x + \frac{1}{2}$ and $-\frac{e^\in-1}{2e^\in+2} \cdot x + \frac{1}{2} - \frac{e^\in-1}{2e^\in+2} \cdot x + \frac{1}{2}$ respectively.

$$\Pr[\hat{x} = t | x] = \begin{cases} \frac{e^\in-1}{2e^\in+2} \cdot x + \frac{1}{2}, & ift = \frac{e^\in+1}{e^\in-1} \\ -\frac{e^\in-1}{2e^\in+2} \cdot x + \frac{1}{2}, & ift = -\frac{e^\in+1}{e^\in-1} \end{cases} \quad (2)$$

We refer to this simply as the Duchi perturbation mechanism and employ it in our approach to inject noise into the data.

2.2 Related Work

With the increasing demand for sharing big data containing individual information, many important researches and works have been carried out in the area of data stream publishing around LDP domains, while the analysis of data stream aggregation under LDP protection has yet to be thoroughly investigated. Ding et al. [13] considered the use of a memetic approach for continuous collection of counter data, but this approach can only deal with constant and unchanging privacy values. They suggested a rounding technique to address this issue, which discretizes the domain and enables each user to report their own number after rounding the true value to the segmentation's midway. However, this solution is only useful in situations when the data fluctuates very little over an extended period of time. Wang et al. [8] proposed a framework that employs a hierarchical approach to segment the data stream and uses an exponential mechanism to find a threshold to truncate the perturbation before the user value to reduce the amount of noise. It does, however, presume that there exist m extra values to aid in the determination of such a threshold, and in their experimental setting, the value of m is quite huge, equivalent to 65536, and their method is not able to refine the estimate to a specific moment in time. Joseph et al. [14] proposed the THRESH algorithm, which identifies the estimate through a voting protocol and updates the estimate only when it becomes sufficiently inaccurate. However, their proposed method has limitations and needs to satisfy that all users' data are repeatedly extracted from a particular distribution, such as the Bernoulli distribution. Duchi et al. [12, 15] proposed a method for mean estimation based on a stochastic response mechanism, which has high data availability but is not applicable to high-dimensional datasets. Subsequently, Nguyên et al. [16] optimized Duchi et al.'s scheme and proposed the Harmony algorithm for data collection and analysis of smart devices, which supports both frequency and mean estimation, and is also capable of completing machine learning tasks with high data availability. Erlingsson et al. [17] shuffled data provided by a user after the use of LDP, without adding additional noise to require tighter privacy constraints. However, it is assumed that each user can modify its underlying Boolean value a maximum of k times in a given time period, and that the user's value can only be altered by 1. Errounda et al. [18] investigated location statistics using the $\omega - event$ LDP mechanism and used an approximation strategy for estimating unperturbed locations, but it may lead to unwanted loss of utility. Bao et al. [19] proposed a related Gaussian mechanism to force $(\in, \delta) - LDP$ to collect streaming data. Utilizing the correlation information between data items, the approach lowers the noise scale by reusing the created noise. However, their proposed method is only applicable to scenarios where the data items exhibit overt autocorrelation. In conclusion, the current research on targeting data stream aggregation under LDP mainly focuses on frequency estimation, which is not comprehensive, and the current solutions continue to have issues with high resource usage or limited data accessibility.

2.3 Definition of the Problem

There is a collection of users, each of whom has a set of private data containing time. At equal intervals, untrustworthy data collectors collect data from the users and analyze and process these values. To ensure privacy, each user hides the real data and reports a noisy data stream, thus satisfying LDP. The formalization is defined as follows:

Problem 1. Given n users, each user $p \in \{1, 2, \cdots, n\}$ has a set of time-inclusive data-value pairs $Y = \{(t_i, y_{p,i})\}_{i=1}^{n}$. Design an algorithm M such that the output satisfies Eq. (3) when the inputs $Y_p = \{y_{p,1}, y_{p,2}, \cdots, y_{p,n}\}$ and is subject to the privacy constraints in Eq. (4), where PP is the set of possible outputs of algorithm M.

$$Y_P' = M\left(Y_p\right) = argmin \frac{1}{n} \sum_{p=1}^{n} \|Y_p - Y_p'\|^2 \tag{3}$$

$$\frac{\Pr\left[M\left(y_{p,1}, y_{p,2}, \cdots, y_{p,n}\right) \in P\right]}{\Pr\left[M\left(y_{p,1}', y_{p,2}', \cdots, y_{p,n}'\right) \in P\right]} = e^{\in} \tag{4}$$

3 Method

We propose a new data aggregation framework that uses a fitted data stream without injecting noise into the original data stream, allowing fewer points to be selected for perturbation, even if the original data stream fluctuates greatly, resulting in significant savings in privacy budget.

3.1 Overall Structure

Figure 2 illustrates the proposed framework, which consists of a data provider and a data collector.

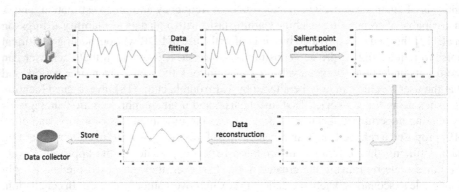

Fig. 2. Local differential privacy based data stream aggregation process

Data Provider:

- Data fitting: we use machine learning algorithms to fit curves over the data stream to obtain a fitted data stream. Our goal is to obtain a smoother data stream that is as close as possible to the original data values, reducing the number of points to be selected while ensuring that data collectors do not have access to the user's real data.
- Salient point perturbation: in order to reduce the privacy budget, we remove redundant data, add noise to only some of the salient points that best represent the direction of the fitted data stream, and report the noisy data to the data collector, thus ensuring that only the user knows the true values and eliminating the dependence on a trusted central server.

Data Collector:

- Data Reconstruction: after receiving the data injected with noise, the data collector computes the missing data values through different methods, reconstructs the data stream, and generates statistical results.

The uniqueness of our work is that it introduces a pseudo data stream against which all subsequent operations are directed. The reason why we can achieve good statistical accuracy based on the fake data stream is that, on the one hand, the fake data stream is close to the real data stream in terms of data values, which ensures the correctness of the report. On the other hand, the pseudo-data stream is much smoother compared to the original real data stream, and the subsequent removal of some redundant points reduces the number of salient points that need to be perturbed. As a result, much less noise is injected into the overall data stream, which improves the statistical accuracy.

We will discuss this in detail in the next sections.

3.2 Data Provider

Data Fitting

The fitted data stream is obtained by utilizing a polynomial function to obtain a fitted curve on top of the original data stream. Let y be a curve connecting all data points in a predefined interval, and let $f(x)$ be a polynomial function of order m, i.e., $f(x) = a_0 + a_1 x + \cdots + a_m x^m$. We need to learn the parameter $\{a_i\}_{i=0}^m$ which minimizes the distance between $f(x)$ and y. Specifically, we use the mean squared error MSE between f(x) and y to express the accuracy of the data fitting as shown below. as shown below.

$$MSE = \underset{(a_0, \cdots, a_p)}{\arg\min} \sum_{i=1}^{n} \|f(x_i) - y_i\|_2 \tag{5}$$

We can consider the timestamps t_i as variables of the functions f and y, and y_i as the corresponding values in y. Therefore, we can fit the curve $f(t)$ according to Algorithm 1, and then get the fitted data streams by calculating $f(t_i)$, $t_i \in \{t_1, \cdots, t_n\}$.

It is worth noting that the degree of the polynomial function affects the fitting accuracy and the selection of the fitted data points, which may lead to errors, but since it cannot be calculated theoretically, we evaluate it in Sect. 4.

Algorithm 1. Data Fitting

Input: user data $Y = \{(t_i, y_{p,i})\}_{i=1}^n$,

Output: fitted data stream F

1: **for** each user p **do**
2: $\quad f(t) = a_0 + a_1 t^1 + \cdots + a_m t^m$
3: $\quad (a_0, a_1, \cdots, a_m) \leftarrow \underset{(a_0, \cdots, a_p)}{\arg\min} \sum_{i=1}^n \|f(t_i) - y_i\|_2$
4: $\quad$ get fitted data stream F $= \{(t_i, f(t_i))\}_{i=1}^n$
5: **end for**

Salient Point Perturbation

After obtaining the fitted data stream, we identify salient points and use the Duchi perturbation mechanism to inject noise into the salient points and send the noisy data to the data collector.

In order to better identify salient points, we first censor the fitted data stream, after removing redundant points, the fitted data stream will be represented by fewer points, but its overall trend will remain unchanged, which ensures the prediction accuracy and reduces the privacy budget.

In general, the redundant points in the data stream can be divided into two categories: the first category is the points whose trend is unchanged during a certain period of time, or the points whose values are unchanged in a consecutive timestamp, and the derivatives of the points can be calculated respectively, when the derivatives are the same positive and the same negative, then the trend is unchanged, and when the derivatives are 0, then the values are unchanged. The formula for calculating the derivative of a data point is as follows:

$$d_i = \frac{y_i - y_{i-1}}{t_i - t_{i-1}} \tag{6}$$

where point (t_{i-1}, y_{i-1}) and point (t_i, y_i) are two consecutive points in time, d_i is the derivative at point i, y_i is the value of the data collected at time t_i, and y_{i-1} is the value of the data collected at time t_{i-1}.

The second category is points that have been fluctuating frequently over a certain period of time, and the maximum interval can be specified as a judgment threshold.

Overall, the steps for removing redundant points and perturbing salient points are as follows:

- Based on the fitted data stream, the derivatives of all points are calculated, and the data points in which the derivative value is equal to 0 are directly deleted, i.e., the points in the first category whose values remain unchanged in consecutive timestamps.
- On this basis, we keep the first point and use the first point as the starting point, and then traverse the remaining data points from the beginning, we can find the data points whose derivatives have the same sign, at this time, we keep these points first.
- Continue traversing backwards until we find the first such turning point, which is characterized by its neighboring data points having derivatives with opposite signs and a time interval to the starting point larger than the threshold we specify $\alpha\alpha$. Keep such a turning point and delete the data points in between from the starting point to that point, i.e., the point with a constant trend for a certain period of time in the first category.
- Taking this turning point as a new starting point, repeat the above steps until the entire data stream sequence is traversed, at which point the last point is the last turning point for which the previous steps are repeated.
- Finally the remaining points in the data stream are the salient points after removing the redundant points, injected with noise and sent to the data collector.

Algorithm 2 shows the process of identifying salient points. In it, lines 1–9 remove all the points whose derivatives are 0 and store the remaining points temporarily in list1; lines 10–21 compare the derivative signs of the data points and remove the redundant points. Where t_s is the time of the starting point, t_f is the time of the previous point where the comparison was made, and t_{in} is the time interval between the two points. The last remaining points along with the start and end points are used as the identified salient points and stored in the noise point set $P^{\wedge'}$; lines 22–24 add the noise under the Duchi perturbation mechanism to the salient points to end up with noisy data points.

Algorithm 1. Perturbation of Salient Points

Input: fitted data stream $F = \{(t_i, f(t_i))\}_{i=1}^{n}$, threshold α

Output: noisy point set P'

1: $i = 1, (t_0, f(t_0)) = (0, 0)$

2: **for** each point $(t_i, f(t_i))$ in F **do**

3:　$d_i = \frac{f(t_i) - f(t_{i-1})}{t_i - t_{i-1}}$

4:　**if** $d_i = 0$ **then**

5:　　delete $(t_i, f(t_i))$

6:　**else**

7:　　store $(t_i, f(t_i))$ in list1

8:　**end if**

9: **end for**

10: store $(t_0, f(t_0))$ in list2, $j = 2, t_f = 0$

11: **for** each point $(t_j, f(t_j))$ in list1, $i = \{j, j+1, \cdots, count(list1)\}$ **do**

12:　$t_{in} = t_f - t_s$

13:　**if** $(d_j > 0 \;\&\&\; d_{j+1} > 0) \;\|\; (d_j < 0 \;\&\&\; d_{j+1} < 0)$ **then**

14:　　delete $(t_j, f(t_j))$

15:　**else if** $t_{in} \leq \alpha$ **then**

16:　　delete $(t_j, f(t_j))$

17:　**else**

18:　　store $(t_j, f(t_j))$ in list2, $t_s = t_f$

19:　**end if**

20:　store $(t_n, f(t_n))$ in list2

21: **end for**

22: k=0

22: **for** each point $(t_k, f(t_k))$ in list2 **do**

23:　$P' \leftarrow Duchi\left(f(t_k), \frac{\epsilon}{|P|}\right)$

24: **end for**

3.3　Data Collector

After receiving the noisy data from the data provider, the data collector reconstructs the noisy data according to the user, essentially connecting the discrete noisy data points and recovering the original data stream by calculating the values of the missing data points. The connection process uses different mechanisms that can cause different errors, one of the common methods is linear estimation, which refers to the use of a straight line to connect two discrete neighboring salient points in a sequential manner, in which the missing data can be found directly.

In linear estimation, two neighboring salient points (x_i, y_i) and (x_{i+1}, y_{i+1}) are known and they form a subinterval function expression as follows:

$$F(x) = \frac{y_{i+1}-y_i}{x_{i+1}-x_i}(x - x_i) + y_i,$$
$$x \in [x_i, x_{i+1}], i = 1, 2, \cdots, n-1 \tag{7}$$

Although the linear estimation method is simple, the reconstructed curves are not smooth. In order to evaluate the errors introduced by different methods, in addition to linear estimation, we additionally use the interpolation method for reconstruction to improve the accuracy and data availability. The interpolation method specifies a function that can take a specified value at a specified point, and a function that satisfies the interpolation condition in terms of derivatives is known as Hermite interpolation. Since the derivative values of each salient point are known, we use the segmented cubic Hermite interpolation method for evaluation.

Let h_i denote the length of the i-th subinterval, i.e., $h_i = x_{i+1} - x_i$, and the slope of the interpolating function $F(x)$ at x_i is $d_i = F'(x_i)$. On the subinterval $[x_i, x_{i+1}]$, set the local variables $s = x - x_i$, $h = h_i$, then we have

$$F(x) = \frac{3hs^2 - 2s^3}{h^3}y_{i+1} + \frac{h^3 - 3hs^2 + 2s^3}{h^3}y_i$$
$$+ \frac{s^2(s-h)}{h^2}d_{i+1} + \frac{s(s-h)^2}{h^2}d_i \tag{8}$$

It can be shown that this is a cubic polynomial with respect to s. Thus with respect to x, it satisfies four interpolation conditions with respect to function values and derivative values:

$$F(x_i) = y_k, F(x_{i+1}) = y_{i+1},$$
$$F'(x_i) = d_k, F'(x_{i+1}) = d_{i+1} \tag{9}$$

Compared to the linear estimation method, the curve reconstructed using the segmented cubic Hermite interpolation method is smoother and at the same time conformal.

4 Experiment

In this study, the experiment consists of four aspects, comparing the effects of different levels of fit, different reconstruction methods on the results, as well as the effects of our proposed method with the work of Li et al. and the baseline method that performs the Duchi mechanism for each data point under different privacy budgets and different data sizes.

4.1 Experimental Setup

Dataset Selection. Our experiments are based on the real publicly available dataset PAMAP (Physical Activity Monitoring for Aging People) [20], which is a physical activity monitoring dataset that records information such as heart rate data collected from nine subjects wearing wearable devices for 18 different activities for a total of more than 10 h. Since one of the subjects had incomplete statistics, we chose eight other subjects and extracted their heart rate data at 10 s intervals, with 3,000 data extracted per person, for a total of 24000 data. For this experiment, we generated four different sizes of data sets: 120K, 240K, 480K, and 600K.

Evaluation Metrics. In order to accurately evaluate our proposed method, Mean Squared Error (MSE) is used as the evaluation metric:

$$MSE = \frac{1}{n} \times \sum_{i=1}^{n} \left(\frac{1}{m} \times \sum_{p_1}^{p_m} x_i' - \frac{1}{m} \times \sum_{p_1}^{p_m} x_i \right)^2 \tag{10}$$

In this dataset, n denotes the length of the data stream, i denotes the moment, m denotes the number of people, $p_1 \sim p_m$ denotes all people, x_i' denotes the estimated heart rate value at the i-th moment, and x_i denotes the actual heart rate value at the i-th moment.

Experimental Methods. Our experiments were run on Windows 11, using Python 3.9 and MATLAB 2020b, and evaluated from different perspectives, testing the effects of the degree of fit, reconstruction method, privacy budget, and data size on the experimental results, and calculating the MSE metrics. In our experiments, the parameter α is uniformly set to 30. To minimize the error, we repeat each experiment 20 times and report the average value as the experimental result.

4.2 Experimental Results

Comparison of Different Levels of Fit

In order to evaluate the fitting effect, we set the timestamp to 40 and conduct experiments on a dataset of size 600K, constantly changing the number of times m of the polynomial function and observing the estimation results, as shown in Fig. 3. Obviously, the size of m clearly affects the estimation accuracy, and this effect becomes more obvious when the privacy budget ϵ is smaller. For example, we choose the smallest privacy budget $\epsilon = 0.1$, and the MSE is only 3 when the polynomial number $m = 5$, but when m increases to 25, the MSE increases to nearly 30.

The reason for this result is that different levels of fitting lead to different subsequent selections of salient points, which in turn lead to different errors. In detail, when the number of selected salient points increases, the privacy budget is interfered with by a more detailed segmentation and the MSE increases, especially when the privacy budget itself is small. And when the number of salient points no longer increases with the polynomial number, the MSE starts to decrease.

Comparison of Different Refactoring Methods

In order to evaluate the different reconstruction methods, we change the dataset size and privacy budget, uniformly use our method to fit and add noise sequentially, execute the linear estimation method and the segmented three times Hermite interpolation method to reconstruct the data, respectively, and observe the estimation results, as shown in Fig. 4.

It can be seen that the results of the two methods are very close to each other. Linear estimation is slightly better than the segmented three times Hermite interpolation method

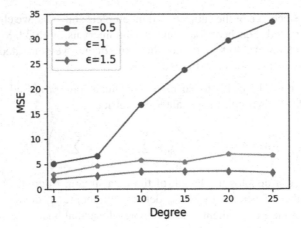

Fig. 3. Effect of different degrees of fit on MSE

because the fitting and injection of noise have deviated from the actual value, while the data stream reconstructed with the interpolation method is smoother, so the deviation is also larger. Therefore, it can be concluded that data reconstruction using linear estimation method is more appropriate.

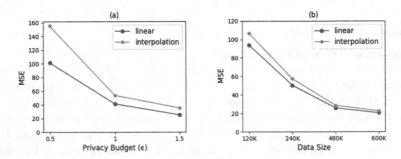

Fig. 4. Effect of different reconstruction methods on MSE

Comparison of Different Privacy Budgets

We choose a dataset of size 600K, execute our proposed method and the work of Li et al. and the baseline method of Duchi perturbation mechanism for each data point respectively, and conduct experiments by adjusting the privacy budget to study the effect of privacy budget on the evaluation metrics, and the results are shown in Fig. 5.

It can be seen that the evaluation metric MSE decreases as the privacy budget keeps increasing. Among the three methods, the baseline method has the largest MSE, while our proposed method is slightly better than the work of Li et al. This is because in our experiments, the original data stream is smoothed by fitting the data stream, which reduces the salient points that need to be selected and saves the privacy budget to some extent.

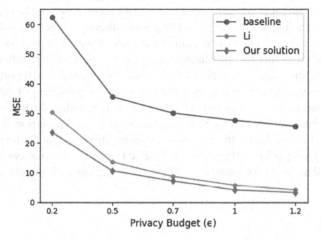

Fig. 5. Impact of different privacy budgets on MSE

Comparison of Different Data Sizes

We specify the privacy budget $\in = 1$, execute our proposed method and the work of Li et al. as well as the baseline method of executing the Duchi mechanism for each data point, respectively, and conduct experiments by adjusting the size of the dataset to study the effect of data size on the evaluation metrics, and the results are shown in Fig. 6.

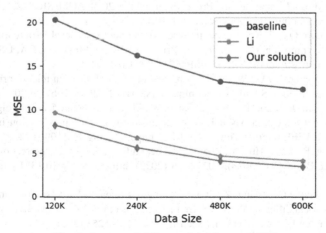

Fig. 6. Impact of different data sizes on MSE

From the figure, it can be seen that the evaluation metric MSE decreases as the amount of data contained in the dataset continues to increase. Among the three methods, the baseline method has the largest MSE, and our proposed method has obvious advantages.

5 Conclusion

For data stream aggregation under LDP, this paper proposes a new mechanism including data fitting, salient point perturbation and data reconstruction. In order to further reduce the privacy budget, instead of adding noise directly to the original data stream, we first generate a smoother fitted data stream and traverse each point based on it, remove redundant points by calculating their derivatives, filter out salient points and add noise according to the Duchi perturbation mechanism, and then report it to the data collector to perform data reconstruction and restore the original data stream. The experiments demonstrate that even though additional noise is introduced when fitting the data stream, it can be compensated by reducing the noise scale. We also evaluate different data reconstruction methods, and extensive experiments show that the linear estimation method slightly outperforms the segmented cubic Hermite interpolation method. All in all, our method achieves the smallest MSE in all settings and effectively improves data availability.

References

1. Papageorgiou, A., Strigkos, M., Politou, E., Alepis, E., Solanas, A., Patsakis, C.: Security and privacy analysis of mobile health applications: the alarming state of practice. IEEE Access **6**, 9390–9403 (2018). https://doi.org/10.1109/ACCESS.2018.2799522
2. Xue, Q., Ye, Q., Hu, H., Zhu, Y., Wang, J.: DDRM: a continual frequency estimation mechanism with local differential privacy. In: IEEE Transactions on Knowledge and Data Engineering, vol. 35, no. 7, pp. 6784–6797 (2023). https://doi.org/10.1109/TKDE.2022.317 7721
3. Kasiviswanathan, S.P., Lee, H.K., Nissim, K., Raskhodnikova, S., Smith, A.: What can we learn privately? SIAM J. Comput. **40**(3), 793–826 (2011)
4. Erlingsson, U., Pihur, V., Korolova, A.: RAPPOR: randomized aggregatable privacy-preserving ordinal response. In: Proceedings of the 2014 ACM SIGSAC Conference on Computer and Communications Security (2014)
5. Dong, W., Luo, Q., Yi, K.: Continual Observation under User-level Differential Privacy. In: 2023 IEEE Symposium on Security and Privacy (SP), San Francisco, CA, USA, pp. 2190–2207 (2023).https://doi.org/10.1109/SP46215.2023.10179466
6. Wang, M., Zhang, X., Li, W., et al.: A review of research on data release-oriented privacy protection techniques. Small Microcomput. Syst. **41**(12), 2657–2667 (2020)
7. Zhang, K., Tian, J., Xiao, H., Zhao, Y., Zhao, W., Chen, J.A.: Numerical splitting and adaptive privacy budget allocation based LDP mechanism for privacy preservation in blockchain-powered IoT. IEEE Internet Things J. (2022). https://doi.org/10.1109/JIOT.2022.3145845
8. Li, Z., Wang, B., Li, J., Hua, Y., Zhang, S.: Local differential privacy protection for wearable device data. PLoS ONE **17**(8), e0272766 (2022). https://doi.org/10.1371/journal.pone.027 2766
9. Wang, T., Chen, J.Q., Zhang, Z., et al.: Continuous release of data streams under both centralized and local differential privacy. In: Proceedings of the 2021 ACM SIGSAC Conference on Computer and Communications Security, pp. 1237–1253 (2021)

10. Warner, S.L.: Randomized response: a survey technique for eliminating evasive answer bias. J. Am. Stat. Assoc. **60**(309) 1965
11. Yang, M., Lam, K., Zhu, T., Tang, C.: SPoFC: a framework for stream data aggregation with local differential privacy. Concurrency Comput. Pract. Exp. **35**(5) (2022)
12. Duchi, J.C., Jordan, M.I., Wainwright, M.J.: Local privacy and statistical minimax rates. In: Paper presented at: 2013 IEEE 54th Annual Symposium on Foundations of Computer Science, pp. 429–438. IEEE (2013)
13. Ding, B., Kulkarni, J., Yekhanin, S.: Collecting telemetry data privately. Adv. Neural. Inf. Process. Syst. **30**, 3571–3580 (2017)
14. Joseph, M., Roth, A., Ullman, J., Waggoner, B.: Local differential privacy for evolving data. NeurIPS. **31**, 2381–2390 (2018)
15. Duchi, J.C., Jordan, M.I., Wainwright, M.J.: Privacy aware learning. J. ACM. **61**(6) (2014). https://doi.org/10.1145/2666468
16. Nguyên, T.T., Xiao, X., Yang, Y., Hui, S.C., Shin, H., Shin, J.: Collecting and analyzing data from smart device users with local differential privacy. arXiv preprint arXiv:160605053 (2016)
17. Erlingsson, Ú., Feldman, V., Mironov, I., Raghunathan, A., Talwar, K., Thakurta, A.: Amplification by shuffling: from local to central differential privacy via anonymity. In: Proceedings of the Thirtieth Annual ACM-SIAM Symposium on Discrete Algorithms, pp. 2468–2479. SIAM (2019)
18. Errounda, F.Z., Liu, Y.: Collective location statistics release with local differential privacy. Future Gener. Comput. Syst. **124**, 174–186 (2021). https://www.sciencedirect.com/science/article/pii/S0167739X21001709
19. Bao, E., Yang, Y., Xiao, X., Ding, B.: CGM: an enhanced mechanism for streaming data collection with local differential privacy. Proc. VLDB Endow. **14**(11), 2258–2270 (2021)
20. Reiss, A., Stricker, D.: Introducing a new benchmarked dataset for activity monitoring. In: 2012 16th International Symposium on Wearable Computers, Newcastle, UK, pp. 108–109 (2012).https://doi.org/10.1109/ISWC.2012.13

BP Neural Network-Based Drug Sale Forecasting Model Design

Yufang He[1], Zhen Gong[1], Dong Han[2], Wenjing Duan[1], Kaiyue Shen[1], Ruoyao Jia[1], and Zheng Xie[1(✉)]

[1] School of Health Management, Changchun University of Chinese Medicine, Changchun 130117, China
xieery@126.com
[2] School of Medical Information, Changchun University of Chinese Medicine, Changchun 130117, China
hd@ccucm.edu.cn

Abstract. Pharmaceutical sales are difficult to show a linear trend due to its inherent high randomness. BP neural network is applied to establish a pharmaceutical sales forecasting model to help enterprises accurately judge the market situation according to the changes in the market environment and thus achieve the expected economic. Benefits. The optimization method is proposed for drug production, thus forming an effective drug sales system model.

Keywords: BP Neural Network · Pharmaceutical Sales · Projections

Pharmaceuticals are essential substances for the prevention, diagnosis, and treatment of human diseases and play an important role in maintaining human health. In order to rationalize the production of drugs, appropriate procurement of drugs and correct introduction of drugs, it is important to study and forecast the sales volume and sales trend of drugs in the process of drug sales. There are many factors that affect the accuracy of drug sales forecasting, such as changes in: pharmaceutical management policies, national macroeconomic conditions new drug launches, unexpected events seasonal changes and the creation of new treatments [1]. Neural network is a method that can solve. The regression problem and appears to be more robust in dealing with higher dimensional input feature space [2]. Artificial neural networks originated in the 1940s. It can be defined as neural network technology is a widely parallel interconnected network of simple adaptive units organized to mimic the interactive responses of the biological nervous system to real-world objects [3].

1 Predicting Drug Sales Using-BP Neural Network

1.1 Sales Volume of Drugs

Sales volume forecasting is the process of extracting the law of change from historical drug sales data and using this law to forecast future sales volume in conjunction with objective conditions and circumstances. For pharmacies, sales forecasting is the front end of their business. Planning and management, playing a role in guiding marketing planning and sales execution, and is an important reference for resource allocation: it helps optimize inventory, release capital and reduce storage costs: it helps improve the transparency of sales: management, guide the reasonable setting of sales tasks, effective control of sales and cooperation channels, and improve the initiative of business initiatives.

1.2 BP Neural Network

The basic idea of the BP algorithm is: the input signal is input from the input layer, and the output signal is obtained from the output layer after the function of the hidden layer, if the actual output signal does not match the expected value. If the actual output signal does not match the desired value, the error is propagated backwards, and the weights and thresholds between the layers are repeatedly modified. Until the desired output is reached, that is, the network global error is minimized. It can be seen that the network learning process of BP algorithm is composed of forward propagation of the signal and backward propagation of the error.

2 Main Algorithms and Model Building for Drug Sales Forecasting

2.1 Sales Forecasting Modeling-Algorithm Using BP Neural Network

Neural networks are trained by two main learning algorithms, namely guided learning algorithms and unguided learning algorithms [4]. In addition, there is a third learning algorithm namely, reinforcement learning algorithm, which can be considered as a special case of supervised learning. In addition, there is a third learning algorithm, the reinforcement learning algorithm, which can be considered as a special case of supervised learning. Supervised learning algorithms need to provide the desired or target output signal. The unsupervised learning algorithm does not need to know the desired output, and during the training process, the neural net end is able to automatically adapt the connection rights so that similar features group the input patterns into aggregates, as long as the input patterns are provided to the neural net end.

The back propagation network in this paper is a supervised learning algorithm, which uses 4 input nodes, 3 hidden nodes and 2 output nodes, and the action function of the nodes is chosen as $f(x) = \frac{1}{1+e^{-x}}$ The learning algorithm is as follows.

Step 1: Choose the initial value of the weight coefficient W (the weight should be chosen as the optimal sales volume for the year, so as to reduce the step of modifying the weight). Repeat the following steps until the target output Y and the actual output Y of each sample. Meet $\left| \frac{y_j - y_j'}{j} \right| < \varepsilon$.

Step 2: Calculate the input and output of each implied layer node for 1 to 4 i_{jk} and output o_{jk}. And the output of the output node y_j, (forward computation process).

Step 3: Calculate the error gradient of each weight based on the error $\frac{\partial E_j}{\partial W_{ij}}$ need to correct the weights $W : W_{ij} = W_{ij}' \Delta W_{ij}$, where $\varepsilon = 0.05, W \in (0.2, 0.3)$. The weights are corrected using the momentum correction method: $\Delta W_{ij}(m) = \eta \sum \delta_{ij} O_{ij} + \alpha \Delta W_{ij}(m - 1)$, instead of modifying the weight coefficients immediately after each sample is given, the error-weight coefficient adjustment and error sum are calculated for each sample effect after all samples are applied $\sum \delta_{ij} O_{ij}$ and the last adjustment amount $W_{ij}(\text{m} - 1)$ related to (ε is the number of iterations, $\eta, \alpha \in (0.1, 0.2)$; the actual value of $\eta = 0.3$ $\alpha = 0.4$).

2.2 Data Normalization Methods for the Input Layer

When training a deep network, the update of the training parameters of the previous layer will lead to changes in the input data distribution of the later layers. Take the second layer of the network as an example: the input of the second layer of the network is calculated from the parameters and input data of the first layer, and the parameters of the first layer keep changing throughout the training process, so it will inevitably lead to changes in the input data distribution of each subsequent layer. As the model parameters are constantly being modified and propagated forward, the input distribution of each layer is constantly changing, i.e. the input data distribution of each layer of the network is always changing. Researchers refer to the change in data distribution in the middle layers of the network during training as the internal covariate shift (ICS), which requires the model to be trained with a small learning rate and careful selection of the initial values of the weights. ICS leads to slower training and the use of saturated non-linear activation functions (e.g. sigmoid, where both positive and negative sides saturate with a gradient of 0) [5].

The solution to this common phenomenon of internal covariance variation is to normalised the input to each layer. The Batch Normalization algorithm consists of the following three steps.

Calculate statistical values: Calculate the statistical values, including the mean and variance, required for normalisation on a small batch of samples. Suppose the input is $X \in R^{m*d}$, where m refers to the size of the current batch, i.e. how many training samples are in the current batch. d refers to the size of the input feature map.

$$E\left(X^k\right) = \frac{1}{m}\sum_{i=1}^{m}X_i^k, \ Var\left[X^k\right] \leftarrow \frac{1}{m}\sum_{i=1}^{m}(X_i^k - E\left[X^k\right])^2 \tag{1}$$

Normalisation: Each element of the input vector is normalised separately as an independent random variable, so that each variable in the vector is independent and there is no covariance matrix. This normalization accelerates convergence even when the variables are correlated, using the following formula, which is an approximate whitening process. For d-dimensional input data $x = (x^{(1)} - x^{(d)})$, normalise each dimension.

$$\overline{X} = \frac{x^k - E[x^{(k)}]}{\sqrt{Var[x^{(k)}]}} \tag{2}$$

Linear Transformation: Normalising the input only may change the properties or distribution that the input would otherwise exhibit, e.g. adding a batch normalisation algorithm to a sigmoid function may cause the input to change from non-linear to linear. To solve this problem, the original distribution can be fitted with a learnable parameter gain γ and bias β.

$$y^k = y^k * \overline{x}^{(k)} + \beta^{(k)} \tag{3}$$

When $y^k = Var[X^k]$ and $\beta^{(k)} = E(X^k)$ gives the same distribution as the input in theory. The linear transformation is added here to make it possible to restore the original input to the BN deliberately added for training purposes.

Based on the above three points, the overall flow of the batch normalisation algorithm is shown below.

Input: Values of x over a mini-batch: $\mathcal{B} = \{x_{1...m}\}$;
Parameters to be learned: γ, β
Output: $\{y_i = \mathrm{BN}_{\gamma,\beta}(x_i)\}$

$$\mu_\mathcal{B} \leftarrow \frac{1}{m} \sum_{i=1}^{m} x_i \qquad \text{// mini-batch mean}$$

$$\sigma_\mathcal{B}^2 \leftarrow \frac{1}{m} \sum_{i=1}^{m} (x_i - \mu_\mathcal{B})^2 \qquad \text{// mini-batch variance}$$

$$\widehat{x}_i \leftarrow \frac{x_i - \mu_\mathcal{B}}{\sqrt{\sigma_\mathcal{B}^2 + \epsilon}} \qquad \text{// normalize}$$

$$y_i \leftarrow \gamma \widehat{x}_i + \beta \equiv \mathrm{BN}_{\gamma,\beta}(x_i) \qquad \text{// scale and shift}$$

2.3 Sales Forecasting Model Using BP Neural Network

BP neural network is a very effective intelligent prediction method, and its topology presents higher accuracy and sensitivity compared with traditional statistical methods [6, 7]. It is more effective in dealing with complex systems, such as impact detection, evaluation system construction, simulation results testing and other research [8, 9]. Neural network adopts smooth activation function, the input signal of the input layer is output by the hidden layer, in the process of signal forward transmission, the network weights are fixed, only when the output layer does not get good output, the error signal reverse transmission, at this time the network weights are adjusted by the error feedback to make the network output close to the actual value. In the process of signal forward transmission, the network weights are fixed, and only when the output layer does not get a good output, the error signal is transmitted in the reverse direction [10]. In this paper, since the sales quantity of a drug in the coming year is unknown, a warehouse of historical sales data of a drug should be built before planning production as a basis for predicting the sales quantity of a drug in the following year. Let the historical sales data (20 years in total) of a certain drug be vector $C(C_1, C_2, C_3 ---, C_{20})$, where C_1 is the sales volume of the current year, C_2 is the sales volume of one year ago, $C3$ is the sales volume of two years ago, and so on, the sales data vector C is specified as vector $SC = (SC_1, SC_2, ----, SC_{20})$, from which four sample vectors are constructed X_j and the corresponding target output Y_j is specified as shown in Table 1.

Table 1. Sample vectors

Sample Sequence	Sample Vector	Target Output
1	SC_2,SC_3,SC_4,SC_5	SC_1
2	SC_3,SC_4,SC_5,SC_6	SC_2
3	SC_4,SC_5,SC_6,SC_7	SC_3
4	SC_5,SC_6,SC_7,SC_8	SC_4

This reasonably establishes the relationship between the first four years and the fifth year of sales, and then the first four years of sample data each sample and the corresponding target-output into the BP neural network for learning, adjusting the weights between the neurons until the output error of all samples reaches the lower limit E, then the learning phase is completed. At this time the mapping function is established between the Y value of the BP neural network output and the X value of the input $Y = W(X)$, and the first four years of samples and the fifth year of samples are consistent with this function. That is the neural network can accurately predict the sales quantity of the first 4 years based on the historical data, so it can predict the sales quantity of the next year. As long as the sales volume of the previous 4 years (SC_1,SC_2,SC_3,SC_4) taken as the sample input, the output of the network with the smallest error from the previous 4 years is the predicted quantity for the next year. The backward propagation network is shown in Fig. 1.

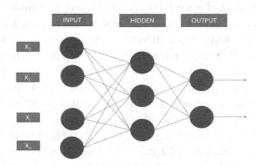

Fig. 1. Reserve propagation network

3   Data Description and Empirical Analysis

3.1   Data Description

The data required for the study were obtained from the official annual reports of a listed pharmaceutical company. In order to make the experimental results representative, the company chosen was a pharmaceutical group in the industry that focuses on the research and development, production and sales of natural drugs, novel formulations and biotechnology products, and has a completed industrial chain. Because BP neural network is a kind of error back-propagation multi-layer feed-forward neural network, which can rely on the data itself to explore its intrinsic connection but also has the characteristics of slow convergence, easy to fall into the local optimum and other disadvantages [11], the data from 2003–2022 were selected. Figure 2 shows the cyclical nature of the company's overall revenue trend over the last 20 years.

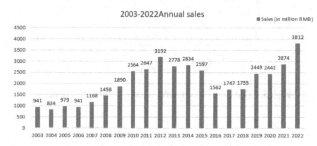

Fig. 2. 2003–2022 Annual sales

3.2 Empirical Analysis

After establishing the BP neural network model, the BP neural network model is trained by entering the relevant code hits in MATLAB R2016A software, taking every 4 years as 1 cycle, the drug sales data of the first 4 years as the input of BP neural, and the current drug sales as the output of BP neural network, using a single-layer BP neural network based on a single layer to build a drug sales prediction model, using the data of the first 18 years. After the training was completed, it was found that the correlation between the modal prediction value and the output layer data was high, which indicated that the training accuracy of the model was good. Figure 3 shows that the correlation degree of the prediction results based on the BP neural network drug sales is the training set R = 0.80773, which indicates that the constructed model better reflects the nonlinear mapping relationship between the sample data. After bringing the data in the table into the model established by the above algorithm for training, it was found that after six times of deep learning the error did not decrease significantly as shown in Fig. 4, so the training was stopped.

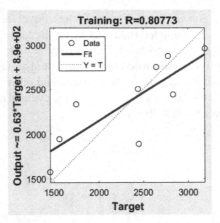

Fig. 3. The degree of correlation of predicted results

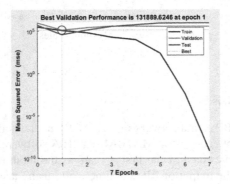

Fig. 4. Number of iterations versus error value

4 Conclusion

The improvement of technology, the number of production entities, the expansion and merger of production entities, the sales volume of medicines, and even the upgrading of the pharmaceutical industry all require financial support, and if the profit margin of a medicine is too small the enterprise cannot complete the initial capital accumulation. Sacrificing the interests of the company for the sake of the public interest is in fact a form of short-term profit-seeking. A stronger pharmaceutical industry can make a significant contribution to China's economic development, and the benefits to the public are obvious.

BP neural network has the characteristics of simple structure, simple training, strong self-adaptive ability, fast convergence and the ability to approximate any complex non-linear function, so it is applied to drug sales forecasting. This is the result of the proposed BP neural network prediction model. In practice, an effective sales forecast should not only include the forecast data, but also update the actual sales data and retain the original forecast data. Each time a company makes a sales forecast it should also consider: the difference between the actual sales and the forecast data for the time period that has already occurred; the difference between the updated data and the original forecast data for the future forecast time period; and the difference between the updated forecast data and the actual sales data of the past.

In practice, it is important to combine quantitative and qualitative analyses flexibly, and it is important to choose the right measure for the company to determine the weighting coefficients when making the combination, and to consider various factors on the basis of the data forecasting model, so as to achieve more accurate forecasting of pharmaceutical sales. Therefore, how to improve the learning algorithm of BP neural network to make it have better prediction ability is a problem that needs to be further explored in the future. The model is useful for pharmaceutical manufacturers to make production plans on the one hand, and for pharmaceutical operators to make sales plans for the next phase based on sales data in the past period on the other hand, so the proposed model has certain practical significance for pharmaceutical manufacturers and operators. With the booming pharmaceutical industry, effective forecasting of end-use sales can also provide data to support decision making and optimisate at each stage of the pharmaceutical supply chain.

References

1. Wang, C., Xiang, B., Xie, S., Tong, X.: Application of BP neural network based on principal component analysis in drug sales forecasting. Pharm. Biotechnol. **16**(04), 385–387 (2009). https://doi.org/10.19526/j.cnki.1005-8915.2009.04.022
2. Forte, R.M.: Predictive analytics - R Language Implementation. Machinery Industry Press, Beijing, pp. 84–107 (2017)
3. Ojas, R.R.: Neural Networks: A Systematic Introduction. Springer - Verlag (1996). https://doi.org/10.1007/978-3-642-61068-4
4. Cai, Z., Xu, G.: Artificial Intelligence and its Applications. Tsinghua University Press, Beijing (2004)
5. Wang, Y.: A study of normalization techniques for deep neural networks. Nanjing University of Posts and Telecommunications (2019). https://doi.org/10.27251/d.cnki.gnjdc.2019.000632
6. Hu, Z., Wu, Y.: Research on analysis and prediction of factors affecting science and technology output - a multiple regression and BP neural network based approach. Sci. Res. **30**(07), 992–1004 (2012). https://doi.org/10.16192/j.cnki.1003-2053.2012.07.004
7. Cheng, J., Xu, J., Wang, Y., Zhang, H., Li, X.: Prediction model of rice temperature in bins based on BP neural network. Modern Electron. **44**(19), 178–182 (2021). https://doi.org/10.16652/j.issn.1004-373x.2021.19.036
8. Tang, L., Liu, Y., Pan, Y., Ren, Y.: Evaluation and zoning of rural regional functions in Pinggu District of Beijing based on BP model and ward's method. Geogr. Sci. **36**(10), 1514–1521 (2016). https://doi.org/10.13249/j.cnki.sgs.2016.10.008
9. He, X., Yang, H., Wang, X.: Rural farm tourism adaptation effects, patterns and their influencing factors - a case study of 17 case villages in Xi'an and Xianyang city. **38**(09), 2330–2345 (2019)
10. Lou, H., Jia, Y., Lu, S.: BP neural network based gas monitoring between sewage grids. Chin. Prod. Safety Sci. Technol. **19**(03), 171–176 (2023)
11. Mi, Y., Li, Z., Zhang, J.: AGA-BP neural network based data resource value assessment in online platform trading environment. Intell. Theory Pract. **43**(01), 135–142 (2020). https://doi.org/10.16353/j.cnki.1000-7490.2020.01.020

Data and Privacy Crisis on Internet Application of Artificial Intelligence

Sen Xu[1](✉), Fei Yan[2], Yu Fu[3], Jincan Xin[1], and Hua Xu[1]

[1] China Telecom Research Institute, Beijing, China
{xusen,xinjc,xuh41}@chinatelecom.cn
[2] China Telecommunications Corporation, Beijing, China
yanfei@chinatelecom.cn
[3] China Unicom Research Institute, Beijing, China
fuy186@chinaunicom.cn

Abstract. The widely use of intelligent applications makes us live with data and algorithms every day. While improving the quality of our lives, its potential misuse of personal information poses a threat to our privacy. People are gradually paying more attention to privacy protection as the increasing number of such incidents indicates the seriousness of the personal information leakage problem and the intensification of privacy risks. This paper analyzes data security issues and metrics for privacy protection in Artificial Intelligence, then some recommendation for privacy protection is proposed.

Keywords: Personal Information · Artificial Intelligence · Privacy Protection

1 Introduction

As the revolution of artificial intelligent, the development and update of technology will bring unprecedented changes to our lives. The update of the Internet, big data, and other related technologies have accelerated the pace of artificial intelligence (AI) applications, quietly bring our production and lifestyle a new vitality. While the development of technology brings opportunities to society, we also cannot ignore its drawbacks and a series of negative consequences, especially in the current era of privacy-free transparency. We have to face and solve the privacy leakage problem when we give up some of our rights in exchange for the convenient services brought by artificial intelligent applications. Finding a balance between the development of AI technology and privacy protection is the key to solving the privacy protection problem in AI applications. Nowadays, AI technology is applied by various industries, which inevitably causes personal privacy leaks for various reasons. Under such circumstances, we must formulate strong measures to deal with such risks to truly and effectively protect personal privacy while using AI applications in exchange for personal data to enjoy economic benefits and social value [1].

In recent years, there have been repeated cases of privacy violations. For example, Facebook leaked users' personal information to Cambridge Analytica for improper purposes without their permission, while it used the browsing behavior of Internet users, among other things for precisely target advertisements. The University of Cambridge Psychometrics Center analyzed the gender, sexual orientation, and extroversion or introversion of each user by which posts and news they read and like. Food delivery companies such as Hungry Food, Dianping, and Meituan used algorithms to push some recommended foods and restaurants to help users to make eating decisions. Self-driving technology makes it easy for people to travel. Google Maps, Baidu Maps, and other similar intelligent navigation systems reduce the time and effort of users to find routes. Moreover, all location information is stored in the form of data collected and used by companies or other subjects. Some smartphone applications even excessively collect and use personal information excessively and illegally, exposing personal private information to the risk of being leaked or stolen. The increasing number of such incidents indicates the seriousness of the personal information leakage problem and the intensification of privacy risks. Therefore, people are gradually paying more attention to privacy protection.

The author in [2] analyzed eight elements of data security in cloud storage systems: data confidentiality, data integrity, data availability, fine-grained access control, dynamic group-based secure data sharing, leak prevention, complete data deletion, and privacy protection. The article also introduced encryption principles of IBE, ABE, homomorphic encryption, and searchable encryption, as well as research directions for new encryption models. Finally, the article summarized data encryption technologies and protection methods.

In the paper "When Machine Learning Meets Privacy: A Survey and Outlook," existing research on privacy protection is divided into three categories: (i) privacy-preserving machine learning, (ii) machine learning-assisted privacy protection, and (iii) privacy protection against machine learning attacks. The article comprehensively reviewed the research status in this field and concluded that private machine learning has recently received the most widespread attention, and in this type of research, many people are trying to use differential privacy criteria for analysis.

PEFL can not only prevent privacy leakage of local gradients but also prevent privacy leakage of shared parameters is present in [3]. PEFL is non-interactive in each aggregation, and can provide a high level of privacy protection even when the adversary colludes with multiple honest entities. Under the consideration of collusion attacks, it is proved that PEFL has post-quantum security and aggregation unaware security. Performance evaluation shows that PEFL has practical accuracy on the MNIST dataset.

This paper analyzes data security issues and metrics for privacy protection in Artificial Intelligence, then some recommendation for privacy protection is proposed. The remaining part of the paper is organized as follows. Section 2 describes the state-of-the-art in terms of privacy protection and deep learning technology. Next, artificial intelligence data security issues is presented in Sect. 3. In Sect. 4 we try to "attack their shield with their son's spear" strategy to address data compliance and privacy protection issues based on artificial intelligence algorithms. Finally, the artificial intelligence data security and privacy protection technology recommendations are provided in Sect. 5.

2   Related Work of Privacy Protection and Deep Learning Technology

2.1   Metrics for Privacy Protection in AI

Privacy preservation in AI refers to the methods and techniques used to protect personal information and data collected by AI systems. This can include methods for de-identifying personal data, reducing data collection, and implementing privacy-enhancing technologies. The goal of privacy preservation in AI is to ensure that the personal data collected by AI systems is not misused or disclosed to unauthorized parties.

There are several metrics for privacy preservation in AI, including data minimization, data de-identification, data encryption, and privacy-enhancing technologies.

Data Minimization: Data minimization refers to the practice of collecting and storing only the minimum amount of personal data necessary to achieve a specific purpose. This reduces the amount of personal data that is stored and processed by AI systems, reducing the risk of privacy violations. To measure data minimization, organizations and individuals can use metrics such as the number of personal data elements collected, the retention period for personal data, and the number of individuals whose personal data is collected [4].

Data De-identification: Data de-identification refers to the process of removing or altering personal information in a manner that makes it impossible to identify an individual. This can be achieved through techniques such as data masking, data anonymization, and data perturbation. To measure data de-identification, organizations and individuals can use metrics such as the percentage of personal data elements that have been de-identified, the level of de-identification achieved, and the success rate of re-identification attacks.

Data Encryption: Data encryption refers to the process of converting personal data into a code to prevent unauthorized access. This helps to protect personal data from theft, unauthorized disclosure, and other privacy violations. To measure data encryption, organizations and individuals can use metrics such as the percentage of personal data elements that are encrypted, the strength of the encryption algorithm used, and the success rate of decryption attacks.

Privacy-Enhancing Technologies: Privacy-enhancing technologies refer to technologies and techniques designed to protect personal data and enhance privacy. This can include technologies such as privacy-preserving data sharing, privacy-preserving data analysis, and privacy-preserving data storage. To measure privacy-enhancing technologies, organizations and individuals can use metrics such as the number of privacy-enhancing technologies implemented, the level of privacy protection provided by each technology, and the success rate of privacy-enhancing technology attacks.

2.2   State of Art of Privacy Protection in AI

Based on the current development of PPDL technology, PPDL methods can be divided into three categories: HE-based PPDL, MPC-based PPDL, and differential privacy-based PPDL. As shown in Fig. 1, the classification of privacy-preserving methods are divided into classical PPDL and hybrid PPDL. Classical privacy-preserving methods do not include any deep learning technology, while hybrid PPDL is a combination of classical privacy-preserving methods and deep learning. Traditional privacy-preserving techniques have become outdated, and hybrid PPDL technology is now more mainstream.

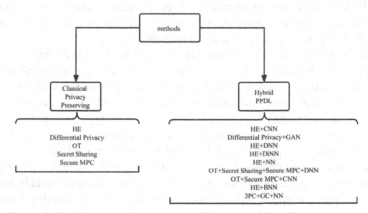

Fig. 1. Classification of Privacy Preservation technology in deep learning

Five metrics were used, including accuracy, runtime, data transmission, client privacy (PoC), and model privacy (PoM) to investigate the measurement standards for PPDL [5].

- Accuracy refers to the correctness of the PPDL model predictions.
- Runtime is the time required for the model to encrypt, transmit data from the client to the server, and perform classification.
- Data transmission (communication volume) is the amount of data transmitted from the client to the server.
- PoC means that neither the server nor any other party knows the client's data.
- PoM means that the client or any other party does not know the model classifier in the server.

3 Artificial Intelligence Data Security Issues

AI and data are interdependent: On the one hand, data provides learning samples for AI. The quantity and quality of the training dataset are the main factors determining the performance of AI models. The richer the training data reserve, the better the performance of the model trained based on it. Thus, AI has a robust data dependency. On the other hand, AI technology can facilitate data collection, storage, and utilization. In a large and disorganized sea of data, AI technology plays an irreplaceable role in searching for target data and analyzing it in depth with precision and efficiency. AI can efficiently collect and acquire users' production and life data and excavate valuable information from them, thus improving the level of data utilization. Therefore, to promote the sustainable development of artificial intelligence, it is necessary to start from three aspects: the quality and security of data, data privacy, and data application security.

3.1 Data Quality Security

Data contamination losing authenticity: Data is the bridge between real space and virtual space. If the data quality is poor, it cannot reflect the actual situation of the real world. The artificial intelligence model built on this basis will have deviations, resulting in deviations or errors in prediction results and the phenomenon of "garbage in, garbage out." At the same time, the development process of AI technology involves the whole life cycle of data collection, transmission, storage, use, and circulation, and there may be data pollution or tampering problems in each link [6].

Data poisoning attacks: Data Poisoning attack technology is adding contaminated data (e.g., wrong or malicious samples) to a training dataset to cause bias in the decision-making of a trained model, thus affecting the validity and usability of the model. AI models are vulnerable to data poisoning attacks during the training process. Attackers can compromise the integrity of training data by implementing malicious behaviors such as label-flipping or adding backdoors, which can undermine the correctness of model decisions. In recent years, poisoning attack of AI models has caused a significant negative impact on several world-renowned companies with severe consequences. For example, Amazon has caused malicious behavior to induce users to commit suicide because its Alexa smart speaker learned bad information from the Internet. Training data integrity has become a significant problem that hinders the development of artificial intelligence. In addition, user data remains at risk of privacy breaches under data poisoning attacks.

3.2 Data Privacy Security

Data over-collection: Data collection must comply with the principle of least necessary. Due to the popularity of intelligent terminal devices, various applications often compulsorily read users' private data, e.g., personal geolocation information, address book, or even run privately in the background without the users' knowledge, collecting data unrelated to the services they provide. In typical application scenarios such as uncrewed vehicles, smart homes, and smart cities, the data mainly rely on deploying various types of sensors or collection terminals in the public environment to collect environmental information in real-time without discrimination or orientation. It is difficult to anticipate the data objects and types to be collected in advance for on-site collection. When performing the on-site collection in open environments, especially in public spaces, it will inevitably lead to over-collection problems due to the expansion of the collection scope. For example, in the driverless scenario of intelligent networked cars, self-driving cars' sensors need to collect street data to support the decision of the intelligent driving system to control the car. However, this kind of indiscriminate street data collection will undoubtedly collect the personal data of pedestrians, including sensitive personal information such as pedestrians' face data, causing the risk of privacy leakage of pedestrians. In addition, it may even capture important data such as the distribution of roadside infrastructure, geographic location information, and military camps, thus posing a risk to national security.

Data breaches raising personal privacy risks: Artificial intelligence in the algorithm model research, may need to study personal privacy data, hackers may steal the model in the data of the training model to reverse the reduction, there may be a risk of privacy disclosure.

3.3 Data Application Security

Data Misuse Raising Social Security Issues: Data misuse refers to the improper use of data in which personal information or data assets are used in a manner unknown to the data subject without informed consent. Driven by the interest in data value, data resources have led to using data with excessive authority, analysis with an excessive agreement, and even creating a "black and gray" interest chain of illegal data trading. These situations have caused great infringement on personal privacy, commercial secrets, and national security. The combination of artificial intelligence and data abuse has led to much redundant information in people's lives, such as artificial intelligence-based harassing phone calls, harassing information, and massive push information [7].

Data Generation Models Raising Social Equity Issues: There is discrimination in machine learning algorithm models in AI, some are conscious, and some are unconscious. The trained models need a massive amount of data, but the data are so ponderous that the algorithms capture the valuable data on the one hand and learn human biases. On the other hand, without good-quality data, the algorithms cannot make fair and just decisions.

Data Models Raise Ethical Issues: The ethical issues arising from applying artificial intelligence technology in labor and employment have also received much attention in recent years. For example, Meituan's "delivery riders trapped in algorithms,"; Amazon's "Amazon logistics staff supervised by AI,"; and 150 employees of the Russian online payment service Xsolla were fired for being detected by the algorithm as "unprofessional and inefficient." With the help of AI, companies have more comprehensive and precise control over their workers, have an absolute advantage in resource scheduling and benefit distribution, and can build a stable organizational form without even hiring a management team. The employed workers will not only be alienated from the labor product, the labor process, and their labor essence but will also become atomized individuals and lose the ability to fight against the employed. Therefore, while artificial intelligence improves the efficiency of work, it also makes people gradually fall into the situation of labor involution. The problems of devaluation of subjective values and alienation of labor goals are also increasingly highlighted.

4 International and Domestic Artificial Intelligence Data Security Technology Practice

Based on artificial intelligence algorithms, fintech/technology companies are trying to "attack their shield with their son's spear" strategy to address data compliance and privacy protection issues. These companies are using technology to solve the challenges posed by technology to ensure secure, compliant, and efficient use of data. For example, the application of algorithms and technologies such as homomorphic encryption and federal learning [8].

Federated learning is a distributed framework for data processing proposed by Google in 2016, which attempts to address data privacy security issues in AI algorithms and model applications. Based on ensuring data privacy security and legal compliance, Federated learning achieves joint modeling to improve the effectiveness of AI models.

Security considerations are added to the joint learning framework to prevent malicious parties from obtaining additional information in joint learning. They are first learned in distributed models and then aggregated into federated models for integrated training and analysis by extracting feature labels from different models. Depending on the data distribution, federated learning can be divided into three main categories: horizontal federated learning, vertical federated learning, and federated migration learning. Big technology companies such as Facebook, Amazon, and Apple employ this technology. Besides, domestic fintech companies and universities are also working on data privacy security technology, and Fourth Paradigm has also applied transfer learning algorithms to its core product "Prophet" platform, which has been applied in the medical field.

5 Conclusion

The artificial intelligence data security and privacy protection technology recommendations are provided as below:

Data collection stage: To authenticate and authorize the data collection operations required for AI, ensure that AI data collectors have globally unique identifiers, clarify role classification and permission control rules, and ensure that collectors can access and can only access the data that their role level can access. To establish a system of AI data security collection, determine and continuously track malicious collection behavior, and prevent data and privacy leakage caused by malicious behavior.

Data processing and preservation stage: To establish data interface specifications to manage the data docking and other operations involved in the processing and storage of data to avoid the risk of privacy leakage during the processing and storage of data. To establish a data storage management system, role rights, and responsibilities requirements for users who access the storage space, access authentication, and authorization, and clarify visitors' identity and access to data. The scope of data use is clarified when using data labeled with privacy-related data. Build a hierarchical storage system to prevent users from inadvertently manipulating privacy data to cause privacy leakage problems. When storing data with large amounts of data, if distributed storage technology is used, a sound network security mechanism should be established to prevent the data from being accessed and stolen by illegal users through network transmission.

Data circulation stage: For each participant in AI data transmission, first of all, basic requirements need to be made regarding basic qualifications as well as organizational management. Secondly, some parties' transmission processes, technology, and content are specified. The transmission process and the transmission content are specified for the provider and the demander of AI data. Before transmission, the identity of the other party must be verified, and the data related to privacy and national strategy must be desensitized; during transmission, different transmission methods are chosen for basic and enhanced requirements to ensure the security of transmission in all aspects; after transmission, a log recording the transmission behavior must be established for audit. It is stipulated that the supply and demand sides can only directly transmit general-level AI data to each other, and the restricted-level data must be transmitted through a qualified AI data hosting platform.

References

1. Eason, G., Noble, B., Sneddon, I.N.: On certain integrals of Lipschitz-Hankel type involving products of Bessel functions. Phil. Trans. Roy. Soc. London **247**, 529–551 (1955)
2. Yang, P., Xiong, N., Ren, J.: Data security and privacy protection for cloud storage: a survey. IEEE Access **8**, 131723–131740 (2020)
3. Hao, M., Li, H., Luo, X., Xu, G., Yang, H., Liu, S.: Efficient and privacy-enhanced federated learning for industrial artificial intelligence. IEEE Trans. Ind. Inf. **16**(10), 6532–6542 (2020)
4. Maxwell, J.C.: A Treatise on Electricity and Magnetism, vol. 2, 3rd edn, pp.68–73. Clarendon, Oxford (1892)
5. Jacobs, I.S., Bean, C.P.: Fine particles, thin films and exchange anisotropy. In: Rado, G.T., Suhl, H. (eds.) Magnetism, vol. III, pp. 271–350. Academic, New York (1963)
6. Elissa, K.: Title of paper if known. unpublished
7. Nicole, R.: Title of paper with only first word capitalized. J. Name Stand. Abbrev. in press
8. Yorozu, Y., Hirano, M., Oka, K., Tagawa, Y.: Electron spectroscopy studies on magneto-optical media and plastic substrate interface. IEEE Transl. J. Magn. Jpn. **2**, 740–741 (1987)
9. Young, M.: The Technical Writer's Handbook. University Science, Mill Valley (1989)

Flexible Sensor Array and Newton Interpolation Algorithm for Molded Dichromatic Sole by Operating Robot

Lizhi Gu[1,2], Shanping Gao[1], Dong Wang[3], and Jinling Song[1,4(✉)]

[1] Key Laboratory of Virtual Manufacturing Technology of Fujian Universities, Quanzhou University of Information Engineering, No. 249, Bodong Road, Fengze District, Quanzhou 362000, Fujian, China
sjl13969@163.com

[2] College of Mechanical Engineering and Automation, Huaqiao University, 668# Jimei Avenue, Jimei District, Xiamen 361021, China

[3] College of Mechanical Engineering, Jiamusi University, 258# Xuefu Avenue, Jiamusi 154007, Heilongjiang, China

[4] Quanzhou Normal University, No.398 Donghai Street, Fengze District, Quanzhou 362000, Fujian, China

Abstract. Based on the principle of system engineering and optimization method, the multi-color soles mold operation robot was constructed instead of manual operation, and the double-layer quality inspection standard of two-color soles was proposed under the conditions of high efficiency and high quality soles production process. According to the working diagram, the sole is divided into 3 areas - (1) arch area, (2) heel area and (3) palm area, and the "plum blossom" lattice is used to determine the detection feature points, so as to achieve the goal of effectively and reliably inspecting the entire sole parts with fewer measurement points. A 24-point sensor array based on 40 code shoes was formed by using the quasi-equidistant distribution design of adjacent inspection points -- the shape of plum pile hierarchy. By adopting the fusion of geometric sequence and arithmetic sequence, using the ultra-small lead/diameter ratio with double lead screw mechanisms for 6 groups of probes to move vertically and 4 groups of probes to rotate and move laterally by rotation of a arm, forming a series of new 24probes of sensor arrays with a variety of new pile measuring points on the intelligent adaptation test of 35–45 code soles. The flexible adjustment topology is drawn accordingly. At the same time, the topological diagram points out the methods, ways, regulations and practical effects of flexible adjustment corresponding to the two directions of different shoe codes. In order to further improve the accuracy and reliability of the inspection, Newton interpolation algorithm was used to realize the high reliability, high flexibility and high accuracy of the workpiece online qualification inspection. Results of simulation and experiment show that the 24-point plum blossom probe distribution test and Newton interpolation test algorithms meet the production requirements completely.

Keywords: plum blossom-like distribution of detected feature points ·
Geometric sequences/arithmetic sequence coalescent adjustment · Sensor array ·
Error norm · Newton interpolation algorithm

H. Jin et al. (Eds.): IAIC 2023, CCIS 2060, pp. 242–262, 2024.
https://doi.org/10.1007/978-981-97-1332-5_20

1 Introduction

Shoes are indispensable basic daily necessities for everyone. China is a shoe-making country, providing about 40% of the world's shoes, has long become the world's largest footwear production center and sales center, especially injection molded soles of leisure and sports styles for the largest amount, about 5 billion pairs per year. Among theses shoes two-color and multi-color soles are the strongest in demand. However This has formed a prominent contradiction between the current and future need and the existing low production and supply capacity.

For two-color or more color soles, not only the mold weight is significantly increased, up to 50 kg or even more. It is necessary to layer and stage injection molding, the process is further complicated, the mold closing and opening need repeated manual operation, and the labor intensity is further increased. The technical difficulty is increased, the quality stability is worse, the finished product rate is further decreased, and the productivity is lower.

In order to fundamentally solve the above problem, efforts have been made and some progresses have been gained. Rawashdeh, Nathir A. et al.[1] created a visual inspection system of glass ampoule packaging defects: effect of lighting configurations. Witek, Maciej [2] aimed at evaluating the inaccuracies of pipe wall flaws sizing derived from an axial excitation magnetic flux leakage technology (MFL), in respect to the defect field checking. Martinez, Pablo et al. [3] proposed an automated vision-based online inspection system for screw-fastening operations in light-gauge steel frame manufacturing with the aim of providing alternatives to current manual quality control activities. Chen, Wei et al. [4] proposed an approach to handling the fine-grained and frequently occurring IaC code errors. The approach extracted code changes from historical commits and clusters them into groups, by constructing a feature model of code changes and employing an unsupervised machine learning algorithm. Adam Hamrol et al. [5] presented an approach to the planning and optimization of quality inspections within a multistage manufacturing process based on quality costs and the value added to the production process by inspections. Inspection errors and the resulting costs of repair and scrapping were taken into account. Huang, Zhiqiang et al. [6] introduced a new acoustic positioning method to solve the problem of space positioning for online inspection robots within the storage tank. Dennis Mosbach et al. [7] presented a new scalable approach to determine a small number of well-placed camera viewpoints for optical surface inspection planning. The initial model was approximated by B-spline surfaces. Aydin Tarik Zengin et al. [8] presented a new approach to measure the sag amount by using sensor data of a power line inspection robot, precisely and reliably with a inspection robot moving on the power line remotely controlled and send sensor data. Long Thanh Cung et al. [9] based on an experimental study of the coupling of a magnetic cup core coil sensor with a metallic layered structure. Ruoxu Ren et al. [10] presented a generic deep-learning-based approach for automated surface Inspection. Marwa Haj Ibrahim et al. [11] put forward an improvement of the optimization phase by reequationting the objective function of the algorithm. A correction matrix associated with the importance of each measurement point, and therefore with the local rigidity of the part during the registration operation, was introduced and incorporated. Gabara Grzegorz and Sawicki Piotr [12] presented the results of testing a proposed image-based point clouds measuring method for geometric parameters determination of a railway track. Urbanic R J and Djuric A M. [13] presented methodologies to predetermine regions of feasible operation for multiple kinematic chain mechanisms, machine tools, industrial robots, and manufacturing cells with a 'boundary of space' model representing all possible positions which may be occupied by a mechanism during its normal range of motion. Ramachandran R K et al. [14] created a novel distributed method for constructing an occupancy grid map of an unknown environment using a swarm of robots with global localization capabilities and limited inter-robot communication with the domain by performing Levy walks. Fu L et al. [15] proposed Maxwell model of stability property based impedance controller with a novel Cartesian admittance or position-based impedance control scheme which was more suitable to be implemented in most common position-controlled robots. Usvyatsov M and Schindler K [16] was of insight into the idea to instead learn a Siamese CNN that acts as similarity function between pairs of training examples with class predictions obtained by measuring the similarities between a new test instance and the training samples. Ge L. et al. [17] Introduced a maintenance robot system with the hardware frame based on gesture recognition achieved through the acquisition of human skeleton information by a Kinect sensor. Liu J.H.et al. [18] put forward a controller design of Two-wheel Self-balancing Robots, and the simulation results show the effectiveness of the proposed method, which was verified also on the DSP platform at the same time.Sprute D.et al. [19] proposed a multi-stage object localization system based on human pointing gestures that considered the whole intelligent environment as interaction partner, accompanying experiment of multi-stage approach. Bolano G and Tanev A [20] proposed an intuitive method for controlling a robot end-effector using human gestures. Vision based techniques were used to track the position of the user's hand, which is directly translated in control signals. The use of a 3D camera sensor allows to easily control the robot tool position in all dimensions. Yang, Pan et al. [21] described a result of Systems Theoretic Accident Model and Processes (STAMP) analysis example of "Fallen Barrier Trap at Railroad Crossing" with an automaton model checker. Jin, Xiating et al. [22] established a deep multimodel RIS (DM-RIS) for surface defect where fast and robust spatially constrained Gaussian mixture model was presented for segmentation proposal and Faster RCNN was utilized for objective location in a parallel structure. Moradi, Saeed et al. [23] proposed a novel approach for automated anomaly detection and localization in sewer CCTV inspection videos with an algorithms employing three-dimensional (3D) Scale Invariant Feature Transform (SIFT) to extract spatio-temporal features in sewer CCTV videos. Taifeng,Li et al. [24] considering of the accuracy and efficiency of free-form surface parts quality inspection, introduced a maximum distance approach for model simplification with normal deviation information, to find the optimal transformation parameters with an improved fruit fly optimization algorithm. Wang, Shanshan et al. [25] proposed an approach consisting of ontology-based models and data-centric algorithms, not only formally representing concepts and relations between concepts involved in predicting whether a contract is efficient, but also organizing multichannel data such as news, marketplace reports and industry databases containing information of factors impacting the unobservable noncontract parameters' fluctuations. Özbilge, Emre. [26] introduced a new expectation-based novelty-detection system with an online recurrent neural network approach that learned the data by inserting new nodes or deleting unused nodes from its structure. Delgado, Pedro et al.[27] proposed multivariate statistical process control based on principal component analysis in solder paste printing process where 100% automatic inspection was already installed. Binbin Zhang et al. [28] looked into a deep neural network, Convolutional Neural Network (CNN), towards a robust method for online monitoring of AM parts. Dejan Jovanović and Bruno Dutertre [29] presented a new model-based interpolation procedure for satisfiability modulo theories (SMT). The procedure uses a new mode of interaction with the SMT solver that we call solving modulo a model. Bednarczyk, B. and Jaakkola, R. [30] considered the family of guarded and unguarded ordered logics, that constitute a recently rediscovered family of decidable fragments of first-order logic (FO), in which the order of quantification of variables coincided with the order in which those variables appear as arguments of predicates meanwhile the complexities of their satisfiability problems were well-established, providing some insight into it. Gianpiero Cabodi et al. [31] aimed at understanding whether and how interpolants could speed up BMC checks, as they represented constraints on forward and backward reachable states at given unrolling boundaries, distinguishing an interpolant generation (learning) phase and a subsequent interpolant exploitation phase in a BMC run with experiments of costs, benefits, as well as invariant selection options, on a set of publicly available model checking problems. Tae-Won Kang et al. [32, 33] put forward a post-processing method called bidirectional interpolation method for sampling-based path planning algorithms, such as rapidly-exploring random tree (RRT) with interpolation to the path generated by the sampling-based path planning algorithm. R. C. Mittal et al. [34] studied the effect of diffusion in coupled reaction-diffusion systems named the Gray-Scott model for complex pattern formation with the help of cubic B-spline quasi-interpolation (CBSQI) method and capture various formates of these patterns. Matthias Schlaipfer and Georg Weissenbacher [35] generalised the parametrised interpolation system which subsumes existing interpolation methods for propositional resolution proofs and enables the systematic variation of the logical strength and the elimination of non-essential variables in interpolants with extension of generalizing two existing interpolation systems for first-order logic and relates them in logical strength. Geng Chen et al. [36] presented a method of inline inspection with an industrial robot (IIIR) for mass-customization production lines. A 3D scanner was used to capture the geometry and orientation of the object to be inspected. As the object entered the working range of the robot, the end effector moved along with the object and the camera installed at the end effector performed the requested optical inspections.

In this study, robot technology, sensing technology, machine vision, optimization theory, exquisite design, contour and geometry detection methods were used to realize online flexible and intelligent inspection of two-color soles by molding parts with operating robot.

2 Error Pattern Analysis and Qualification Evaluation of Soles

2.1 Analysis of Sole Working Drawing

Molded part——two-color sole quality problem is the regional filling problem. Because the mold cavity in the injection molding process is injected with plastics or rubber in the form of "quasi-Newton fluid" full and overflow, in the cooling condensation process obeying the rule of " expansion by heating and condensation by cooling", from the injection port and accessories of the liquid material to further fill the cavity, followed by release of the formation of the sole. In this way, there will be some residual flash material around the stripping and parting of the parts, but it will not affect the body of the soles. On the other hand, due to insufficient filling, the soles shrink locally, resulting in "missing". Therefore, the inspection of the qualified parts only tests the "insufficient filling and insufficient degree".

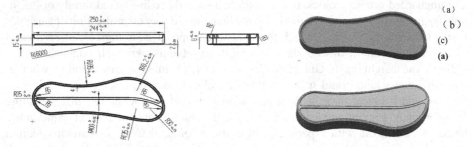

Fig. 1. The working drawing of the dichromatic sole

2.2 Analysis of Characteristics of Sole Molded by Robots in Production Line

The two-color sole molding production line is carried out on the 24-station rotary work-bench, and two injection molding machines are uniformly distributed on the rotary diameter of the workbench, respectively, to undertake the injection molding of A-color and B-color TPR materials. First of all, A-color is injected and molded. After some while the first part of the sole is inspected and checked. If the inspection is qualified then, B-color TPR is injected and molded, too. After a period of cooling and curing, the molded part is inspected. If qualified for B-color injection molded sole, after cooling and curing then it is thought of as qualified totally for the sole. The above process is carried out in the continuous rotation of the table to complete two tests (Fig. 2 and 3).

3 Principle and Method of the Inspection of the Sole

3.1 Physical Model/Geometry Model/Mathematical Model

In order to effectively inspect the soles, it is necessary to transform the soles from physical model to mathematical model. The physical model is shown in Figs. 1 (b) and (c). If the sole is further quantified and described in detail, the geometric model is given in Fig. 1 (a), although it is a flat view.

Figure 1 (a) not only specifies the final geometry of the molded soles, but also determines the tolerance range of the molded soles. When the contour of the part is within this tolerance zone, the part is considered qualified, otherwise it is not qualified.

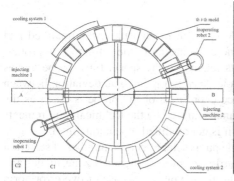

Fig. 2. Robot operating line with 24 positions

Fig. 3. Operating robot with sensing array

The engineering problem we intend to solve is whether the molded product is qualified, so we need to meet the specific mathematical model of the inspection requirements.

We select a certain number of measuring points according to certain rules -- select measuring points and express them with a matrix, as shown in Fig. 4 and Eq. (1). When these selection points are qualified, the product is considered qualified.

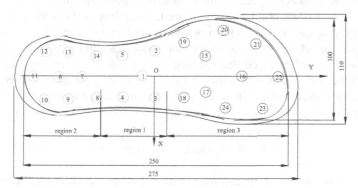

Fig. 4. Matrix expression of selected checking points

$$C_1 = \begin{bmatrix} 5 & 0 & 2 \\ 0 & 1 & 0 \\ 4 & 0 & 3 \end{bmatrix}; C_2 = \begin{bmatrix} 12 & 13 & 14 \\ 11 & 6 & 7 \\ 10 & 9 & 8 \end{bmatrix}; C_3 = \begin{bmatrix} 19 & 0 & 20 & 0 & 21 \\ 0 & 15 & 0 & 0 & 0 \\ 0 & 0 & 0 & 16 & 22 \\ 0 & 17 & 0 & 0 & 0 \\ 18 & 0 & 24 & 0 & 23 \end{bmatrix} \qquad (1)$$

3.2 Inspection Method

Using triple model construction and double error norm algorithm, the qualified and unqualified products are checked and screened. According to the setting of 24 probes, and distributed in the sole to pick up points relative to the space. Each probe moves vertically and linearly until it reaches the sole detection point. The deviation between the termination coordinate and the reference coordinate of the probe is taken as the criterion. If the deviation is within the tolerance required by the technical requirements of the sole, it is considered qualified, and if it is not qualified, it is unqualified.

According to the surface profile of the sole parts, select a certain number of selection points, through the establishment of standard size model, tolerance zone model and measured model, the sensor probe to measure the deviation, the use of two error norm calculation and discrimination to determine whether the parts qualified.

In order to further improve the detection accuracy, polynomial interpolation method is used.

3.3 Determination of Selected Checking Points and Plum Blossom Layout

According to the surface profile of the sole parts, select a certain number of selection points, through the establishment of standard model, zero-tolerance model and measured model, the sensor probe to measure the deviation, the use of two error norm calculations and discrimination to determine whether the parts are qualified.

Using triple model construction and double error norm algorithm, the qualified and unqualified products are checked and screened. According to the setting of 24 probes, and distributed in the sole to pick up points relative to the space. Each probe moves vertically and linearly until it reaches the sole detection point. The deviation between the termination coordinate and the reference coordinate of the probe is taken as the criterion. If the deviation is within the tolerance required by the technical requirements of the sole, it is considered qualified, and if it is not qualified, it is unqualified. When the norm d is used to evaluate or measure the conformity, it can be expressed in terms of.

$$d = \left| a_{ij} - a_{\text{standard}\,ij} \right| \leq + \frac{es - ei}{2} \tag{2}$$

In Eq. (2), norm d will be presented in different forms according to the error nature and measurement method, which indicates the criterion that the finished part is qualified when a batch of parts is processed by the adjustment method. According to the principle and method of "three model construction", the selection and matrix representation of detection feature points are used. If the elements are technical indicators on workpiece processing requirements, equation Eq. (2) can be used as the basis for workpiece processing qualification, and the color test.

For molded soles, after the first color is formed, it is necessary to test, when the error is large, not only has a direct impact on the second color injection molding, but also has a direct and significant impact on the overall quality of the sole. How to complete the selection of detection feature points, according to the sole parts diagram, the sole is divided into 3 areas - (1) arch area and (2) foot area, (3) palm area, using the "plum blossom" lattice to determine the feature points, in order to achieve the goal of expressing the system features with fewer points.

In the arch area (1) 5 points, in the foot area (2) 9 points, in the palm area (3) 10 points, a total of 24 points. The detection feature points determined by the "plum" lattice are expressed by matrix of qualification criteria, then C1, C2 and C3 matrices are used to represent the distribution of detection points in the three sub-regions, and the above list of selected test points is stored in the database.

4 Flexible Sensor Array and Its Optimal Design

4.1 Relationship Between the Codes and the Sizes of the Shoe

The change of shoe code is mainly the change of length, but still a little change in the direction of width. The relationship between shoe code and the shoe size is shown in Table 1.

Table 1. Some of the Chinese Codes and counterparts of Europe

Code(China)	35	35.5	36	36.5	37	37.5	38	38.5	39	39.5	40
Europe	22.5		23		23.5		24		24.4		25
Length	225	227.5	230	232.5	235	238.5	240	242.5	245	247.5	25.
width	90	91	92	93	94	95	96	97	98	99	100

The base is taken as the standard one, and its length and width are observed.
For a length of a code shoe

$$L_{\text{code}} = \frac{n_{\text{code}} + 10}{2} \times 10 \tag{3}$$

And for a shoe of code 40

$$L_{40} = \frac{n_{\text{code}} + 10}{2} \times 10 = \frac{40 + 10}{2} \times 10 = 250 \tag{4}$$

The length change for every one code change is

$$\Delta L_{\text{code}} = \frac{(n+1)_{\text{code}} + 10}{2} \times 10 - \frac{n_{\text{code}} + 10}{2} \times 10 = 5 \tag{5}$$

The width change for every one code change is

$$\Delta W_{\text{code}} = \left[\frac{(n+1)_{\text{code}} + 10}{2} \times 10 - \frac{n_{\text{code}} + 10}{2} \times 10 \right] \times \frac{2}{5} = 2 \tag{6}$$

4.2 Flexible Sensor Array

In order to effectively detect whether the molded sole is qualified or not, a set of pressure sensors are used, and the probe corresponds exactly to the selected test point, and a threshold value is set. When the detected pressure after the pressure sensor contacts the selected test point of the measured sole, the point is judged to be qualified; If all the selected points are qualified (or the number of qualified points is set to meet the requirements), the molded sole is judged to be qualified.

The above analysis is based on the sole of code 40. In order to make the detection system have good flexibility, that is, adapt to the detection requirements of different shoe sizes without increasing the number of selected measurement points, it is necessary to design an adjustable pressure sensor probe combination - flexible sensor array.

According to Fig. 4 and Eqs. (3), (4), (5), (6), the commonly used shoe codes 35–45 can be taken into account, and a set of sensor array can be properly adjusted to achieve the detection target.

4.3 Optimal Design of Sensor Array

4.3.1 Target Requirement

(1) The standard array of code 40 soles covers the sole area evenly with a fewer number of sensors;
(2) The sensor array is flexible, and through appropriate adjustment, the sensor array meets the requirements of codes 35–45 of sole;
(3) Simple structure, easy operation, reliable and convenient adjustment;
(4) After adjustment, it can still basically meet the requirements of (1).

The above 4 points are described as follows:
Let N be the number of sensors in the sensor array and S be the area of the sole parts. The area for which a sensor is responsible is

$$S_u = \frac{S}{N} \tag{7}$$

If a sensor probe is located at pprobej, the distance to the nearest probe $p_{probej-1}$ or $p_{probej+1}$ is

$$a \le r_j \le b \tag{8}$$

where a——the minimum distance between two adjacent types of value points;
b——the maximum distance between two adjacent value points.

According to the "plum matrix", draw a circle with p_{probej} as the center of the circle and r_j as the radius to form a regular pentagon of the outer circle, see Fig. 5, and the distance between adjacent two points on the circumference is

$$\frac{1}{2}\sqrt{10 - 2\sqrt{5}}r_j \tag{9}$$

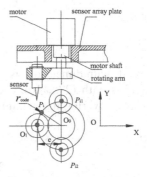

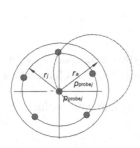

Fig. 5. Plum lattice and distance between two adjacent SCPs

Fig. 6. Transverse adjustment of the sensor array

4.3.2 Longitudinal Adjustment of the Sensor Array

Considering the rationality of the sensor array before and after adjustment, the sensor probe distribution is relatively uniform, and the longitudinal adjustment quantity in the process of retracting and retracting follows the law of geometric sequence.

Longitudinal maximum shoe size adjustment of Group 1, 11.5; Group 2, 6.0; Group 3, 3.0

The ultra-small lead/diameter ratio is adopted to move 6 sets of probes longitudinally to achieve longitudinal adjustment of the sensing array.

Take the common ratio, $\varphi = 1.26$ according to the shoe size change to take the proportional series to adjust.

$$\Delta y_{P_{ij}} = f_j(N_{\text{code}}(i)) = \frac{\Delta_{j\max}}{\varphi^{(9-i)}} + N_{\text{code}}(40), i = 0, 1, 2, \ldots\ldots, 9; \; j = 1, 2, 3 \quad (10)$$

where $\Delta_{j\max}$——maximum adjustment value of the j-th group,

$$\Delta_{j\max} = \begin{cases} 11.5, j = 1 \\ 6.0, j = 2 \\ 3.0, j = 3 \end{cases} \quad (11)$$

If the shoe of code 40 is taken as the standard shoe, then $N_{\text{code}}(40) = 0$.

When $i = 9$, the longitudinal adjustment reaches the greatest value, $\Delta_{j\max} = 11.5$.

4.3.3 The Transverse Adjustment of the Sensor Array

The maximum adjustment is ± 10, and unilateral value is 5 mm. Based on the code number, the arrangement of the transverse adjustment runs with arithmetic sequence. Transverse maximum shoe size adjustment of Group 1 is 5.0; Group 2 is 2.5.

Figure 6 shows shows the transverse adjustment of a sensor probe in the sensor array. The rotating sensor is in the P_{i1} position. When the angle, α, is turned counterclockwise, then the probe is turned to the P_i position with the radius of e, and the change in the X direction is

$$\Delta x_{P_i} = e \times \sin \alpha \tag{12}$$

Shoe code with dimensions is as follows

code 45 is with (275,110), code 40 with (250,100), and code 35 with (225,90), respectively,. Covering the maximum and minimum sizes. Therefore, take $e = 5$, is desirable.

When $\alpha = \frac{\pi}{2}$, the transverse adjustment reaches the maximum value:

$$2\Delta x_{P_i} = 2e = 10 \tag{13}$$

4.3.4 The Optimal Design and Application of the Sensing Array

(a) Construction and access of the database, with the shoe code function

$$N_{\text{code}}(X_j) = \left\{ 35 + \frac{1}{2}(j-1) \right\}, j = 1, 2, \ldots\ldots, 21 \tag{14}$$

(b) Matching shoe size and bidirectional adjustment parameters; (c) Real-time testing to obtain 24 sets of basic data; (d) Linear interpolation test and polynomial interpolation test; (e) Adjustment process parameters retest; (f) Automated system inspection.

The above is the situation when adjusting to large shoe size. Similarly, information and data can be obtained when adjusting to small shoe size, see Fig. 7.

(1) Adapt to the mold operation of multi-color soles, based on the principle of system engineering and optimal design method, propose and determine the first color inspection and second color inspection of sole parts, and cooperate with other process links to form the automation and intelligence of the entire process;
(2) Sole quality inspection methods and technology research, to achieve automatic screening of qualified products and unqualified products.

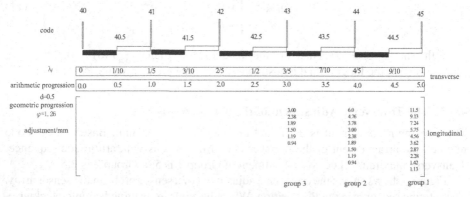

Fig. 7. The adjustment diagram of sensor array

The ultra-small lead/diameter ratio screw mechanism is designed and adopted. The working diameter is Φd, the number of threads is n, and the lead is L. General lead/diameter ratio $\geq 1/8$. We take

$$\frac{L}{\varphi d} = \frac{1}{18} \tag{15}$$

In this way, when the screw turns for a revolution, the distance the nut moves along its axis is

$$S = n \times L \tag{16}$$

The lead screw turns a small angle, $\Delta\alpha$, and the nut moves with distance

$$\Delta S = \frac{1}{360}(n \times L) \times \Delta\alpha \tag{17}$$

If take $n = 1$, then we have

$$\Delta S = \frac{1}{360} \times \frac{\varphi d}{20} \times \Delta\alpha \tag{18}$$

or

$$\Delta S = \frac{1}{180} \times \frac{L}{\varphi d} \times \Delta\text{arc} \tag{19}$$

where Δarc——infinitesimal.

From Eqs. (16) or (17) or (18), it can be seen that when the lead/diameter ratio is smaller, a smaller and more accurate axial displacement can be obtained by turning a certain angle or a certain arc length.

The lead screw is made into a double helices and matched with a nut respectively. When the double helix lead screw mechanism is driven by a motor, the two nuts will move in the same direction or in the opposite direction simultaneously, so as to realize the synchronous and accurate adjustment of the two groups of probes.

So. The length adjustment is guaranteed effectively. Now the probe positions in Region 2 and Region 3 are adjusted, meanwhile the probe positions in Region 1 remain in the the previous ones.

The first adjustment group with the numbers of transducers, 10, 11, 12 (in Region 2), and 21, 22, 23 (in Region 3) Adjustment is

$$\Delta y_{ij} = \lambda_j(y_i - y_0) - \Delta_1 \tag{20}$$

where, λ_j——adjusting factor for the j-th group of the adjustment, $\lambda_j \in [0, 1]$

y_i——coordinate in y direction in the extreme position in adjustment. For the first adjustment group

$$|(y_i - y_0)| = 12.5$$

where y_0——coordinate in y direction for the standard original position

Take $\lambda_j = 1$, $\Delta_1 = 1$, then

$$\Delta y_{ij} = (y_i - y_0) - \Delta_1 = 11.5$$

Similarly, for the second adjustment group, with transducer numbers 9,6,13(in region 2) and 20,16,24 (in region 3), the maximum adjustments for group 1 and group 2 are as follows, respectively.

$$\Delta y_{ij} = (y_i - y_0) = 6.0$$

For the third adjustment group with the numbers of transducers, 8,7,14 (region 2);15,17 (Region 3)

$$\Delta y_{ij} = (y_i - y_0) = 3.0$$

The topological diagram of three groups for length adjustment as follows (Fig. 8).

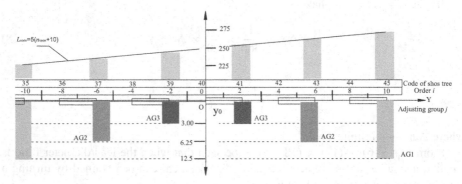

Fig. 8. Topological diagram of three groups for length adjustment

5   Test and Result

5.1   Inspection End-effector and Testing System

In the equation, norm d will be presented in different forms according to the error nature and measurement method, which indicates the criterion that the finished part is qualified when a batch of parts is processed by the adjustment method. According to the principle and method of "three model construction", the selection and matrix representation of detection feature points are used. If the elements in Eq. (2) are technical indicators on workpiece processing requirements, Eq. (2) can be used as the basis for workpiece processing qualification.

For molded soles, after the first color is formed, it is necessary to test, when the error is large, not only has a direct impact on the second color injection molding, but also has a direct and significant impact on the overall quality of the sole. How to complete the selection of detection feature points, according to the sole parts diagram, the sole is divided into 3 areas - (1) arch area and (2) foot area, (3) palm area, using the "plum blossom lattice" to determine the feature points, in order to achieve the goal of expressing the system features with fewer points.

In the arch area (1) 5 points, in the foot area (2) 9 points, in the palm area (3) 10 points, a total of 24 points. The detection feature points determined by the "plum" lattice are expressed by matrix of qualification criteria, then C1, C2 and C3 matrices are used to represent the distribution of detection points in the three sub-regions, and the above list of selected test points is stored in the database. Figure 10 gives the inspection system (Fig. 9).

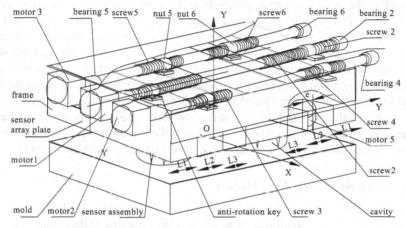

Fig. 9. Flexible sensor array and optimal design

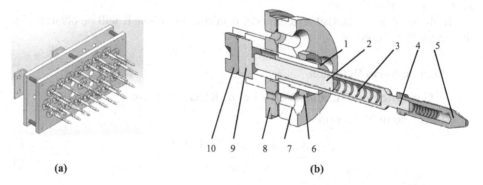

(a)   (b)

Fig.10. End-effector of the capture - inspection of the sole with probe array

The detection probe assembly is composed of a probe, a detection needle, a sensor spring, a thimble, a pressure sensor, a compression cap, and a sensor housing. The detection probe assembly is fixed on the sensor fixing plate through the sensor lower adjustment cover, the sensor upper cover and the locking nut. When the probe contacts the surface of the sole, the detection needle is firmly connected with the probe thread, so that the detection needle moves slightly upward, the detection needle external compression sensor spring, and the pressure is transmitted to the pressure sensor through the thimble, and the pressure sensor gives a signal back. For different sole surfaces, in order to make the detection end effector device more passable, when the sole mold is changed, the detection needle can be adjusted to make the probe contact with the corresponding position of the standard sole surface; At the same time, for different positions on the surface of the same sole, in order to completely contact the probe and the surface of the sole, the contact length between the detection needle and the probe can be adjusted.

In order to improve the detection accuracy, 24 detection points are arranged on the surface of the sole, and 24 detection probe holes are designed on the sensor fixing plate for installing the detection probe group. In order to suit different shoe sizes, as well as different positions of the shoe surface, the quality requirements are different, the sensor adjustment cover is designed for different shoe size length direction adjustable. The detection probe group is installed on the sensor fixing plate through the sensor adjustment cover. When the detection device is near the predetermined position of the surface of the sole product, the probe contacts the surface of the sole product, compresses the spring, the spring is compressed by the reaction force, feeds back to the pressure sensor, transmits the feedback data to the background. After the background processing of the feedback data and the background triple model construction data, and the norm test algorithm, it can distinguish the surface of the sole product after injection and molding.

5.2 Testing Result with Three Models

Several tables are used for the databases with respect to the three models (Table 2).

In the X 'Y' Z '−O' coordinate system, the sole of the shoe is the origin of Z '. Then take Z ' $= 14.925$ and determine the origin of XYZ minus O coordinates. After such conversion, the tolerance zone of the first color A is distributed symmetrically, so is the second color (Table 3).

In the same way, the real tested model database for colour B will be obtained, but the detail of it is omitted.

5.3 Double Norm Detection

5.3.1 Construction of Double Norm 1 and Detection of A-color Sole

Construct a norm in terms of

$$d_1 = |X_1 - X_{1i}| \le \varepsilon, i = 1, 2, 3......24 \tag{21}$$

$$d_{1j} = |X_{1sj} - X_{1wj}| = \sqrt{(x_{1sj} - x_{1wj})^2 + (y_{1sj} - y_{1wj})^2 + (z_{1sj} - z_{1wj})^2} \le \delta_1, j = 1, 2, 3......24 \tag{22}$$

where, X_1 ——qualification standard 1 characteristic parameter matrix;

X_{1i}—— the actual parameter matrix of the measured workpiece.

Only both that 90% of the above selected points meet the first requirements and that the points that do not meet the first requirements but do satisfy the second requirement are in positions they are not adjacent to each other will they be considered qualified. As a result, for the first requirement

$$24 \times 90\% = 21.6$$

Table 2. Perfect database model (both colour A and B)

Order	region	Space			Order	code	Space		
		X	Y	Z(A,B)			X	Y	Z(A,B)
1	(1)	0.0	−11.5	(−0.075,7.925)	15	(3)	−35.0	48.0	(−0.075,7.925)
2	(1)	−40.0	0.0	(−0.075,7.925)	16	(3)	0.0	78.0	(−0.075,7.925)
3	(1)	40.0	0.0	(−0.075,7.925)	17	(3)	28.0	48.0	(−0.075,7.925)
4	(1)	40.0	−60.75	(−0.075,7.925)	18	(3)	43.5	28.0	(−0.075,7.925)
5	(1)	−40.0	−60.75	(−0.075,7.925)	19	(3)	−48.0	28.5	(−0.075,7.925)
6	(2)	0.0	−85.0	(−0.075,7.925)	20	(3)	−51.0	60.0	(−0.075,7.925)
7	(2)	0.0	−50.0	(−0.075,7.925)	21	(3)	−48.0	90.0	(−0.075,7.925)
8	(2)	47.5	−40.0	(−0.075,7.925)	22	(3)	0.0	115.0	(−0.075,7.925)
9	(2)	−48.5	−68.5	(−0.075,7.925)	23	(3)	47.5	92.0	(−0.075,7.925)
10	(2)	46.5	−95.5	(−0.075,7.925)	24	(3)	45.5	60.0	(−0.075,7.925)
11	(2)	0.0	−115.0	(−0.075,7.925)					
12	(2)	−46.5	−95.5	(−0.075,7.925)					
13	(2)	−48.5	−68.5	(−0.075,7.925)					
14	(2)	−47.5	−40.0	(−0.075,7.925)					

There are 22 selected points that are supposed to be qualified, and the three non-adjacent points may be appropriately relaxed to the error a little bigger than the allowable error for the first requirement.

If the whole sole is divided into three sub-regions, then for the three points with the second requirement there are only 2 points allowed in, sub-region 1, 0 in sub-region 1, and 1 in sub-region 3. The conditions to be qualified are as follows.

$$d1_k = |X1 - X1_k| \le \varepsilon_2, k = j + 1, j = 15, 16, 17, \ldots\ldots, 23, 24; k \ne 19, k \ne 24 \tag{23}$$

The error of the test point of the first color A part of the sole can be slightly larger, because the second color B can fill and compensate the deviation of color A at A certain level. We take

$$\varepsilon_1 = \varepsilon_2 = \pm 0.1$$

Table 3. Real tested data for colour A

Order	(region code)	Space(standard)			Space(experiment)			
		X	Y	Z	Z1	Z2	Z3	Z_{MEAN}
1	region(1)	0.0	−11.5	(0.075)	0.068	0.072	0.059	0.068
2	(1)	−40.0	0.0	(0.075)	0.058	0.063	0.057	0.059
3	(1)	40.0	0.0	(0.075)	0.046	0.054	0.055	0.052
4	(1)	40.0	−60.75	(0.075)	0.024	0.036	0.033	0.031
5	(1)	−40.0	−60.75	(0.075)	0.018	0.021	0.023	0.021
6	region(2)	0.0	−85.0	(0.075)	0.004	0.007	0.006	0.006
7	(2)	0.0	−50.0	(0.075)	0.017	0.022	0.019	0.019
8	(2)	47.5	−40.0	(0.075)	0.038	0.043	0.036	0.038
9	(2)	−48.5	−68.5	(0.075)	0.044	0.055	0.052	0.050
10	(1)	46.5	−95.5	(0.075)	0.068	0.062	0.059	0.063
11	(1)	0.0	−115.0	(0.075)	0.053	0.057	0.054	0.055
12	(2)	−46.5	−95.5	(0.075)	0.066	0.070	0.061	0.066
13	(2)	−48.5	−68.5	(0.075)	0.069	0.061	0.073	0.068
14	(2)	−47.5	−40.0	(0.075)	0.056	0.052	0.058	0.055
15	region(3)	−35.0	48.0	(0.075)	0.048	0.042	0.044	0.045
16	(3)	0.0	78.0	(0.075)	0.038	0.032	0.031	0.034
17	(3)	28.0	48.0	(0.075)	0.018	0.012	0.016	0.016
18	(3)	43.5	28.0	(0.075)	0.004	0.002	0.003	0.003
19	(3)	−48.0	28.5	(0.075)	−0.008	−0.002	−0.005	−0.005
20	(3)	−51.0	60.0	(0.075)	−0.021	−0.019	−0.018	−0.019
21	(3)	−48.0	90.0	(0.075)	−0.033	−0.032	−0.038	−0.034
22	(3)	0.0	115.0	(0.075)	−0.047	−0.041	−0.050	−0.046
23	(3)	47.5	92.0	(0.075)	−0.038	−0.032	−0.039	−0.036
24	(3)	45.5	60.0	(0.075)	−0.008	−0.002	−0.005	−0.005

5.3.2 Detection of B-Color Sole

Similar to Eq. (6), we have

$$d_2 = \left| X_2 - X_{2j} \right| \le \delta_2, j = 1, 2, 3 ......24 \tag{24}$$

where, X_2—— Spatial coordinate matrix of colour B for the standard model of the sole.
X_{2j}——Real coordinate matrix of the selected points.
Then

$$d_{2j} = \left| X_{2sj} - X_{2wj} \right| = \sqrt{\left(x_{2sj} - x_{2sj}\right)^2 + \left(y_{2sj} - y_{2sj}\right)^2 + \left(z_{2sj} - z_{2sj}\right)^2} \le \delta_2, j = 1, 2, 3 ......24 \tag{25}$$

If the selected points obey all the requirements of sole parts,by 100%, then it is considered qualified. We take

$$\delta_1 = \delta_2 = \pm 0.075$$

6 Newton Interpolation Algorithm.

6.1 Comparison Among Interpolations

6.1.1 Linear Interpolation

Suppose that the two type value points are given in terms of z0(x0, y0, z0), z1(x1, y1, z1),take the real numbers $\lambda \in [0, 1]$,then

$$z = z_0 + \lambda(z_1 - z_0) \tag{26}$$

Such interpolation with threshold known,the mid point between the two type value points filled;And if the error rule is nonlinear,the interpolated error will be considerable.

6.1.2 Polynomial Interpolation——Lagrangian Interpolation

For function $y - f(x)$,with $n + 1$ points,$x_0,x_1,x_2,\ldots\ldots, x_n$ and values of the function.

$$y_k = f(x_k), \quad k = 0, 1, 2 \ldots\ldots, n,$$

$$y = f(x) \approx \sum_{k=0}^{n} \prod_{\substack{i=0 \\ i \neq k}}^{n} \left(\frac{x - x_i}{x_k - x_i} \right) y_k, \tag{27}$$

If the error rule is nonlinear, the interpolation precision is higher. However, every time the interpolation point is changed, it needs to be re-calculated with the previous value points.

6.2 Newton Interpolation

Similar to the above expression, for function $y = f(x)$ with $n + 1$ points,$x_0,x_1,x_2,\ldots\ldots,x_n$, and values of the function.

$$y_k - f(x_k), \quad k = 0, 1, 2 \ldots\ldots, n,$$

$$\begin{aligned} y = f(x) = f(x_0) + (x - x_0)f(x_0, x_1) + (x - x_0)(x - x_1)f(x_0, x_1, x_2) + \\ +, \ldots\ldots, +(x - x_0)(x - x_1)\ldots\ldots(x - x_{n-1})f(x_0, x_1, \ldots\ldots, x_n) \end{aligned} \tag{28}$$

Increment value point, the previous calculation is still valid, only need to calculate the increased part. Through experimental observation, linear interpolation, second-degree lagrangian interpolation and second-degree newton interpolation are analyzed and compared, and polynomial interpolation error is small. Especially in the same precision, newton interpolation equation is relatively simple, convenient to calculate and has obvious advantages, see Fig. 11.

Newton Interpolation not only provides the higher accuracy probably up to one number rate, but also calculates errors with ease.

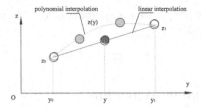

Fig. 11. Comparison of interpolations

7　Conclusion

We have proposed and designed a multi-color sole molding operation robot instead of manual work,especially the dichromatic sole production, which can realize the two-color sole molding with new process under automatic and intelligent conditions. Complete the basic and important processes of opening and closing mold, first color flash residual cleaning, first color mold opening and inspection, second color mold closing, second color flash residual cleaning, second color release and inspection, and picking and final inspection.

The quality inspection standard of two-color soles of molded parts under the conditions of robot operation production is put forward and applied with Code 40 as the inspection basis. According to the engineering drawing, the sole is divided into 3 areas, as (1) arch region, (2) heel region, and (3) palm region. With the goal of quasi-equidistance between two adjacent points, the detection feature point "plum" lattice is optimized to achieve the goal of effectively and reliably inspecting the entire sole parts with fewer measurement points.

Optimal design of flexible sensor array for part inspection. As the core of the end effector for the inspection of the sole parts of the robot, corresponding to the detection feature points, a 24-point sensor array based on 40 size shoes is designed using the quasi-equidistant distribution of adjacent inspection points.

By adopting the fusion of isometric series and arithmetic series, the ultra-small lead/diameter lead screw mechanism is used to move 3 groups of probes vertically and the arm is rotated and moved horizontally to realize the 2-dimensional flexible adjustment of the sensor array and the intelligent adaptive inspection of 35–45 code soles.

In order to further improve the accuracy and reliability of the inspection, Newton interpolation algorithm was used to carry out high-precision inspection on the basis of the two-dimensional flexible adjustment of the sensor array. The efficiency of the inspection was 99.9%, the accuracy was 99.99, and the high reliability, high flexibility and high accuracy of the workpiece was realized.

References

1. Rawashdeh, N.A., Abu-Khalaf, J.M., Khraisat, W., Al-Hourani, S.S.: A visual inspection system of glass ampoule packaging defects: effect of lighting configurations. Int. J. Comput. Integr. Manuf. **31**(9), 848–856 (2018)
2. Witek, M.: Validation of in-line inspection data quality and impact on steel pipeline diagnostic intervals. J. Natural Gas Sci. Eng. **56**, 121–133 (2018)

3. Martinez, P., Al-Hussein, M., Ahmad, R.: Intelligent vision-based online inspection system of screw-fastening operations in light-gauge steel frame manufacturing. Int. J. Adv. Manuf. Technol. **109**(3–4), 645–657 (2020)
4. Chen, W., Wu, G., Wei, J.: An approach to identifying error patterns for infrastructure as code. In: Proceedings of 29th IEEE International Symposium on Software Reliability Engineering Workshops, ISSREW 2018, 16 November 2018, pp 124–129 (2018)
5. Hamrol, A., Kujawińska, A., Bożek, M.: Quality inspection planning within a multistage manufacturing process based on the added value criterion. Int. J. Adv. Manuf. Technol. **108**, 1399–1412 (2020)
6. Huang, Z., et al.: Research on spatial positioning of online inspection robots for vertical storage tanks. Ind. Robot **47**(2), 187–195 (2020)
7. Mosbach, D., Gospodnetić, P., Rauhut, M., Hamann, B., Hagen, H.: Feature-driven viewpoint placement for model-based surface inspection. Mach. Vision Appl. **32**, 1–21 (2021)
8. Zengin, A.T., Erdemir, G., Akinci, T.C., Seker, S.: Measurement of power line sagging using sensor data of a power line inspection robot. IEEE Access **8**, 99198–99204 (2020)
9. Cung, L.T., Nguyen, N.H., Joubert, P.Y., Vourch, E., Larzabal, P.: A model-based approach for inspection of aeronautical multi-layered structures by eddy currents. Int. J. Comput. Math. Electr. Electron. Eng. **38**(1), 382–394 (2019)
10. Ren, R., Hung, T., Tan, K.C.: A generic deep-learning-based approach for automated surface inspection. IEEE Trans. Cybern. **48**(3), 929–940 (2017)
11. Haj Ibrahim, M., Aidibe, A., Mahjoub, M.A., Tahan, A., Louhichi, B.: A novel approach to the inspection of deformable bodies by adapting the coherent point drift algorithm and using a clustering methodology. Int. J. Adv. Manuf. Technol. **103**, 409–422 (2019)
12. Gabara, G., Sawicki, P.: A new approach for inspection of selected geometric parameters of a railway track using image-based point clouds. Sensors **18**(3), 791 (2018)
13. Urbanic, R.J., Djuric, A.M.: Methods for evaluating the functional work space for machine tools and 6 Axis serial robots. SAE Int. J. Mater. Manuf. **9**(2), 465–473 (2016)
14. Ramachandran, R.K., Kakish, Z., Berman, S.: Information correlated levy walk exploration and distributed mapping using a swarm of robots. IEEE Trans. Rob. **36**(5), 1422–1441 (2020)
15. Fu, L., Wu, R., Zhao, J.: On the stability of maxwell model based impedance control and cartesian admittance control implementation. In: 4th IEEE International Conference on Advanced Robotics and Mechatronics, Osaka, Japan, pp. 793–798 (2019)
16. Usvyatsov, M., Schindler, K.: Visual recognition in the wild by sampling deep similarity functions. In: 2019 IEEE International Conference on Robotics and Automation (ICRA), Montreal, Canada, pp. 2341–2347 (2019)
17. Ge, L., Wang, H.J., Xing, J.S.: Maintenance robot motion control based on Kinect gesture recognition. J. Eng.-JOE **23**, 8794–8796 (2019)
18. Liu, J.H., Cao, L., Wang, C.L.: Control of two-wheel self-balancing robots based on gesture recognition. Wearable Sensors Rob. **399**, 525–538 (2017)
19. Sprute, D., Rasch, R., Aljoscha Pörtner, A., Sven Battermann, S., Konig, M.: Gesture-based object localization for robot applications in intelligent environments. In: 14th International Conference on Intelligent Environments (IE), Shanghai, China, pp. 48–55 (2018)
20. Bolano, G., et al.: Towards a vision-based concept for gesture control of a robot providing visual feedback. In: IEEE International Conference on Robotics and Biomimetics (ROBIO), Kuala Lumpur, Malaysia, pp. 386–392(2018)
21. Yang, P., Karashima, R., Okano, K., Ogata, S.: Automated inspection method for an STAMP/STPA - fallen barrier trap at railroad crossing - fallen B. Procedia Comput. Sci. **159**, 1165–1174 (2019)
22. Jin, X., et al.: DM-RIS: deep multimodel rail inspection system with improved MRF-GMM and CNN. IEEE Trans. Instrument. Meas. **69**(4), 1051–1065 (2020)

23. Moradi, S., Zayed, T., Nasiri, F., Golkhoo, F.: Automated anomaly detection and localization in sewer inspection videos using proportional data modeling and deep learning-based text recognition. J. Infrastruct. Syst. **26**(3), 04020018 (2020)

24. Li, T., Gao, L., Pan, Q., Li, P.: Free-form surface parts quality inspection optimization with a novel sampling method. Appl. Soft Comput. **62**, 550–570 (2018)

25. Wang, S., Chen, K., Liu, Z., Guo, R.Y., Chen, S.: An ontology-based approach for supply-chain quality control: From a principal–agent perspective. J. Inf. Sci. **45**(3), 283–303 (2019)

26. Özbilge, E.: Experiments in online expectation-based novelty-detection using 3D shape and colour perceptions for mobile robot inspection. Robot. Auton. Syst. **117**, 68–79 (2019)

27. Delgado, P., Martins, Cristina, Braga, Ana, Barros, Cláudia., Delgado, Isabel, Marques, Carlos, Sampaio, Paulo: Benefits of multivariate statistical process control based on principal component analysis in solder paste printing process where 100% automatic inspection is already installed. In: Gervasi, O., Murgante, Beniamino, Misra, Sanjay, Stankova, Elena, Torre, Carmelo M., Rocha, Ana Maria A C., Taniar, David, Apduhan, Bernady O., Tarantino, Eufemia, Ryu, Yeonseung (eds.) ICCSA 2018. LNCS, vol. 10961, pp. 351–365. Springer, Cham (2018). https://doi.org/10.1007/978-3-319-95165-2_25

28. Zhang, B., Jaiswal, P., Rai, R., Guerrier, P., Baggs, G.: Convolutional neural network-based inspection of metal additive manufacturing parts. Rapid Prototyping J. **25**(3), 530–540 (2019)

29. Jovanović, D., Dutertre, B.: Interpolation and model checking for nonlinear arithmetic. In: Silva, A., Leino, K.M. (eds.) CAV 2021. LNCS, vol. 12760, pp. 266–288. Springer, Cham (2021). https://doi.org/10.1007/978-3-030-81688-9_13

30. Bednarczyk, B., Jaakkola, R.: Towards a model theory of ordered logics: expressivity and interpolation. In: SEFM, pp. 382–387 (2017)

31. Cabodi, G., Camurati, P., Palena, M., Pasini, P., Vendraminetto, D.: Interpolation-based learning as a mean to speed-up bounded model checking (short paper). In: Cimatti, A., Sirjani, M. (eds.) SEFM 2017. LNCS, vol. 10469, pp. 382–387. Springer, Cham (2017). https://doi.org/10.1007/978-3-319-66197-1_25

32. Mittal, R.C., Rohila, R.: A numerical study of the Burgers' and Fisher's equations using barycentric interpolation method (2022)

33. Kang, T.W., Kang, J.G., Jung, J.W.: A bidirectional interpolation method for post-processing in sampling-based robot path planning. Sensors **21**(21), 7425 (2021). https://doi.org/10.3390/s21217425

34. Mittal, R.C., Kumar, S., Jiwari, R.: A cubic B-spline quasi-interpolation algorithm to capture the pattern formation of coupled reaction-diffusion models. Eng. Comput. **38**, 1375–1391 (2022)

35. Schlaipfer, M., Weissenbacher, G.: Labelled interpolation systems for hyper- resolution, clausal, and local proofs. J. Autom. Reason. **57**, 3–36 (2016)

36. Chen, G., Yang, J., Xiang, H., Ou, D.: Inline inspection with an industrial robot (IIIR) for mass-customization production line. Sensors **20**(11), 3008 (2020). https://doi.org/10.3390/s20113008

Element Extraction from Computer Science Academic Papers for AI Survey Writing

Fan Luo and Xinguo Yu[✉]

Central China Normal University, Wuhan, China
xgyu@ccnu.edu.cn

Abstract. With the exponential growth of research papers, text summarization tools have emerged. However, existing text summarization tools merely extract existing sentences or words based on their frequency and may not be particularly well-suited for papers. To address this gap, this study develops a model based on DistilBERT, primarily focusing on information extraction and dataset labeling and augmentation techniques. The model's central objective is entity recognition, aiming to identify two specific entities from the full text of research papers. The model takes these critical segments of papers as input and aims to identify the research problems and content contained within them. In response to the limitations of existing datasets, this research augments a dataset with over 4000 full-text arXiv computer algorithm papers through manual annotations.

The developed model demonstrates exceptional performance on several evaluation metrics, including accuracy, precision, F1 score, and recall. For comparative experiments, we employed several baseline models based on BERT. These results demonstrate the effectiveness of the proposed model. As part of a comparative experiment, we trained our models using three different dataset training methods. Additionally, to evaluate our dataset's quality and underline the importance of full-text data, we manually annotated a random selection of 4000 papers from the ARXIV Data dataset, extracting only their titles and abstracts. As a result, Our proposed model outperforms all the baseline models, achieving an accuracy of 0.823 and an F1 Score of 0.798 and models trained on the proposed full-text annotated dataset outperform those trained on other datasets.

Keywords: Information Extraction · Dataset Labeling and Augmentation · Automating Literature Review · Text Mining

1 Introduction

Computer algorithm researchers, particularly during the process of conducting literature reviews, often face the daunting task of investing substantial time in comprehending diverse computer algorithm papers. Their goal is to extract valuable

© The Author(s), under exclusive license to Springer Nature Singapore Pte Ltd. 2024
H. Jin et al. (Eds.): IAIC 2023, CCIS 2060, pp. 263–274, 2024.
https://doi.org/10.1007/978-981-97-1332-5_21

information, including research problems and algorithmic enhancements. This manual and labor-intensive process has prompted the development of various text processing tools like FastRead and EDAM, designed to streamline this endeavor. However, these existing text processing tools have limitations, especially when confronted with the intricate structure of computer algorithm papers [1].

Currently, existing text processing tools fall short of meeting the specific needs of computer algorithm researchers, primarily due to their inability to effectively extract essential elements crucial for this field, which includes research problems and research content (encompassing improved algorithms and novel algorithm applications) [2].

Looking ahead, the research direction focuses on creating a system that simplifies survey writing while providing insights into the critical papers in the field and the relationships between them - a visualized literature system. This article presents an approach to constructing a visualized literature system that revolves around the extraction of essential research elements, namely research problems and research content, which includes improved algorithms and novel algorithm applications. These elements are transformed into binary vectors, serving as representations of entire papers, facilitating vector space clustering and similarity analysis. This method enables the discovery of related papers, the recognition of paper relationships, and the identification of key contributions within the realm of computer algorithms.

The integration of the above segments results in a coherent narrative highlighting the limitations of current text processing tools and the promising future of visualized literature systems in facilitating the work of computer algorithm researchers.

Connecting to the following discussion, it is crucial to emphasize that the pivotal step in achieving this vision of a visualized literature system is the extraction of essential paper elements. However, the contemporary challenge lies in the scarcity of datasets directly offering titles, abstracts, introductions, and conclusions, along with research problems and research content (algorithms) from papers [3].

To tackle this limitation, this paper introduces the creation of a specialized dataset known as arxiv-TAIC-RPAL, tailored explicitly for computer algorithm papers. Within this dataset, research problems and research content are meticulously annotated and enriched. The research adopts the DistilBERT model for training an element extraction model specifically designed for computer algorithm papers. This model emphasizes the extraction of research problems and research content from the titles, abstracts, introductions, and conclusions of computer algorithm papers, effectively addressing the pressing needs of computer algorithm researchers.

The proposed solution not only serves as a time-saving tool for researchers but also empowers them to efficiently extract vital information. This enhancement significantly bolsters their capacity to conduct comprehensive literature reviews and stay abreast of the latest developments in the field. Consequently, this work makes a notable contribution to the advancement of automated literature analysis and information extraction within the realm of computer algorithm papers.

Furthermore, the long-term significance of this model lies in its potential to represent entire papers as binary vectors. These vectors facilitate the construction of a paper visualization system, enabling researchers to quantify relationships between articles, discover pivotal contributions, and navigate the landscape of academic literature with greater efficiency and precision.

2   Related Work

The content of this paper focuses on the Information Extraction task, extracting elements from academic papers in the field of computer science, which belongs to the essential part of the task of automating literature review. An Automating Literature Review involves five main steps: creating literature review rules, searching the literature collection, selecting and filtering relevant papers, extracting information from the papers, and integrating the information. The focus of this paper's model lies in the fourth step, which is extracting information from the literature.

Additionally, it is essential to provide an overview of the dataset proposed in this paper.

2.1   Information Extraction

Information Extraction from literature is a challenging task due to variations in the structure and formats of published papers. Attempting to create a fully automated extraction tool to handle all possible cases would be impractical. Instead, a more realistic approach is to categorize and develop different information extraction tools based on the subject matter and different arrangements of literature. This tailored approach will better address the specific challenges posed by different fields and document formats [4].

In a research conducted in 2018, text mining and artificial intelligence models were employed to enable algorithms to automatically identify chapter titles corresponding to different paragraphs. Furthermore, the data extraction was carried out using the most frequent word phrases within paragraphs. However, this method's performance was limited to specific structured documents, and it exhibited poor performance for unstructured or differently structured papers [5].

Relation extraction is a subset of information extraction, primarily focusing on structurally extracting information, particularly relationships between entities from textual data. It can be categorized into traditional relation extraction and relation extraction enhanced with Seq2Seq methods [6].

Early relation extraction methods followed a pipeline approach. These methods typically started by recognizing named entities in the text and then, using separate models, such as tables, classified the relationships between every pair of entities. These models primarily relied on convolutional neural networks (CNN) and long short-term memory networks (LSTM) [7].

More recent models have embraced transformer-based architectures, often building upon BERT or ALBERT models and incorporating entity-aware components. This has significantly enhanced the performance of relation extraction. The key advantage of these models is their ability to effectively capture semantic relationships between entities in text [8].

Relation Extraction models based on tables or pipelines demonstrate excellent performance for single-entity relations. However, when dealing with multiple relations between entities, the computational complexity increases exponentially due to the need to infer all possible relations. To address this issue, Seq2Seq-based models were introduced [9]. These models incorporate a recursive decoder that adjusts future decoding based on prior entity predictions, enabling them to handle cases with multiple relationships between entities and even situations where relations might be incompatible [10].

However, this approach has its own set of challenges. For example, the extracted entities are ordered alphabetically, which is not an ideal solution. Furthermore, these models can only predict known relations and cannot handle positional relationships effectively.

In this context, since the goal is to extract questions and research content from the text, and these elements may not necessarily exist in traditional predefined relationship forms, it aligns more closely with the realm of information extraction. The inputs and outputs can be precisely defined, whereas outputs usually contain information extracted from the inputs. Therefore, this article adopts a semi-supervised approach, where both the input and output of the dataset are labeled, facilitating direct training of DistilBERT.

2.2 Datasets

In a previous study, data augmentation and enrichment were conducted to address the task of sentence classification, focusing on the integration of medicine-related and algorithmic research papers from the PubMed baseline database. The abstract sections of approximately 200,000 published papers were divided into categories such as Background, Methods, Results, and Conclusion. To meet the requirements of machine learning training, the dataset was further split into training, validation, and testing sets. Each set comprised different sections extracted from the abstract of a paper, identified by its corresponding PMID. The evaluation of the model's performance utilized four metrics: Precision, Recall, F1-score, and Runtime [11].

3 Datasets

The goal of this study is to build an information extraction model based on computer algorithm research papers' titles, abstracts, introductions, and conclusions, in order to extract information about the research questions and research content (algorithms). However, finding a dataset that perfectly aligns with these requirements, containing both the title, abstract, introduction, conclusion, and the research questions and content (algorithms) for a single article, is extremely rare. Due to the specialized nature of the research questions, datasets usually only include the first four elements (title, abstract, introduction, and conclusion), without readily available research questions and content (algorithms) (Fig. 1).

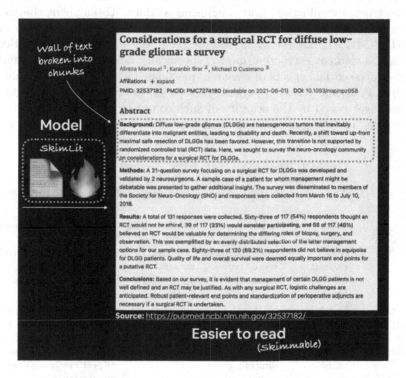

Fig. 1. The PubMeb RCT 200 Dataset [11]

The arXiv dataset consists of textual data containing research paper titles, abstracts, and full content, typically provided in the form of text files. It serves as a valuable resource for researchers, allowing them to train and evaluate natural language processing and text mining algorithms, such as information extraction, text classification, text summarization, and topic modeling. The dataset covers a wide range of academic disciplines and is particularly comprehensive in the field of computer science [12].

Therefore, as a result, we have augmented an existing arXiv dataset through manual inspection. By referencing the original full-text articles, we added the research questions and research content (algorithms) to the dataset.

Our approach involves leveraging the existing data available from the arXiv dataset and manually curating the missing elements. This combination ensures that the dataset becomes a valuable resource for training and evaluating our element extraction model. It allows us to create a comprehensive and more complete dataset, essential for developing an accurate and robust model capable of extracting the desired information from the different sections of research papers.

Finally, a dataset of over 4000 arXiv papers was extracted, focusing on the structure and impact of algorithmic research papers. This dataset included the title, abstract, introduction, and conclusion sections. To augment the dataset, additional information was manually added, resulting in the creation of the arXiv-TAIC-RPAL dataset. The resulting dataset provides a foundation for training the element extraction model, specifically designed to understand and capture the essential aspects of research questions and algorithmic content from computer algorithm research papers' various sections. Our approach addresses the challenges posed by the scarcity of such complete datasets and contributes to advancing the capabilities of element extraction in specialized domains.

Table 1. The arXiv-TAIC-RPAL Dataset Instances

Title	Abstract	Introduction	Conclusion	Research Problem	Algorithm
learning numeral embedding	word embedding is an essential building block ...	word embeddings, the distributed vector representations ...	in this paper, we propose two novel numeral embedding methods	Learning Numeral Embedding	representing the embedding of a numeral
tensor-tensor products for optimal representation and compression	in this era of big data, data analytics and machine learning ...	much of real-world data is inherently multidimensional ...	we have demonstrated theoretically ...	determine if tensor-based approximation techniques can outperform metrication	tensor-tensor products, Tensor-SVDs
simultaneous identification of tweet purpose and position	tweet classification has attracted considerable ...	over the past few years, microblogs have become ...	in this paper, we study the problem to identify tweet purpose and position	Simultaneous identification of tweet purpose and position ...	multi-label classification method with post-processing, referred to as RAkEL

As shown in Table 1, this is a partial excerpt of the dataset, and it is evident that the dataset covers the entire corresponding sections of the research papers. Furthermore, the research questions and research content (algorithms) are well-parsed and clearly presented.

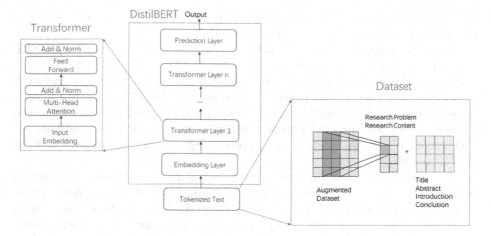

Fig. 2. The Proposed DistilBERT Structure

4 Methods

For this task of training a model to extract the research problem and research content (algorithm) from the title, abstract, introduction, and conclusion of a research paper, DistilBERT is chosen as a suitable model. DistilBERT is a lightweight pre-trained language model based on the Transformer architecture. It is a variant introduced by Hugging Face, aiming to compress large BERT models into smaller and more efficient versions, allowing faster execution on devices with limited computing resources. The design inspiration of DistilBERT originates from the concept of Knowledge Distillation, which involves "distilling" knowledge from a large model to a smaller one. By learning from a large BERT model and leveraging its knowledge, DistilBERT achieves a significant reduction in parameters and computational cost while retaining high performance. Knowledge distillation is a pivotal stage in DistilBERT training, involving an objective function defined as (Fig. 2):

$$\mathcal{L}_{\mathrm{KD}} = \alpha \cdot \mathrm{CE}(P, T) + \beta \cdot \mathrm{CE}(P_T, T) \tag{1}$$

In this equation, $\mathcal{L}_{\mathrm{CE}}$ corresponds to the cross-entropy loss function, P signifies the predicted probability distribution of DistilBERT, Q represents the predicted probability distribution of a larger model (e.g., BERT), P_T stands for the temperature-adjusted predicted probability distribution of DistilBERT, Q_T denotes the temperature-adjusted predicted probability distribution of the larger model, and α and β are weighting parameters for the loss [13].

As a pre-trained language model, DistilBERT undergoes pre-training on a large-scale dataset, enabling it to acquire rich language representations. These learned representations can be utilized to extract key information from text, aiding in better comprehension and processing of information extraction tasks.

Moreover, the adoption of DistilBERT is particularly suitable for this study, considering the usage of a small dataset for training.

Now, we proceed with the formal training process, using the DistilBERT model from the transformers library, including the DistilBertTokenizer and DistilBertModel modules. To measure the model's performance, we employ the accuracy score, recall score, and f1 score functions from the sklearn metrics library. To split the arxiv-TAIC-RPAL dataset into training, testing, and validation sets, we utilize the train test split function from sklearn model selection module. For text preprocessing, we leverage the preprocessing module from sklearn and the NLTK module.

The specific training process is as follows: First, we extract the required fields ('Title', 'Abstract', 'Introduction', 'Conclusion', 'Research problem', 'Algorithm') from the local CSV file, which is the arxiv-TAIC-RPAL dataset. Next, we proceed with data cleaning and preprocessing, which involves removing punctuation and special characters, converting text to lowercase, and replacing dirty or empty data with 'N/A'. Since DistilBERT supports a maximum input length of 512, we also divide the text into multiple segments.

Then, we load the DistilBERT model and tokenizer to define the model architecture. During the data processing stage, we convert the data into the format required by the DistilBERT model and then encode the labels into numerical codes. Subsequently, we convert the encoded labels into tensors, create padding tensor lists, and establish data loaders. We define the batch size and ensure that input sequences are padded to the same length. Finally, we define the loss function and optimizer to begin the formal training process. After training, we evaluate the model on the test set, and finally, save the trained model to the local storage.

This comprehensive training process ensures that the model is equipped to efficiently extract essential elements from computer papers. It not only contributes to saving researchers valuable time but also advances the capabilities of information extraction, ultimately leading to enhanced literature analysis in the domain of computer algorithms.

Table 2. DistilBERT Model Training Parameters

Training Parameter	Value
Max Sequence Length	512
Epoch	50
Batch Size	16
Dropout	0.1
Learning Rate	2e−5
Classifier	Linear
Optimizer	Adam

5   Results and Discussion

Machine learning involves numerous evaluation metrics, which allow us to horizontally compare the performance of different models. In this study, the evaluation metrics employed include Accuracy, Precision, Recall, and F1 score as shown in Table 3.

By utilizing these evaluation metrics, the model's performance can be accurately assessed. Results are listed in Table 3:

Table 3. The Performance of Trained DistilBERT-based Model

Accuracy	Precision	Recall	F1 Score
0.823	0.810	0.785	0.798

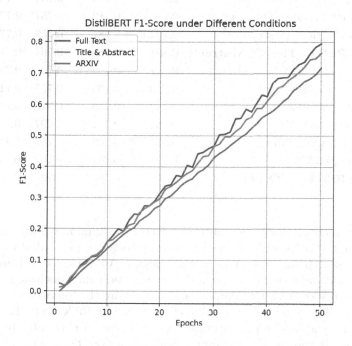

Fig. 3. DistilBERT F1-Score under Different Datasets

In the preceding sections, we presented the results of training the DistilBERT model on our proposed dataset. To conduct comparative experiments, we included several baseline models, such as BERT, RoBERTa, SBERT, XLNET, ALBERT, and Electra. These models were evaluated on the same dataset to provide a comprehensive comparison. Furthermore, to assess the quality of our dataset, we intentionally incorporated the ARXIV data, comprising over 24,000 computer science papers, each containing only titles and summaries. To maintain

Table 4. Performance Metrics Comparison for Models and Datasets

Model	Accuracy	Precision	Recall	F1 Score
BERT (Full Text)	0.818	0.805	0.780	0.793
RoBERTa (Full Text)	0.817	0.803	0.779	0.791
SBERT (Full Text)	0.821	0.807	0.783	0.796
XLNET (Full Text)	0.812	0.799	0.774	0.787
ALBERT (Full Text)	0.815	0.802	0.777	0.790
Electra (Full Text)	0.816	0.804	0.779	0.792
DistilBERT (Full Text)	0.823	0.810	0.785	0.798
BERT (Title & Abstract)	0.795	0.785	0.760	0.773
RoBERTa (Title & Abstract)	0.793	0.783	0.758	0.771
SBERT (Title & Abstract)	0.798	0.788	0.763	0.776
XLNET (Title & Abstract)	0.789	0.779	0.754	0.767
ALBERT (Title & Abstract)	0.792	0.782	0.757	0.770
Electra (Title & Abstract)	0.794	0.784	0.759	0.771
DistilBERT (Title & Abstract)	0.801	0.784	0.764	0.774
BERT (ARXIV)	0.742	0.732	0.707	0.719
RoBERTa (ARXIV)	0.740	0.730	0.705	0.717
SBERT (ARXIV)	0.748	0.738	0.713	0.725
XLNET (ARXIV)	0.732	0.722	0.697	0.709
ALBERT (ARXIV)	0.735	0.725	0.700	0.712
Electra (ARXIV)	0.737	0.727	0.702	0.714
DistilBERT (ARXIV)	0.750	0.732	0.715	0.723

experimental consistency, we manually selected 4,000 papers from this dataset, focusing on the titles and summaries. Subsequently, we annotated these papers for their research questions and methodologies. Additionally, within our proposed full-text dataset, we excluded the introduction and conclusion sections to test the quality of the paper dataset and underscore the importance of full-text information. The summarized results are provided in Table Two (Table 4).

From the graphical representation, we can discern a noticeable performance enhancement in the initial ten training epochs. However, beyond the tenth epoch, although there is a general upward trend, there is a significant degree of fluctuation, ultimately resulting in performance stabilization. The final results, presented in the table, indicate that the trained model demonstrates a reasonable level of performance in terms of the evaluation metrics, albeit not reaching the high values achieved in other tasks.

Analysis of the table reveals that DistilBERT consistently outperforms other models across all three dataset scenarios. It excels when trained on our self-introduced Arxiv-RPAL dataset, and its performance remains impressive when applied to other datasets. These findings underscore the effectiveness of our proposed model in various data conditions, verifying its reliability.

Moreover, the comparative analysis reveals that models trained on our full-text dataset consistently outperform those trained on datasets with only titles and abstracts, reinforcing the value of full-text data. This suggests that the utilization of full-text datasets can yield more precise extraction results.

Furthermore, when tested under conditions where only the titles and summaries are available, our dataset still outperforms other models. This demonstrates the high quality of our proposed dataset, possibly attributed to its representation of a wide range of computer science papers.

In summary, the information extraction model exhibits promising results, particularly in terms of accuracy. However, there remains room for improvement in precision and recall. The fluctuating performance observed after the initial improvement highlights the potential benefits of further fine-tuning and optimization. Additionally, given the subjective nature of evaluating the extracted information, manual inspection and verification will be vital to ensure the quality and reliability of the results.

6 Conclusion and Future Work

This research addresses the practical challenges faced by computer algorithm researchers. It begins by expanding and curating a comprehensive Dataset that comprises more than 4000 full-text articles from arXiv. This Dataset is enriched with manually added annotations, focusing on research problems and research content (algorithmic details), both of which are integral to this study. Subsequently, the research leverages the DistilBERT model to train an element extraction system. This model excels in extracting research problems and research content from the full text of algorithm papers, demonstrating commendable performance across various evaluation metrics.

In terms of future research directions, several key priorities emerge. Firstly, there is a need to expand the dataset further, as the current volume of data samples presents limitations to the model's scope and applicability. Future datasets should also encompass interdisciplinary papers on computer algorithms to enhance their generalizability.

Moreover, ongoing efforts will be directed toward optimizing the model's architecture and training it with the expanded Dataset to further enhance its performance. These steps are essential for creating a more robust and versatile element extraction system, capable of serving a broader array of applications and research domains.

This research effort stands as a significant step forward in alleviating the challenges faced by computer algorithm researchers. Developing a model that can efficiently extract critical research elements, known as research problems and research content, lays the foundation for comprehensive clustering and similarity analysis of papers. This, in turn, facilitates the discovery of related papers, the elucidation of paper relationships, and the identification of key contributions within the field of computer algorithms. In conclusion, this paper provides an extensive methodology and outlines the contributions of this study, shedding light on the critical future directions that will drive advancements in the field.

References

1. Felizardo, K.R., Carver, J.C.: Automating systematic literature review. In: Contemporary Empirical Methods in Software Engineering, pp. 327–355. Springer, Cham (2020). https://doi.org/10.1007/978-3-030-32489-6_12
2. McNabb, L., Laramee, R.S.: How to write a visualization survey paper: a starting point. In: Eurographics (Education Papers), pp. 29–39 (2019)
3. Loza, V., Lahiri, S., Mihalcea, R., et al.: Building a dataset for summarization and keyword extraction from emails. In: LREC, pp. 2441–2446 (2014)
4. Jonnalagadda, S., Goyal, P., Huffman, M.: Automating data extraction in systematic reviews: a systematic review. Syst. Rev. 4(1), 78 (2015)
5. Aliyu, M.B., Iqbal, R., James, A.: The canonical model of structure for data extraction in systematic reviews of scientific research articles. In: 15th International Conference on Social Networks Analysis, Management and Security (SNAMS 2018), pp. 264–271 (2018)
6. Cabot, P.L.H., Navigli, R.: REBEL: relation extraction by end-to-end language generation. In: Findings of the Association for Computational Linguistics, EMNLP 2021, pp. 2370–2381 (2021)
7. Kenton, J.D.M.W.C., Toutanova, L.K.: BERT: pre-training of deep bidirectional transformers for language understanding. In: Proceedings of NAACL-HLT, vol. 1, p. 2 (2019)
8. Nayak, T., Ng, H.T.: Effective modeling of encoder-decoder architecture for joint entity and relation extraction. In: Proceedings of the AAAI Conference on Artificial Intelligence, vol. 34, no. 05, pp. 8528–8535 (2020)
9. Yamada, I., Asai, A., Shindo, H., et al.: LUKE: deep contextualized entity representations with entity-aware self-attention. arXiv preprint arXiv:2010.01057 (2020)
10. Zhang, R.H., Liu, Q., Fan, A.X., et al.: Minimize exposure bias of Seq2Seq models in joint entity and relation extraction. arXiv preprint arXiv:2009.07503 (2020)
11. Blloshmi, R., Conia, S., Tripodi, R., et al.: Generating senses and RoLes: an end-to-end model for dependency-and span-based semantic role labeling. In: IJCAI, pp. 3786–3793 (2021)
12. Dernoncourt, F., Lee, J.Y.: PubMed 200k RCT: a dataset for sequential sentence classification in medical abstracts. arXiv preprint arXiv:1710.06071 (2017)
13. Gehrke, J., Ginsparg, P., Kleinberg, J.: Overview of the 2003 KDD cup. ACM SIGKDD Explor. Newsl. 5(2), 149–151 (2003)

An Emotion Recognition Method in Conversations Based on Knowledge Selection and Fuzzy Fingerprints

Yuanzheng You and Yongli Wang[✉]

Nanjing University of Science and Technology, NanJing 210000, China
yongliwang@njust.edu.cn

Abstract. In the context of large-scale pre-trained language models such as BERT or RoBERTa, they exhibit outstanding performance in natural language processing tasks, but they suffer from a lack of interpretability and the potential introduction of irrelevant information when incorporating external knowledge into sentiment recognition tasks (ERC). This paper presents an innovative ERC sentiment analysis model that combines large-scale pre-trained language models and knowledge pruning techniques while introducing a fuzzy fingerprint module. The workflow of this model consists of several steps: firstly, through the knowledge pruning module, irrelevant external knowledge that does not align with the sentiment of words is removed to enhance the accuracy of sentiment analysis. Next, each utterance in the dialogue as well as the previous dialogue is transformed into contextual embedding representations. Finally, these representations are fed into the fuzzy fingerprint classification module for sentiment recognition. Experimental results demonstrate that this model outperforms other models significantly in sentiment classification tasks.

Keywords: RoBERTa · Emotion Recognition in Conversations · Emotion Recognition

1 Introduce

Conversational Emotional Recognition (ERC) refers to the key process for recognizing and understanding a speaker's emotional states during a dialogue or conversation. This process involves recognizing a wide range of emotions expressed by the speakers in the conversation, such as happiness, anger, sadness, joy, anxiety, satisfaction, and also includes the evolving emotional states as the conversation progresses. The main goal of ERC is to extract emotional information from text dialogues to better understand the emotional context of the conversation and the speakers' emotional experiences. This is essential for enhancing interactions between computers and humans and improving the performance of emotion-aware systems.

Emotion Recognition in Conversation (ERC) r is about recognizing emotions for every expression in a dialogue. However, machines face certain difficulties in recognizing the emotions in speech when compared to humans. These difficulties partly stem from the

H. Jin et al. (Eds.): IAIC 2023, CCIS 2060, pp. 275–286, 2024.
https://doi.org/10.1007/978-981-97-1332-5_22

lack of common-sense knowledge in machines. While existing methods are gradually integrating knowledge into ERC, they still struggle to adapt knowledge to different utterances and contexts automatically. Fuzzy fingerprinting technology has been used successfully to classify interpretable text, but like most other techniques, its performance lags behind large pre-trained language models such as BERT and RoBERTa. These models significantly outperform others on various natural language processing tasks, but have problems with interpretation.

We propose an approach that combines knowledge elimination and a fuzzy fingerprint RoBERTa model to create a more simplified and interpretable classifier based on a large language model. Firstly, we employ a Knowledge Elimination (KE) module to remove external knowledge. We then insert the expressions into the dialogue and the previous conversation is converted into a pre-trained RoBERTa model to achieve contextual embedding of the expressions. Finally, these representations are fed into the corresponding fuzzy fingerprint classification module. We validate our approach on the widely used ERC DailyDialog benchmark dataset and the results show that we achieve state-of-the-art performance with a lighter model.

2   Related Work

2.1   Emotion Recognition in Conversation

Over the past few decades, the challenge of emotion recognition has been considered as a part of text or content classification tasks. Recent research has started to incorporate external knowledge into Emotion Recognition in Conversation (ERC) models. For instance, some researchers have introduced the COSMIC method, which utilizes bidirectional Gated Recurrent Units (GRU) to model the psychological states of speakers [1]. Another approach is the SKAIG model, which aims to capture the structure and psychological interactions between statements [2]. In addition, there is a KET model that uses graphical attention networks (GATs) to dynamically adjust conceptual knowledge weights [3]. Xie et al. proposed an AI-Net model that uses a self-attention-based self-adaptation module to simulate explicit interactions between expressions and knowledge [4].

While the incorporation of external knowledge enriches the representation of utterances, it poses a common challenge in ERC, as some of this knowledge may be ineffective or even play a crucial role in emotion detection. Consequently, some studies have introduced knowledge elimination modules into their models to remove irrelevant external knowledge.

2.2   Pre-Trained Language Models

A very popular category of large pre-trained models is the family of transformer-based language models. It is worth noting that BERT (Bilateral Encoder Representations from Transformers) has attracted great attention in this area [5]. BERT is a multi-layer, bidirectional transformer encoder that has undergone extensive pre-training for expensive language modeling and subsequent sentence prediction tasks.

RoBERTa (A Robustly Optimized BERT Pretraining Approach) can be viewed as an evolutionary successor to BERT [6]. It uses a similar architecture to BERT, but undergoes more extensive pre-training with larger amounts of data, longer training times and larger batches, eliminating the task of predicting the next BERT sentence. RoBERTa consistently outperforms BERT on many tasks with the same amount of data.

These pre-trained language models (PLMs) are adapted to specific tasks after the pre-training phase, eliminating the need to retrain the entire model. The optimization process includes supervised PLM training on the target dataset and adjusting the weights in the accompanying classification module to match the characteristics of the dataset, including the set of classification labels. During this process, the weight adjustments in the PLM aim to achieve optimal performance on the target dataset.

2.3 Fuzzy Fingerprint

Fingerprint recognition is a well-known and widely used technology in forensics. In computer science, fingerprinting is the process of mapping large amounts of data into a more compact block of information (fingerprint) that uniquely identifies the original data in all practical applications. This research introduced the concept of fuzzy fingerprinting (FFP) as a technique for solving one-to-many problems in tasks such as web user identification [8] and text author identification [7]. Fuzzy fingerprints (FFP) are created based on feature frequencies. For example, text classification creates a fingerprint for a specific category from all texts associated with that category. The frequency of each word in each text is used to create a fingerprint for that category. Fuzzy fingerprints are also used in many text classification tasks, such as topic detection of social media tweets [9] or cyberbullying detection [10]. To achieve a certain level of interpretability, a fuzzy fingerprint module was added to the model. This module maintains the performance level of RoBERTa while increasing its interpretability. This initiative enables a better understanding of the model's decision-making process and improves its transparency.

First, a fuzzy fingerprint library is created. Then, the training set is deal by calculating the first K feature lists to every category. Fuzzy fingerprint detection is subsequently performed. To determine the category of unknown instances, such as text T, the process involves first calculating the fingerprint Φ_T of T, which consists of K elements. Then, the fingerprint of T is compared with the fingerprint Φ_j of all classes which belong to the fingerprint library. If the fingerprint of the text is most similar to Φ_j it is classified into category j. The similarity between fingerprints, $sim(\Phi_T, \Phi_j)$, is computed as follows:

$$sim(\Phi_T, \Phi_j) = \sum_{v \in S_T \cup S_j} \frac{min(\mu_v(\Phi_T), \mu_v(\Phi_j))}{N} \quad (1)$$

The term $\mu_v(\Phi_x)$ represents the degree of membership related to the rank of element v in the fingerprint x. The function, which is determined on fuzzy AND operations, where we utilize the minimum value or the Gödel t-norm.

3   ROBERTA-Based Emotion Model with Knowledge Selection and Fuzzy Fingerprint

3.1   Task Definition

In this research, we define a task involving a conversation $U = \{u_1, u_2, \ldots, u_N\}$, where each utterance u_i is composed of a sequence of words, denoted as $u_i = (w_{i1}, w_{i2}, \ldots, w_{iT_i})$. Each utterance u_i corresponds to a specific emotion from a predefined set of emotions, denoted as $emotion_i$. The vector representation of the i-th utterance in the embedding layer of the conversation is represented as $E(u_i) \in R^{T_w * d_u}$. Here T represents the number of utterances in the conversation and d_u is the dimensionality of the utterance vectors.

Additionally, we have T sets of external knowledge $K = \{\{k_1\}, \{k_2\}, \ldots \{k_T\}\}$ corresponding to each utterance, where $K \in R^{T * T_w * T_d}$. We also have corresponding weights $G = \{\{g_1\}, \{g_2\}, \ldots \{g_T\}\}$, with $G \in R^{T * T_w * T_d}$. The end goal of the task is to match each expression in the conversation with the correct emotion.

In this task, the goal of the model is to determine the appropriate emotional label for each utterance based on the vector representation of the utterance, external knowledge, and associated weights. This is crucial for the task of emotion analysis within a conversation.

3.2   Word-Level Knowledge Elimination

For conceptual knowledge, knowledge retrieval in ConceptNet should not rely solely on confidence values (g) between concepts. Instead, we should place more emphasis on the specificity of a particular task and select the appropriate knowledge adaptively, rather than introducing knowledge with fixed weights across tasks.

We consider conversational emotion recognition (ERC), where the main goal is to predict different emotional labels for each expression. In this context, the emotional properties of external knowledge play an inevitable role in knowledge integration. For example, consider a statement marked "positive." and contains the word "smile". When we use the knowledge of "smile," it must be clearly associated with a positive thing. However, in reality, this strongly positive word "smile" may appear in a negative context, associated with certain negations, as in Fig. 1.

In such cases, blindly eliminating the knowledge associated with these negations can hinder the learning of knowledge representations. To solve this problem, we propose a Word-Level Knowledge Eviction (KE) module that randomly removes emotion-incompatible knowledge from words in an expression as follows:

$$W_{(w,i)}^{(j,p)} = \alpha_i^{(j,p)} * g_i^{(j,p)} \tag{2}$$

$$\alpha_i^{(j,p)} = 1_{\left[r \geq \pi \text{ or } s(w_j) = s\left(k_{(i)}^{(j,p)}\right)\right]}, p = 1, 2, \ldots T_d \tag{3}$$

$k_i \in R^{T_w * T_d}$, k_i represents the knowledge associated with the i-th utterance, where T_w is the maximum number of words while T_d is the maximum knowledge dimensions.

$k_{(i)}^{(j,p)}$ is the knowledge corresponding to the j-th word w_j in the current i-th expression. $g_i \in R^{T_w * T_d}$, g_i is a vector representing the initial weights of the knowledge k_i. $g_i^{(j,p)}$ is the initial weight of $k_{(i)}^{(j,p)}$. $W_{(w)} \in R^{T * T_w * T_d}$, $W_{(w)}$ represents the new weight matrix, and $W_{(w,i)}^{(j,p)}$ represents the updated weight for $k_{(i)}^{(j,p)}$ after word-level knowledge elimination. $\alpha_i^{(j,p)}$ represents the result of word level knowledge elimination. If t $\alpha_i^{(j,p)}$ is zero, it shows that the knowledge is eliminated.

l represents the pointer function. s is a function for extracting emotional content from words. It processes both single words and multi-word combinations based on the SenticNet emotional lexicon [11] and the natural language analysis tool Stanford NLP [12].

3.3 Context-Sensitive Embedded Discourse Representation

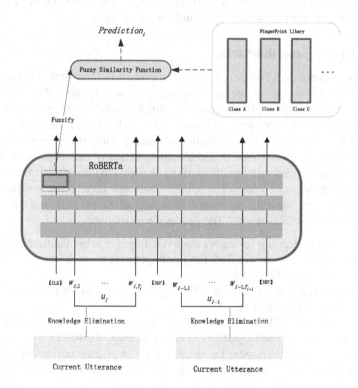

Fig. 1. Model architecture. In this example input, the expression, u_i, and its conversational context, u_{i-1}, , are passed to the RoBERTa encoder, whose last layer token [CLS] is passed via the fuzzy fingerprint module.

In ERC (Emotion Recognition in Conversation), the most common approach involves generating context-independent representations for each utterance using pre-trained language models (PLMs). Subsequently, a classification module, which may include gating mechanisms and graph neural networks, is used to model the obtained representations in the context. However, in recent research [13], we introduced a context-based method for generating context-dependent representations. These representations not only capture the current utterance, but also take into account a range of previous utterances within the conversation. This contextual approach allows us to eliminate the need for complex classification modules. Instead, simply add a fully connected linear softmax layer to a variant like RoBERTa for optimal performance.

In the pooling process of the embedding layer, we opted for the representation from the last layer L, specifically the first embedding corresponding to the [CLS] token used for classification, as shown in Eq. 4:

$$\text{pooled}_i = RoBERTa_{L,[CLS]}(\text{input}_i) \tag{4}$$

We then introduce this embedding into the fully connected linear layer using softmax to maximize the probability of correct labeling. In this study, we adapt RoBERTa on a fully connected layer, then remove it and replace it with a fuzzy fingerprint classification module, as described in the next section and shown in Fig. 1.

3.4 Fuzzy Fingerprint RoBERTa

In contrast to traditional FFP (Fuzzy Fingerprint), where the size of the utterance interval is finite but unknown (the number of all words present), the size of the fingerprint in RoBERTa FFP is limited to 768. Therefore, RoBERTa FFP can be seen as a discrete fuzzy set consisting of 768 output elements, each with a relevance membership value calculated based on its activation strength and ranking. Only the membership values of the top K objects are greater than 0, and the composition of the set is ordered based on the ranking of the membership functions of the elements to meet practical requirements.

Hence, the process of creating a specific emotion fingerprint can be succinctly described as "generating (the emotion's) training set by ranking and fuzzifying the activations of RoBERTa's output." The process is as follows:

1) Create a fuzzy fingerprint for each emotion (category) using the training data.
2) Initialize a 2D vector of size 768 for the fingerprint of each emotion, where each position's value is initialized to 0.
3) Perform context-based fine-tuning of RoBERTa by sequentially providing all training examples for a given emotion.
4) Each example's RoBERTa output consists of a real-valued vector of size 768. Add real output values to the fingerprint. Therefore, once RoBERTa passes all training examples for a given emotion, the fingerprint of a given emotion consists of a 2D vector, with each entry containing an index and the cumulative value of the RoBERTa scores for all examples.
5) Sort the fingerprint vector by the accumulated values (in descending order).
6) Decrease a 2D vector to a single dimension by removing columns with cumulative values (only ranking matters). Thus, for each emotion, you obtain a 768-dimensional vector that represents the ranking of RoBERTa's output based on activations.
7) For fingerprinting, only the first K outputs of RoBERTa and not all 768 outputs of RoBERTa are used for classification purposes. The value of K is selected based on the validation set.

8) Obtain the FFP by applying the following fuzzification function to the top-K-sized vector:

$$\mu_i = 1 - \frac{a \times i}{K}, \forall a \in [0, 1] \tag{5}$$

And represents the element index of the classification vector, K is the size of the fingerprint and the slope adjustment function. This feature assigns broader membership relationships to parent elements. Other functions were tested abpve the validation set, but the function delivered the best performance.

Once you have obtained the possible fingerprints, you can proceed with the classification of a given utterance sequence u_i:

1) Pass u_i through RoBERTa.
2) Create an emotional fingerprint for u_i using the same procedure (i.e., classify the activations of the output vector, select the first K elements, and apply the blurring function mentioned above).
3) Utilize the fingerprint similarity function to compare the fingerprint of u_i with each emotion fingerprint. Select the emotion with the highest similarity.

4 Experiment

4.1 Dataset

DailyDialog [17] is a web resource created for practicing everyday English dialogues. Table 1 provides key statistics about this resource. In terms of emotion recognition, this dataset includes six basic Ekman emotions [18], which are anger, disgust, fear, happiness, sadness, surprise, and neutral emotions. The distribution of these emotion labels is shown in Table 2, and these labels are defined based on Yanran's publicly available split.

Table 1. Statistics for the DailyDialog Dataset

Dialogues num	Turns/Labels num	Avg. Turns/labels per dialogue
13118	102979	7.9

Table 2. DailyDialog Proportions of Each Emotion Label in the DailyDialog Dataset

Anger	Disgust	Fear	Happy	Sad	Surprised	Neutral
1.0%	0.3%	0.2%	12.5%	1.1%	1.8%	83.1%

4.2   Training Details

To get contextually relevant embedded dialogue representations, the RoBERTa-base model from Hugging Face [14] was used, and training was performed using cross-entropy loss. The Adam optimizer was used with the initial learning rate set to $1e^\wedge(-5)$ and $5e^\wedge(-5)$ for the encoder and decoder respectively, with a class decay rate of 0.95 after each training epoch. Layer freezing was performed in the first epoch and the batch size was set to 4, with a slope of 1.0. Training will stop early if the F1 macro point on the validation set exceeds the threshold for 5 consecutive epochs, with a maximum of 10 epochs allowed. The checkpoint used in the experiment is the one that achieves the highest F1 macro score in the validation set. The fuzzy fingerprint module was proposed and implemented by the authors. The DailyDialog dataset has class imbalance issues, not only because the neutral class is dominant but also due to relative imbalances among different categories. To improve overall performance for each category, the F1 points was chosen for model picking. Regarding the choice of parameter "a," experimental observations led to the selection of 0.8 as an appropriate value.

4.3   Results and Analysis

The reported results are the averages of 5 runs, with each run using a different random seed, and these seeds have been documented for meaningful experimental comparisons. The use of these averages is because the same experiment can yield significantly different macro F1 scores under different random seeds, which aligns with our findings in improving state-of-the-art models. This approach is also consistent with the practice of using the averages of 5 runs in some ERC task research [2, 3, 15, 16].

Comparison with Other Models. Figure 2 shows the evolution of the F1 score with different K fingerprint sizes. Other models use all RoBERTa output, which is equivalent to K = 768.

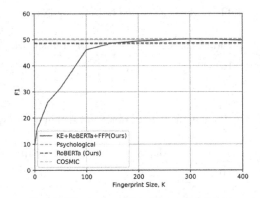

Fig. 2. F1 changes with fingerprint size K.

First, we examine the impact of FFP K size on performance. From Table 3 and Fig. 2, it can be seen that our proposed method performs on par with using 768 RoBERTa output (achieving state-of-the-art performance) when K is larger than 150. Hence, it can be concluded that for achieving high performance, there's no need to use all RoBERTa outputs, and once the size of K is determined, a smaller and computationally less expensive model can be used. This also suggests that we can train smaller base models to fit this task.

Furthermore, it can be seen from the graph that performance peaks at K = 300, with a corresponding F1 score of 50.32, This indicates that our method has a significant advantage of 0.51 points compared to the F1 score without the fuzzy fingerprint module.

Table 3. F1-SCORE changes with fingerprint size K

K	1	5	10	25	50
F1	9.66	16.12	18.34	26.03	31.56
K	100	150	200	300	400
F1	46.13	48.68	49.44	50.32	49.78
F1	Without fuzzy fingerprint module				49.81

Table 4. Compare with state-of-the-art works

Other works	macro-F1
RoBERTa [1]	47.63
RoBERTa + Dialogue RNN [1]	48.96
COSMIC [2]	49.54
Contextual RoBERTa [13]	49.76
Psychological [2]	50.38
Contextual RoBERTa + FFP	50.32

Compare with State-of-the-Art Works. Our method was compared with other RoBERTa-based methods, which allows for a fair comparison as the use of PLMs significantly improves performance compared to other acoustic feature extraction methods. We compared our method with COSMIC [1], RoBERTa implemented by the authors of COSMIC, RoBERTa with dialogue RNN, and a psychological model [2], all of which were described in Sect. 2. The results in Table 4 are averages of 5 runs.

Our method not only outperforms simple RoBERTa but also demonstrates superior performance when compared to more complex models like Dialogue RNN and COS-MIC. Although the psychological model's performance is slightly better than ours, it employs RoBERTa Large, a substantial common-sense knowledge base (COMET [19]), and complex dialogue graphs as a session-level encoder, requiring the use of graph transformers. In contrast, our model, after knowledge elimination, utilizes RoBERTa base with context-based fine-tuning and a concise Fuzzy Fingerprint classification module.

FFP Interpretability Example. The method proposed in this paper leverages FFP to enhance the interpretability of PLMs. Below are three examples to support this viewpoint. We chose to use small fingerprints for visualization and examples. Table 5 presents the FFP emotions generated when $K = 10$. It's worth noting that the output index 578 appears at the top of five categories, which can potentially be used for multi-label classification to enhance ERC performance since 578 does not appear in the Neutral FFP, which is the majority of categories and the eventual outcome of most misclassifications.

Table 6 provides three examples, including statement fingerprints, category fingerprints, and their similarity (consider $n = 1$). The first example illustrates that there are many common elements between the statement and the FFP category, which certainly support correct classification.

The next two examples demonstrate an interesting misclassification scenario and how FFP can help explain these errors. For the sentence, (You haven't given me the file I asked for), FFP clearly shows a strong similarity to anger, with significant overlap between some fingerprints. However, according to the dataset labels, this expression should be neutral. This type of misclassification cannot be explained by FFP. However, for native English speakers, this sentence might be considered to carry a clear negative sentiment. Therefore, the current emotion label may reflect anger more than neutrality, possibly due to annotation errors or implicit knowledge that the annotators have about the utterance, which cannot be explicitly captured from the text. The same information can be expressed in a more polite and neutral way, such as, "Please provide me with more information about this matter." In this case, FFP clearly indicates a neutral sentiment, even though both sentences convey the same information with a different tone.

Table 5. Class fingerprints

$FFP_{Neu}=$	{(198,1),	(632,0.9),	(529,0.8),	(701,0.7),	(388,0.6),	(315,0.5).	(402,0.4),	(58,0.3),	(562,0.2),	(101,0.1)}
$FFP_{Ang}=$	{(6,1),	(651,0.9),	(192,0.8),	(278,0.7),	(639,0.6),	(561,0.5),	(98,0.4),	(613,0.3),	(168,0.2),	(296,0.1)}
$FFP_{Dis}=$	{(578,1),	(565,0.9)	(19,0.8),	(149,0.7),	(315,0.6),	(56,0.5),	(549,0.4),	(5,0.3),	(488,0.2),	(427,0.1)}
$FFP_{Fear}=$	{(578,1),	(297,0.9),	(643,0.8),	(391,0.7)	(725,0.6),	(336,0.5),	(609,0.4),	(358,0.3),	(273,0.2),	(6,0.1)}
$FFP_{Hap}=$	{(578,1),	(569,0.9),	(372,0.8),	(582,0.7)	(753,0.6),	(302,0.5),	(730,0.4),	(549,0.3),	(460,0.2),	(121,0.1)}
$FFP_{Sad}=$	{(359,1),	(578,0.9),	(5,0.8),	(148,0.7),	(3,0.6),	(79,0.5),	(541,0.4),	(388,0.3),	(502,0.2),	(413,0.1)}
$FFP_{Sur}=$	{(678,1),	(578,0.9)	(84,0.8),	(565,0.7),	(516,0.6),	(518,0.5),	(645,0.4),	(363,0.3),	(354,0.2),	(601,0.1)}

5   Conclusion and Future Work

In this paper, we proposed a method that combines the Knowledge Elimination (KE) module with the Fuzzy Fingerprint RoBERTa model. Specifically, the KE module is used to eliminate external knowledge, and then we obtain contextually relevant embedded utterance representations from RoBERTa's pretrained language model. These representations are then fed to the adaptive FFP classification module. We validated our

Table 6. Classification examples

Text:	Sorry, sir. I've tried my best								
$FFP_{Sanole} =$	{(5,1),	(363,0.9),	(144,0.8),	(538,0.7)	(82,0.6),	(5,0.5),	(218,0.4),	(406,0.3),	(387,0.2),	(428,0.1)}
Class Similarity:	$Neu = 0$	$Ang = 0$	$Dis = 0.2$	$Fear = 0$	$Hap = 0$	$Sad = 1.5$	$Sur = 0$		
Text:	You still have not given me those files I 've asked you for								
$FFP_{Sample} =$	{(7,1),	(576,0.9),	(287,0.8),	(582,0.7)	(255,0.6),	(65,0.5),	(118,0.4),	(187,0.3),	(432,0.2),	(402,0.1)}
Class Similarity:	$Neu = 0$	$Ang = 1$	$Dis = 0$	$Fear = 0.1$	$Hap = 0$	$Sad = 0$	$Sur = 0$		
Text:	Please tell me something about the event								
$FFP_{Sample} =$	{(315,1),	(624,0.9)	(508,0.8),	(196,0.7),	(91,0.6),	(405,0.5),	(192,0.4),	(687,0.3),	(377,0.2),	(534,0.1)}
Class Similarity:	$Neu = 1.6$	$Ang = 0$	$Dis = 0$	$Fear = 0$	$Hap = 0$	$Sad = 0$	$Sur = 0$		

approach on the ERC DailyDialog dataset, and the results show that our architecture significantly competes with state-of-the-art methods on this dataset, while adding some interpretability. In future, we design to further explore the interpretability potential of our method and investigate size reduction techniques based on fuzzy fingerprints.

References

1. Ghosal, D., Majumder, N., Gelbukh, A., et al.: Cosmic: commonsense knowledge for emotion identification in conversations. arxiv preprint arxiv:2010.02795 (2020)
2. Li, J., Lin, Z., Fu, P., et al.: Past, present, and future: conversational emotion recognition through structural modeling of psychological knowledge. In: Findings of the Association for Computational Linguistics: EMNLP 2021, pp. 1204–1214 (2021)
3. Zhong, P., Wang, D., Miao, C.: Knowledge-enriched transformer for emotion detection in textual conversations. arxiv preprint arxiv:1909.10681 (2019)
4. Xie, Y., Yang, K., Sun, C.J., et al.: Knowledge-interactive network with sentiment polarity intensity-aware multi-task learning for emotion recognition in conversations. In: Findings of the Association for Computational Linguistics: EMNLP 2021, pp. 2879–2889 (2021)
5. Devlin, J., Chang, M.W., Lee, K., et al.: Bert: pre-training of deep bidirectional transformers for language understanding. arxiv preprint arxiv:1810.04805 (2018)
6. Liu, Y., Ott, M., Goyal, N., et al.: Roberta: a robustly optimized bert pretraining approach. arxiv preprint arxiv:1907.11692 (2019)
7. Homem, N., Carvalho, J.P.: Authorship identification and author fuzzy "fingerprints". In: 2011 Annual Meeting of the North American Fuzzy Information Processing Society, pp. 1–6. IEEE (2011)
8. Homem, N., Carvalho, J.P.: Web user identification with fuzzy fingerprints. In: 2011 IEEE International Conference on Fuzzy Systems (FUZZ-IEEE 2011), pp. 2622–2629. IEEE (2011)
9. Rosa, H., Batista, F., Carvalho, J.P.: Twitter topic fuzzy fingerprints. In: 2014 IEEE international conference on fuzzy systems (FUZZ-IEEE), pp. 776–783. IEEE (2014)
10. Rosa, H., Carvalho, J.P., Calado, P., et al.: Using fuzzy fingerprints for cyberbullying detection in social networks. In: 2018 IEEE International Conference on Fuzzy Systems (FUZZ-IEEE), pp. 1–7. IEEE (2018)
11. Cambria, E., Liu, Q., Decherchi, S., et al.: SenticNet 7: a commonsense-based neurosymbolic AI framework for explainable sentiment analysis. In: Proceedings of the Thirteenth Language Resources and Evaluation Conference, pp. 3829–3839 (2022)

12. Manning, C.D., Surdeanu, M., Bauer, J., et al.: The Stanford CoreNLP natural language processing toolkit. In: Proceedings of 52nd Annual Meeting of the Association for Computational Linguistics: System Demonstrations, pp. 55–60 (2014)

13. Pereira, P., Moniz, H., Dias, I., et al.: Context-dependent embedding utterance representations for emotion recognition in conversations. arxiv preprint arxiv:2304.08216 (2023)

14. Wolf, T., Debut, L., Sanh, V., et al.: Huggingface's transformers: state-of-the-art natural language processing. arxiv preprint arxiv:1910.03771 (2019)

15. Shen, W., Chen, J., Quan, X., et al:. Dialogxl: all-in-one xlnet for multi-party conversation emotion recognition. In: Proceedings of the AAAI Conference on Artificial Intelligence, vol. 35, no. 15, pp. 13789–13797 (2021)

16. Shen, W., Wu, S., Yang, Y., et al.: Directed acyclic graph network for conversational emotion recognition. arxiv preprint arxiv:2105.12907 (2021)

17. Li, Y., Su, H., Shen, X., et al.: Dailydialog: a manually labelled multi-turn dialogue dataset. arxiv preprint arxiv:1710.03957 (2017)

18. Ekman, P.: Basic emotions. Handb. Cogn. Emot. **98**(45–60), 16 (1999)

19. Bosselu, A., Rashkin, H., Sap, M., et al.: COMET: commonsense transformers for automatic knowledge graph construction. arxiv preprint arxiv:1906.05317 (2019)

The Security and Privacy Concerns on Metaverse

Kejiang Liu[1], Tengfei Zheng[1], Tongqing Zhou[1], Chang Liu[1], Fang Liu[2(✉)], and Zhiping Cai[1(✉)]

[1] College of Computer, National University of Defense Technology, Changsha 410073, Hunan, China
zpcai@nudt.edu.cn
[2] School of Design, Hunan University, Changsha 410082, Hunan, China
fangl@hnu.edu.cn

Abstract. As the next development of the Internet, metaverse aims to construct an immersive, hyper-realistic, self-sustaining virtual space. In this space, humans can work, social and entertainment. Over the years, with the continuous development and emergence of various technologies, the realization of metaverse has become possible, and has attracted the attention of researchers. However, the privacy and security problems existing in the Internet world are no exception in metaverse, and even more serious problems will arise in metaverse. In this paper we comprehensively summarize the security and privacy threats in metaverse. Specifically we proposes a three-stage metaverse framework according to the user usage process, including: Ambient Perception, Avatar-Enabled Playing and Virtually Feedback. Meanwhile we point out and summarized the security and privacy issues in metaverse under three-stage framework and the according solutions to key challenges.

Keywords: Metaverse · Privacy and Security · Countermeasure

1 Introduction

Metaverse, as the name shows, consists of the words meta and verse, is a computer-generated world with a consistent value system and an independent economic system linked to the physical world [1]. In 1992, Neil Stephenson created the word 'metaverse' in the science fiction Snow Crash [2], where people can have their virtual avatars. In this novel, human beings enter the avatar in metaverse through wearable devices in the physical world to carry out various social activities. In metaverse, people worldwide can be gathered in the same scene regardless of distance obstacles, such as real-time teleconference, virtual 'live' concerts, and on-the-spot teaching. Furthermore, the virtual world can also simulate the taste, touch, and smell of the real world, giving people a fully immersive experience. Generally, metaverse is considered as a virtual environment blending physical and digital, facilitated by the convergence between the Internet and Web technologies, and Extended Reality (XR) [3].Generally, the development of

H. Jin et al. (Eds.): IAIC 2023, CCIS 2060, pp. 287–305, 2024.
https://doi.org/10.1007/978-981-97-1332-5_23

metaverse consists of three successive phases from a macro perspective [3] : (i) digital twins (ii) digital natives and (iii) co-existence of physical-virtual reality. The first phase is to generate large-scale and high-fidelity digital twins of physical world in the virtual world.The properties of the physical counterparts [4] will reflect by the digital twins,such as weight,length, motions.The connection between the virtual and physical twins is tied by their data [5]. The next phase is mainly about native content creation. Avatars representing real users can create in a digital world. This creation can be related to the physical world, or even exist only in the virtual world. In this phase, the intersection between the physical world and the virtual world is created, and the virtual world is gradually separated from the physical world. In the final phase, metaverse developed into a self-sustaining and independent world. In this phase, metaverse coexist with the physical world in a highly independent manner. The avatars can experience heterogeneous activities in real-time characterized by unlimited numbers of concurrent users theoretically in multiple virtual worlds [5]. Driven by real-world demand and the prospect of building metaverse viability, metaverse has recently attracted increasing attention around the world, with many tech giants such as Facebook, Microsoft, Tencent and NVIDIA announcing their investment in metaverse. In particular, Facebook has renamed itself "Meta" and is committed to shaping the meta world of the future.Despite the potential of metaverse, traditional issues such as security and privacy still deserve our attention. From the management of massive data streams, the ever-present analysis of user behavior, the discrimination of artificial intelligence algorithms, to the security of human wearable devices, a variety of security funnels and privacy risks will emerge in metaverse.It can be seen from the three phases of metaverse development that metaverse was built on the most novel technologies and systems available. The related technologies was shown in Fig. 1. So it is inevitable that metaverse will inherit the vulnerabilities and flaws of these technologies and systems. What's more, metasverse is highly integrated with various technologies, and when these technologies collide, existing security problems can be magnified and have more serious consequences in the virtual world. And metaverse can breed new threats nonexistent in physical and cyber spaces can breed such as virtual stalking and virtual spying [6]. In order to ensure the user's immersive experience in metaverse, the data involved in metaverse is more sensitive. For example, users upload facial data, voice recognition data, physical movement, body features and so on by wearing sensors.This opens new horizons for crimes on private big data [7].A hacker could also use the connections between physical devices as an entry point to break into a user's device, and even threaten critical infrastructures such as power grid systems, high-speed rail systems, and water supply systems via advanced persistent threat (APT) attacks [8].

Fig. 1. A collection of key technologies in metaverse

At present the topic of metaverse has attracted the attention of many researchers. There have been a number of survey papers on different aspects of metaverse. Lee et al. [3] discussed and analyzed eight key technologies for the construction of metaverse.Yang et al. [1] explored the application of artificial intelligence technology and blockchain technology in metaverse. Dionisio et al. [9] describes five phases of virtual world development and specifies four features of a viable metavers. Nevelsteen [10] classified the technology of virtual world and defined what is virtual world. Ning et al. [11] investigated the development of metaverse from the aspects of national policies, social activities, development projects and construction. The above articles elaborated the concept of metaverse and the technology used in metaverse, explained the difference between metaverse and virtual world.However users' avatars and physical identities in metaverse are interrelated, and sensitive in- formation in the real world may be maliciously stolen through the terminal. Users are not independent individuals in metaverse. Like the physical world, users will interact with other users. However, due to the differences between users, conflicts between users will inevitably occur, and even malicious harassment and harm users. We notice that there is a lack of literature on privacy and security in metaverse.

In this paper we comprehensively summarize the security and privacy threats in metaverse. First, this paper proposes a three-stage metaverse framework according to the user usage process, including: Ambient Perception, Avatar-Enabled Playing and

Virtually Feedback. Second we point out and summarized the security and privacy issues in metaverse under three-stage framework. Finally, we propose some solution to the key challenges and list some future research directions in metaverse. Table 1 shows the comparison between our work and related works.

Table 1. A comparison between this work and related works

Year	Refs.	Contribution
2009	[12]	Survey about the retailing evolution from traditional way to metaverse
2018	[7]	Survey on privacy issues and countermeasures to surveillance in metaverse
2020	[13]	Survey about applications of education in metaverse
2020	[14]	Survey about applications of social goods in metaverse
2021	[3]	Review the state-of-the-art technologies of building metaverse and opportunities based on the view
2021	[11]	Describe metaverse from national policies, industrial projects, supporting technologies, and social metaverse
2022	[15]	Survey on the hardware, software, content components and some applications in metaverse
2022	[16]	Survey on the fundamentals, security, and privacy of metaverse
Now	**Ours**	Comprehensive survey about the privacy and security problems about metaverse from the three-stage framework we proposed, and future research directions in the protection of metaverse

2 Overview of Metaverse

2.1 Framework of Metaverse

Driven by the understanding of the fundamental architecture of metaverse, we present a three-stage man-machine interactive process of metaverse. In the first stage, metaverse focuses on the perception of data in the physical reality world. Based on the data perceived from the real world, virtual skyscrapers sprang up, and avatars are built and controlled by human users. The second stage describes a parallel and lively virtual environment wherein humans control their avatars to complete various activities, such as social contact, travel, medical treatment, content creation, and economic activities. Finally, the sense perception of avatars about metaverse such as sight, hearing, taste, and touch are fed back to the human body in the physical world. The man-machine interactive framework we proposed is shown in Fig. 2.

Ambient Perception. Metaverse in the past only allow user inputs with traditional devices such as the keyboards and mice duo, which can not follow the movements of avatar and give immerse user experience.

The ambient perception stage bridges physical users and metaverse by the data collected from the real world using the sensing devices. On one hand, metaverse should perceive all kinds of slight but useful data, such as visual data, auditory data, tactile data, and somatosensory. In the film Ready Player One, the man wears HMD, gloves, glasses, headsets, and a special suit so that all user actions can be captured. On the other hand, real-time 3D rendering-related technologies like MR/AR/VR play significant roles in user interaction.

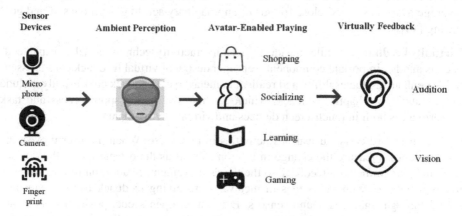

Fig. 2. The three-stage process framework for metaverse

As shown in VR, the essential component device is the head- mounted display [17] (HMD), which can be worn on the head with a display optic. The perception devices can detect the user's input signal data and input it into metaverse through various sensors. In order to give a more immersive, realistic experience in metaverse, the perception devices capture a lots of user information. These pieces of information are from biometrical data of users to spatial data, including surroundings such as bystanders' physical space (rooms) [18, 19]. Based on different functions and purposes, in addition to the traditional mouse and keyboard, perception devices have also expanded with many new forms, such as data gloves for hand posture perception, data clothes for obtaining body posture, and voice interactive microphones.

The development of perception devices such as somatosensory and brain-computer interface have promoted the 3D rendering-related technologies and metaverse into diversification.

Avatar-Enabled Playing. Avatar, which is the digital representation of users in meta-verse, serves as a communication entity in metaverse. Any authorized users can utilize their imagination and create their own highly diversified avatars through open and creative exploration to mimic the users' real-life appearances. For example, Facebook and Bitmoji Avatars provide users rights for editing their personalized image.

At the same time, avatars play a key role in shaping social interaction inside meta-verse. Through the application of 3D rendering-related technologies and computer vision technologies, avatars in 3D virtual environments will be controlled through human pose tracking and eyes tracking to capture and reflect the physical users' behaviors in real-time.

It is well-known that metaverse is an independent virtual world for social gatherings. As for avatars' social activities in metaverse, research studies have also attempted to leverage avatars as users' close friends or an imaginary self to govern oneself and goal setting [14].

Virtually Feedback. While considering user interactivity technologies, the virtual feedback is another important component. A proper design of virtual feedback cues is essential for enhancing the usability and realism in metaverse. Feedback cues usually include visual, audio and haptic feedback, which can improve user responsiveness and task performances both in touchscreen devices and virtual environments.

The user will carry out various activities in metaverse. When the user moves, the motion sensor measure the change in motion as well as the orientation of the devices [20] and give immediate feedback to the user's movement. In addition to this, the user needs to interact with the scenes in metaverse, receiving feedback from images and sound through light and sound sensors. Even in an open scene, position sensors are used to locate the users and tell the users where they are. It's like the most common one position sensors used in Internet of Things (IoT) devices are magnetic sensors and Global Positioning System (GPS) sensors [20].

Meanwhile, new enriched feedback models will be explored, which can be reflected in a new form such as smell and taste. Typical virtual feedback devices are not only limited in screens, but also include a stereo headset with sound feedback, data gloves with force feedback, and large-screen stereo reality systems.

3   Security and Privacy: Threats on Metaverse

In this section we discuss the security and privacy challenges in metaverse through the three stage we proposed in the previous section. Fig. 3 describes the taxonomy of security and privacy threats on metaverse.

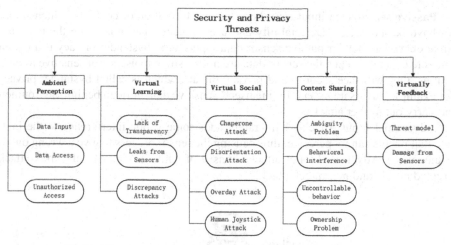

Fig. 3. The taxonomy of security and privacy threats on metaverse

3.1 Threats in Ambient Perception

Security and Privacy of Sensor. In modern society, user information is a very sensitive issue related to security and privacy. Because of the immersion in metaverse and the needing for multi-sensor, the meta universe needs to collect more dimensions and more detailed user information, which makes the harm caused by information leakage greater. An attacker may attack sensitive data in the following two ways as shown in Fig. 4.

1) Data Input. Firstly, we focus on the challenges in ensuring security and privacy of sensor data input, which can be categorized into the protection of input data and the defense of adversarial inputs. Vulnerable sensor data inputs can be divided into two categories based on users' intention. The one is captured from environments and usually is non-user-intended inputs, which is called passive sensor data inputs [18]. On the contrary, the other one which is intentionally provided by users can be defined as active sensor data inputs [18]. The protection of input data are mainly focused on ensuring sensitive information privacy while capturing the input data. However, adversarial inputs [18] are external attacks which have posed security threats to data sensor process rather than the data itself. Based on the above categories, we will describe the security and privacy attack threats on them.

   Active sensor data inputs represent users' gesture input and all types of user inputs that invoke commands to finish some functions. In this process, data leakage and denial of service attack are two main threats to hinder active sensor data inputs. Confidentiality and Integrity of data are compromised.

Passive sensor data inputs are those latent information or contexts, which should not have been detected. Detectability and user content unawareness are the two main threat characteristics for passive sensor data inputs. And bystander privacy is a typical concrete example. When the sensor data are not sensitive to users, but sensitive to other entities, they will face security and privacy threats, which is called bystander privacy. For example, adversaries can infer the users' sensitive data through benign environment and carry out further attacks.

Adversarial Input Attacks deliberately adds some subtle interference to the input data that people cannot perceive, thus causing the sensor process to give a wrong output to metaverse. Adversarial input attacks consist of white-box attack, blackbox attack, targeted attack, and non-target attack.

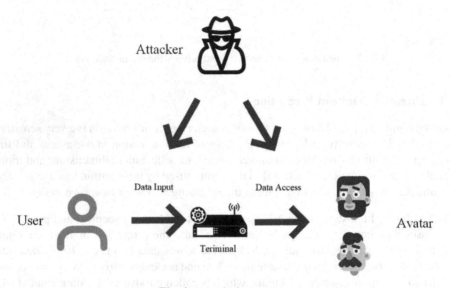

Fig. 4. Attacks in two ways

2) Data Access. When sensors have collected enough data, it goes to the data access process. Data from multiple sensors need to be aggregated, processed, and stored. Once all kinds of applications have collected and stored the sensor data, users may have no control over their own sensitive data, which raises various security and privacy risks. Consequently, data in aggregation, processing, and storage stage are threatened by all kinds of attacks, which will be paid great attention to in this section. The problem of user privacy leakage such as information disclosure, linkability, detectability, and identifiability are seriously threatened in all three stages.

In **data aggregation** stage, data transmission and aggregation will be more open and diversified. At the same time, the traditional method of physical area isolation cannot effectively ensure the security of long-distance transmission. Electromagnetic leakage and eavesdropping will be prominent security threats.

In **data processing** stage, one typical attack is Denial of Service attack. Any interference with services that makes services less available or loses availability is called denial of service. The most common DDoS attacks are computer network bandwidth attacks and connectivity attacks, which poses great threats on remote cloud data processing stage.

In **data storage** stage, users can not know the exact storage location of data, which means that inherent security threats such as tampering, unauthorized access, data leakage, and spoofing often appears. Unauthorized access means that attacker impersonates as a valid user and then gains unauthorized access to sensor data, which will tamper sensor input data and trigger confidential user privacy disclosure panic.

Security and Privacy of Sensor Devices. Immersive virtual reality system, actual MR/AR/VR device access, and the input or output interfaces also need to be concerned. Attacks against immersive VR systems are known as immersion attacks, which can be classified as chaperone attack, disorientation attack, human joystick attack, and Overlay attack. Chaperone attack creates an illusion by modifying virtual environment boundaries, and disorientation attack can cause dizziness and confusion for immersive VR users. Meanwhile, any attack that controls the movement of immersive users to a predetermined physical location without the user's knowledge is called human joystick attack, and any attack that superimposes unwanted images/video/content on the user's VR view is called overlay attack.

The main threats for device access are identity spoofing and unauthorized access. Identity spoofing technology often utilize user-trusted familiar names or addresses to spoof and confuse identification services. Unauthorized access means that attackers impersonate as a valid user and then gain unauthorized access to sensor device. All physical interfaces are easy to be attacked such as external inference or shoulder-surfing attacks and thus threats such as spoofing, tampering ,or denial of service may appear. Meanwhile, three-dimensional input interfaces allow users to interact with metaverse by body movements or gestures, which have transcended traditional input interfaces such as keyboards and mouses. However, this new form of input is detected through visual capture while sensitive information will also be captured. Consequently, latent privacy risks such as detectability and content unawareness arise and need to be pay attention to.

3.2 Threats in Avatar-Enabled Playing

Risks In Virtual Learning. Virtual Learning, also called Social Virtual Reality based Learning Environments (VRLEs), can be methods of motivating learners' understanding of certain events. VRLEs constructs a sharable learning virtual environment based on network and provides immersive and interactive experience for users. Furthermore, in order to build VRLEs, there are four components of a VRLEs requires [21]: knowledge space, communication community, active action, and facility toolkit.

Some attacks have opportunities to damage VRLEs because of virtual learning's special features. For example, the geographically distributed user settings of metaverse system have caused Distributed Denial of Service [22] (DDoS) attacks, which will influence the data collection process, information real-time visualization and VR session building. Meanwhile, some major threats to VRLEs include: Intermittent network discrepancy attacks, lack of transparency [23], tracking, leaks from sensors. Elevation of Privilege [24] (EOP) give attackers elevated access, which will modify the learning environment contents and cause a bad influence. Moreover, attackers can login with illegal tokens can access sensitive private information, which is called impersonation attacks [25].

In literature [26], they proposed a novel risk assessment framework, which utilizes attack trees to calculate a risk score for varied virtual learning threats and evaluates internal and external vulnerabilities.

Issues of Virtual Social Activities. Virtual Social Activities consist of virtual objects such as avatars and created virtual environments. Avatars is a kind of self-presentation in online social places such as metaverse. Metaverse is a representative of multiple users to establish their self, experience and interact with another avatar. For metaverse users, it is inevitable to sacrifice their personal information for a better virtual metaverse experience. At the same time, metaverse platform did not clear the boundaries of private information.There are possibilities of the attacks that virtual environments will get hacked and they can be accessed, modified, or changed by the attacker and can be able to destroy the virtual environment.Due to the inter-connectivity of the edge network and the core cloud in virtual environments, virtual social activities are faced with Immersive attacks [27], which may cause physical damage to users and influence user's immersive metaverse experience. Authors in [27] have classified immersive attacks into the following categories by the attack's outcome:

- **Chaperone Attack**: modifies the virtual environments' boundaries and makes the virtual space smaller or larger.
- **Disorientation Attack**: causes dizziness to any immersed metaverse user.
- **Human Joystick Attack**: realizes immersed users' physical movement by attacking virtual environment.
- Overlay Attack: overlays any unexpected images or videos on users' view,which are fixed in virtual space and cannot be removed.

Meanwhile, the literature [28] have proposed a novel framework to quantify the security and privacy problems caused by immersive attacks

Content Sharing Conflicts. Content in metaverse is an extension of the physical world, rather than isolated digital content, and metaverse incorporates the feel of the real world. According to the characteristics of metaverse, users do not always conduct activities in metaverse in isolation, but interact with the ecology of other users. While metaverse shows great potential for development, the computer security and privacy community has begun to identify and address serious security, privacy, and safety risks that they present. Many of the existing problems focus on the risks that individual users may face from malicious or flawed applications on their own devices. For example, preventing the application from displaying content that user does not want to see. Previous work has almost exclusively considered individual users, yet the risks come not only from flawed or malicious applications, but also from other users. Users may exhibit different sharing behaviors in the metaverse, some virtual content is private and some is public, and violating user's expectations for shared content can hurt users. For example, users may interact with sensitive virtual content without realizing that other users can see it, or they may inadvertently reveal private information that is shared with them but not shared with others nearby. These risks present unique challenges for content-sharing in metaverse. Immersion in the metaverse opens up exciting possibilities, but also potential risks. The virtual objects behave like physical ones, but attackers may deliberately violate victim expectations, such as by having an obstacle pop up in the victim's path. Authors in [29] enumerate many kinds of security and privacy issues that can arise between multiple users:

Ambiguity in Physical World and Metaverse: Avatars behave like physical objects, and what the user sees in metaverse causes them to simultaneously adjust their poses in the physical world. For example, a user may see an obstacle on the ground in metaverse. And in the real world, there is a high probability that he will make a move over the obstacle to avoid tripping himself. The users make avatars behave in a certain way in metaverse, and that action also affects their behavior and reactions in the physical world.

Interfering the Behavior of Others: When multiple users use the same program, participants some- times try to interfere with the activities of other participants. Take gaming activities as an example, where participants are stealing control of their resources or causing damage to other participants' resources. The tension between these users creates challenges around object ownership, visibility, and control.

Obscurity of Other Users' Actions: When virtual content is not shared between multiple AR/VR users, or when a non-AR/VR user interacts with an AR/VR user, multiple people may see different views of the same physical space.When a user behaves, the participants around him cannot determine whether such behavior will harm him or not. Imagine a user suddenly stops and stares at someone, who knows whether he's checking email or has some ill intentions about someone else.

Ownership of Virtual Objects and Physical Spaces: There is no doubt that metaverse allows multiple users to share virtual content.Determining the precise nature of this sharing raises questions such as: what content created by a given user is shared with whom, and how can those other users interact with this content [29]? Without any restrictions, all users can create and expose virtual content to other users, and freely view and interact with other users' creations. However, it is this potentially unlimited interaction that is the focus of our attention.

3.3 Threats in Virtually Feedback

The increasing sense of immersion and blurred boundaries between virtual world and reality in metaverse have caused some security problems. Malicious or vulnerable applications can produce harmful output, such as virtual objects that prevent users from obtaining important information. Or sensory overload caused by flickering vision, harsh audio or strong tactile feedback signal will cause physiological harm to users.The space of possible threats becomes amplified when networked sensor devices can share mixed-reality experiences with each other or retrieve content from the cloud [30, 55].

At present, there are many researches on input privacy and security, but less attention is paid to output security. For the feedback of the virtual world, there are roughly three kinds of threat content model, which have different forms of expression in different outputs. The three models are 1) Obscuring real-world context, 2) Attacking another metaverse application by obscuring or preventing the user from interacting with its content, 3) Distracting or disrupting the user.

The above threat models can be posed by a general set of potential attackers, including an adversarial or bug-ridden application, another user that is connected to the current user over a network, or a cloud-based entity that controls the virtual content sent to the device.In metaverse, users also inevitably have to wear various devices, headphones, and virtual sports equipment for various social activities. Imagine that hackers hack into the user's wearable device and issue abnormal instructions to the wearable device. The user is attacked without even knowing it, but it cannot be prevented. This is a very terrible thing.

4 Countermeasures and Challenges

4.1 Countermeasures for Perception Threats

Secure and Privacy-Preserving Data.

1) Data Input: Input Sanitization techniques act as an intermediary layer between physical world and virtual world and achieves a disinfection function by removing sensitive information. Input sanitization policies are user-defined and the sensitivity permissions are controlled by user's preferences, which means that users can judge and determine whether there exists sensitive information such as faces or ID card numbers and choose to delete them. For example, PREPOSE [34] is atypical intermediary layer which achieves secure gesture detection and recognition by providing the necessary gesture event to metaverse and thus controls least privilege access to metaverse in virtual world. For adversarial defense, the broad approaches are adversarial detection, input reconstruction, and adversarial training. Adversarial detection is achieved by a trained classifier to distinguish adversarial inputs and normal ones. Input reconstruction converts adversarial samples into clean data by encoding. Adversarial training means training adversarial samples to make model robust, which has a obvious improvement on white-box attacks. Public Key Infrastructure (PKI) provides cryptographic protection for sanitization and data transmission.

2) Data Access: In data aggregation stage, the goal is to obtain aggregate data information without revealing personal privacy. Many privacy-preserving approaches such as k-anonymity and differential privacy [35] are applied between the users and data aggregation services. K-anonymity utilizes generalization and concealment technologies to ensure that records are non-inferred from other K-1 records. Differential privacy provides randomness to data to ensure maximize query results and no leakage of personal privacy.After collection and aggregation, here goes to the data processing services. During this processing stage, external attackers have opportunities to directly access users' data. Multi-party secure computing [36], Federated Learning [37] are all effective privacy computing methods. Homomorphic encryption [38]

technology allows services to implement queries and computations over encrypted data and has no influence on processing results. Paillier encryption algorithm is a typical Homomorphic encryption technology and can be applied into data processing services to ensure data privacy. Data manipulation and perturbations utilize Generative Adversarial Network to deceive an adversarial data collector, which aims at recognizing facial identification to provide identity privacy. How to achieve secure data storage is also the problem we are focusing on. Advanced Encryption Standard (AES) has been widely used on separate databases. Meanwhile, researchers are also considering some more powerful methods such as disaster recovery backup and database firewall.

Seamless Authentication for Sensor Devices. In device level, there are two general aspects that need to be protected, which are device access and input or output interfaces, respectively. In order to ensure secure authentication for device access, all kind of methods are utilized or combined. Password, multi-password, PIN, dynamic keys have been recommended as a security enhancement for sensor device access. Meanwhile, for special sensor devices that can capture users' physiological features, fingers, hand, head,or breath features have achieved a high accuracy on authentication. Multi-models combines two or more models, which will create a huge leap as well. In literature [39], authors have investigated how to create seamless and secure authentication in VR. Through comparing different screen sizes, input modalities and password types for detecting attackers in physical world, they get a conclusion that both PINs and patterns are suitable for VR and can provide some level of security. When there is no common central authentication entity in AR, traditional authentication protocols appear powerless. In order to solve this problem, literature [40] has combined facial recognition and wireless localization to build a trusted and secure AR authentication system, which is called "Looks Good To Me" and is capable of defending against Man-in-the-Middle Attacks, User Impersonation Attacks, and Denial-of-Service Attacks.

There are optical or visual cryptography methods for physical interfaces protection. Active camouflaging [41] techniques are typical optical protection strategy, which allow interfaces to hide sensitive information in order not to be captured by detection devices. EyeDecrypt [42] utilizes visual cryptography techniques to encrypt and decrypt physical interfaces, and the private key is only kept by users.

Remaining Challenges. Potential risks in current sensor input or output devices should be investigated whether it has been detected and abused by adversaries. A formal evaluation mechanism needs to be formulated in the future to complete the evaluation and supervision system.

4.2 Countermeasures for Virtual Playing Threats

Secure and Privacy-Preserving Metaverse. Applications based on the privacy and security problems that users will encounter in their social activities in metaverse, the literature [43] has determined three design schemes for the privacy and security disclosure in social virtual reality. These schemes will inspire developers/designers to design a more secure and socially supported future of metaverse technology.

User Education: The technical characteristics in the metaverse will make user information (voice, avatar presentation, motion trajectory) recognized or collected by others. But users may not knowhow this information is collected without their knowledge. At the sametime, different users have different privacy problems, so an effective solution is to provide users with corresponding guidelines to describe the potential harm caused by lack of privacy and security awareness. Before entering the metaverse, users can receive corresponding privacy and security protection training to show users which behaviors may disclose their privacy and the possible consequences of such information disclosure. Such a scheme helps to arouse users' attention to their own information protection, help to reduce the burden of the platform, and help protected users not to disclose too much information.

Platform Embedded Voice Modulators: In virtual social activities, voice chat is one of the functions most used by users, but this function will disclose recognizable voice data. Therefore, the platform provides a corresponding voice modem to process the user's voice and keep it anonymous to other users. Through this method, the sensitive characteristics of users such as gender and age can be hidden.

Non-identifiable Avatars: In order to help users not suffer discrimination by their gender, age, race, skin color, nationality and other information in the physical world, metaverse service providers can help provide a variety of unrecognized and inconsistent avatar templates for users to use. Through this way users cannot infer users' information in the physical world from these modeled avatars.

Secure and Privacy-Preserving Content Sharing. Content sharing in metaverse can be divided into two dimensions: The initiator and receiver of content.The three standard access control permissions are read, write and execute. The expansion of read and write to metaverse produces the user's control over various permissions. In order to achieve the above conditions, literature [44] lists the corresponding control structure should be included in the metaverse-software designing. Permission management: Using the classic access control work [45] to track and enforce per object and per user permission. Two-party sharing consent: The sharer and receiver of digital content are required to determine each other before the completion of sharing activities. Clutter management: Support users to temporarily or permanently remove objects from their field of view. Kimberly Ruth et al. introduced a new visual state in the virtual world, called ghosting. Ghost objects only show their location in space without sensitive content, no matter what angle you watch it. Furthermore, a user with whom a ghost object is shared receives from the sharer only the data needed to instantiate the ghost, rather than the full object data. This further insulates the private content [44]. Devine Maloney et al. put forward the idea of targeting sharing content in the virtual world. Such hierarchical privacy settings can also be applied to metaverse. For example, set up such a function that only the friends and some contacts selected by the user can view their position in metaverse. This will allow users to choose to have different levels of social circles and participate in different levels of social activities.

Remaining Challenges. Users of metaverse come from different countries, with different gender, age, and cultural differences, which influence how they perceive and understand security and privacy issues in social virtual reality. Therefore, in the hierarchical privacy setting or in the training guidance of users entering metaverse, how to provide unique solutions for different users according to their huge personal differences is a more difficult problem at present. Second, in a huge social platform full of immersion like metaverse, disclosure of personal privacy issues is an inevitable trade-off, and users show conflicting attitudes toward privacy issues. Some people are not very concerned about the security concerns posed by privacy concerns, while others are cautious about the content and scope of the information they share. A more in-depth and comprehensive understanding of these issues is required to design a safer and more satisfying metaverse platform.

4.3 Countermeasures for Feedback Threats

Secure and Privacy-Preserving Feedback. A solution to the potential security and privacy risks to users posed by sensory feedback from metaverse is enhancing existing sensor management systems. Xu et al. proposed a sensor management system named Semadroid, which provides monitoring and logging functions to users to authorize the use of sensors by applications. At the same time, the system can create simulated data to verify how the sensor is used from unauthorized applications, thereby preventing malicious behavior. Literature [46] proposes a framework of 6thSense, which adopts a context-aware framework to detect possible threats to users in devices. The framework is based on the observation that for all users of the device, certain sensors are always active. Based on this result, 6thSense builds a context-aware model of each user's normal activity, and uses machine learning algorithms to determine whether the sensor's current context is malicious or harmful to the user. One proposed countermeasure to immune electromagnetic emanation attacks is to use a single inverter ring oscillator (SIRO) [47]. Lebeck et al. recently introduced an OS-level framework called Arya that constrains the visual output of AR applications based on context-specific policies that must be explicitly specified in a conditional fashion using a set of policy primitives [49]. Based on Lebeck's work, Surin Ahn et al. propose a method for generating adaptive policies to preserve visual output using deep reinforcement learning in AR systems [50]. The method utilizes a local fog computing node, which automatically learns appropriate policies to filter out potentially malicious content generated by the application or make the user produce uncomfortable content to ensure a good user experience.

Remaining Challenges. Although many solutions were listed in the previous section, they also have the following limitations.

1) Most of the proposed security mechanisms for IoT devices are anomaly detection frameworks at the application level which are not suitable for detecting sensor-based threats at the system level [51, 52].
2) In metaverse, a large number of devices must be connected to each other. Considering that these devices are basically small devices with limited resources, the security of these devices becomes difficult to manage, and it will be very difficult to implement complex security mechanisms.

3) In the solution of how to use the user's autonomous decision-making to construct the corresponding device usage policy, if the user uses the sensor without any restrictions, the system will blindly regard the program as safe.

4) A key feature in metaverse is modifying the user's view of the world,however, this is a very threatening feature. However, the current design of solutions against such threats is very difficult. Future metaverse platforms must consider and solve such problems.

5   Summary and Future Research Directions

In this part, we will discuss the following future research directions ofmetaverse.

A. **Blockchain in Metaverse.** With the development of metaverse, the same virtual economic ecosystem as the physical world will inevitably emerge in metaverse, and blockchain technology is regarded as the basis for building this ecosystem. However, different sub-verses can be built on heterogeneous blockchains to meet different needs.An example is the exchange of different cryptocurrencies such as Bitcoin and Ethereum. Cross-chain governance is essential to ensure the security and legitimacy of digital asset related activities (e.g., asset trading) across different sub-metaverses built on heterogeneous blockchains [16].

B. **Cloud-server in Metaverse.** Metaverse must be an area where cloud server can provide a lot of power. Metaverse itself needs computing, storage, machine learning, and so on, which are all inseparable from cloud computing. The five key characteristics of cloud computing include on-demand,self-service, ubiquitous network access, location-independent resource pooling, rapid elasticity, and measured service, all of which are geared toward using clouds seamlessly and transparently. Attackers may target servers or users through various attacks to ensure the smooth operation of the system under such a large-scale service load. At the same time, the architecture design and risk assessment of cloud services also need to update the design scheme applicable to metaverse.

C. **Governance in Metaverse.** In analogy to the social norms and regulations in the real world,content creation, data processing, and virtual economy in metaverse should align with the digital norms and regulations [53]. We need to consider supervising the regulator of metaverse. The improper behavior of the supervisors may lead to system paralysis. How to implement a dynamic and effective punishment/reward mechanism for this kind of supervision is the focus of our research. Similarly, the cost of committing a crime in metaverse is low, and it is difficult to track this crime, which is related to the anonymous avatar inmetaverse. How to reconstruct and track such cyber-crime is a challenge for researchers.

6   Conclusion

In this paper, we investigate the privacy and security issues in metaverse. We divide metaverse into three stages: Ambient Perception, Avatar-Enabled Playing and Virtually Feedback. Starting from the existing technical prototype, we study the security and privacy issues under the technical framework of metaverse, and elaborate on the key issues in detail. We also sorted out and summarized the corresponding solutions for the security and privacy problems in these three stages. In addition, we also look forward to the future research direction in metaverse.

Acknowledgements. This work is supported by the National Natural Science Foundation of China (No. 62172155, 62072465, 62102325), the Natural Science Foundation of Hunan Province (No. 2022JJ40564) and the Science and Technology Innovation Program of Hunan Province (Nos. 2022RC3061, 2021RC2071).

References

1. Yang, Q., et al.: Fusing blockchain and AI with metaverse: A survey. IEEE Open Journal of the Computer Society **3**, 122–136 (2022)
2. Stephenson, N.: Snow crash: A novel. Spectra (2003)
3. Lee, L.H., et al.: All one needs to know about metaverse: a complete survey on technological singularity, virtual ecosystem, and research agenda. Computers and Society (2021)
4. Mohammadi, N., Taylor, J.E.: Thinking fast and slow in disaster decision-making with Smart City Digital Twins. Nature Comput. Sci. **1**(12), 771–773 (2021)
5. Grieves, M., Vickers, J.: Digital twin: Mitigating unpredictable, undesirable emergent behavior in complex systems. Transdisc. Perspect. Complex Syst. New Find. Approaches, 85–113 (2017)
6. Leenes, R., Fischer-Huebner, S., Duquenoy, P., Zuccato, A., Martucci, L.: Privacy in the metaverse: regulating a complex social construct in a virtual world. In: FIDIS (2007)
7. Falchuk, B., Loeb, S., Neff, R.: The social metaverse: Battle for privacy. IEEE Technol. Soc. Maga. **37**(2), 52–61 (2018)
8. Hu, P., Li, H., Fu, H., Cansever, D., Mohapatra, P.: Dynamic defense strategy against advanced persistent threat with insiders. In: International Conference on Computer Communications (2015)
9. Dionisio, J.D.N., Burns, W.G., Gilbert, R.L.: 3d virtual worlds and the metaverse: current status and future possibilities. ACM Comput. Surv. **45**, 1–38 (2013)
10. Nevelsteen, K.J.L.: Virtual world, defined from a technological perspective and applied to video games, mixed reality, and the Metaverse. Comput. Animat. Virtual Worlds **29**(1), e1752 (2018)
11. Ning, H., et al.: A survey on the metaverse: the state-of-the-art, technologies, applications, and challenges. IEEE Internet Things J. **10**, 14671–14688 (2023)
12. Bourlakis, M., Papagiannidis, S., Li, F.: Retail spatial evolution: paving the way from traditional to metaverse retailing. Electron. Commer. Res. **9**, 135–148 (2009)
13. Díaz, J., Saldaña, C., Avila, C.: Virtual world as a resource for hybrid education. Int. J. Emerg. Technol. Learn. (iJET) **15**(15), 94–109 (2020)
14. Duan, H., et al.: Metaverse for social good: a university campus prototype. In: Proceedings of the 29th ACM International Conference on Multimedia (2021)
15. Park, S.M., Kim, Y.G.: A metaverse: taxonomy, components, applications, and open challenges. IEEE Access **10**, 4209–4251 (2022)
16. Wang, Y., Su, Z., Zhang, N., Xing, R., Liu, D., Luan, T.H., Shen, X.: A survey on metaverse: fundamentals, security, and privacy. IEEE Commun. Surv. Tutor. **25**, 319–352 (2022)
17. Janin, A.L., Mizell, D.W., Caudell, T.P.: Calibration of head-mounted display for augmented reality applications. In: Virtual Reality Annual International Symposium, 1993. IEEE (1993)

18. de Guzman, J.A., Thilakarathna, K., Seneviratne, A.: Security and privacy approaches in mixed reality: a literature survey. ACM Comput. Surv. **52**, 1–37 (2019)
19. de Guzman, J.A., Seneviratne, A., Thilakarathna, K.: Unravelling spatial privacy risks of mobile mixed reality data. In: Proceedings of the ACM on Interactive, Mobile, Wearable and Ubiquitous Technologies (2021)
20. Sikder, A.K., Petracca, G., Aksu, H., Jaeger, T., Uluagac, A.S.: A survey on sensor-based threats to internet-of-things (iot) devices and applications. Cryptography and Security (2018)
21. Pan, Z., Cheok, A.D., Yang, H., Zhu, J., Shi, J.: Virtual reality and mixed reality for virtual learning environments. Comput. Graph. **30**(1), 20–28 (2006)
22. Lau, F., Rubin, S., Smith, M., Trajkovic, L.: Distributed denial of service attacks. In: Systems Man and Cybernetics (2000)
23. Schwarcz, D.: Transparently opaque: understanding the lack of transparency in insurance consumer protection. Social Sci. Res. Netw. (2013)
24. Shostack, A.: Elevation of privilege: drawing developers into threat modeling. In: Genetics Selection Evolution (2014)
25. Barbeau, M., Hall, J., Kranakis, E.: Detecting impersonation attacks in future wireless and mobile networks. In: Burmester, M., Yasinsac, A. (eds.) MADNES 2005. LNCS, vol. 4074, pp. 80–95. Springer, Heidelberg (2006). https://doi.org/10.1007/11801412_8
26. Gulhane, A., et al.: Security, privacy and safety risk assessment for virtual reality learning environment applications. In: Consumer Communications and Networking Conference (2019)
27. Casey, P., Baggili, I., Yarramreddy, A.: Immersive virtual reality attacks and the human joystick. IEEE Trans. Depend. Secure Comput. **18**(2), 550–562 (2019)
28. Valluripally, S., Gulhane, A., Hoque, K.A., Calyam, P.: Modeling and defense of social virtual reality attacks inducing cybersickness. IEEE Trans. Depend. Secure Comput. **19**(6), 4127–4144 (2021)
29. Lebeck, K., Ruth, K., Kohno, T., Roesner, F.: Towards security and privacy for multi-user augmented reality: foundations with end users. In: IEEE Symposium on Security and Privacy (2018)
30. Schmalstieg, D., Hollerer, T.: Augmented Reality: Principles and Practice. Addison-Wesley Professional, Boston (2016)
31. Hasan, R., Saxena, N., Haleviz, T., Zawoad, S., Rinehart, D.: Sensing-enabled channels for hard-to-detect command and control of mobile devices. In: Computer and Communications Security (2013)
32. Deshotels, L.: Inaudible sound as a covert channel in mobile devices. In: WOOT'14 Proceedings of the 8th USENIX conference on Offensive Technologies (2014)
33. Subramanian, V., Uluagac, S., Cam, H., Beyah, R.: Examining the characteristics and implications of sensor side channels. In: International Conference on Communications (2013)
34. Figueiredo, L.S., Livshits, B., Molnar, D., Veanes, M.: Prepose: privacy, security, and reliability for gesture-based programming. In: 2016 IEEE Symposium on Security and Privacy (SP), pp. 122–137. IEEE (2016)
35. Samarati, P.: Protecting respondents identities in microdata release. IEEE Trans. Knowl. Data Eng. **13**(6), 1010–1027 (2001)
36. Yan, F., Zhang, H., Zhao, B.: A secure multi-party computing model based on trusted computing platform. In: 2009 Ninth IEEE International Conference on Computer and Information Technology, vol. 2, pp. 318–322. IEEE (2009)
37. Li, T., Sahu, A.K., Talwalkar, A., Smith, V.: Federated learning: challenges, methods, and future directions. IEEE Signal Process. Mag. **37**(3), 50–60 (2020)
38. Gentry, C.: Fully homomorphic encryption using ideal lattices. In: Symposium on the Theory of Computing (2009)
39. George, C., et al.: Seamless and secure vr: adapting and evaluating established authentication systems for virtual reality. In: NDSS (2017)

40. Gaebel, E., Zhang, N., Lou, W., Hou, Y.T.: Looks good to me: authentication for augmented reality. In Proceedings of the 6th International Workshop on Trustworthy Embedded Devices, pp. 57–67 (2016)
41. Pearson, J., et al.: Chameleon devices: investigating more secure and discreet mobile interactions via active camouflaging. In: Proceedings of the 2017 CHI Conference on Human Factors in Computing Systems, pp. 5184–5196 (2017)
42. Forte, A.G., Garay, J.A., Jim, T., Vahlis, Y.: EyeDecrypt — private interactions in plain sight. In: Abdalla, M., De Prisco, R. (eds.) SCN 2014. LNCS, vol. 8642, pp. 255–276. Springer, Cham (2014). https://doi.org/10.1007/978-3-319-10879-7_15
43. Maloney, D., Zamanifard, S., Freeman, G.: Anonymity vs. familiarity: self-disclosure and privacy in social virtual reality. Virtual Reality Softw. Technol. (2020)
44. Ruth, K., Kohno, T., Roesner, F.: Secure multi-user content sharing for augmented reality applications. In: Usenix Security Symposium (2019)
45. Lampson, B.W.: Protection. ACM SIGOPS Oper. Syst. Rev. **8**(1), 18–24 (1974)
46. Sikder, A.K., Aksu, H., Uluagac, A.S.: 6thsense: a context-aware sensor-based attack detector for smart devices. In: Usenix Security Symposium (2017)
47. Zafar, Y., Har, D.: A novel countermeasure enhancing side channel immunity in FPGAs. In: 2008 International Conference on Advances in Electronics and Micro-Electronics (2008)
48. Giuseppe, P., Lisa, M.M., Ananthram, S., Trent, J.: Agility maneuvers to mitigate inference attacks on sensed location data. In: IEEE Conference Proceedings (2016)
49. Lebeck, K., et al.: Securing augmented reality output. In: 2017 IEEE Symposium on Security and Privacy (SP). IEEE (2017)
50. Ahn, S., Gorlatova, M., Naghizadeh, P., Chiang, M., Mittal, P.: Adaptive fog-based output security for augmented reality. In: ACM Special Interest Group on Data Communication (2018)
51. Wang, X., Yang, Y., Zeng, Y., Tang, C., Shi, J., Xu, K.: A novel hybrid mobile malware detection system integrating anomaly detection with misuse detection (2015)
52. Sundarkumar, G.G., Ravi, V., Nwogu, I., Govindaraju, V.: Malware detection via api calls, topic models and machine learning. In: Conference on Automation Science and Engineering (2015)
53. Almeida, V., Filgueiras, F., Doneda, D.: The ecosystem of digital content governance. IEEE Internet Comput. **25**(3), 13–17 (2021)
54. Woo, G., Lippman, A., Raskar, R.: Vrcodes: unobtrusive and active visual codes for interaction by exploiting rolling shutter. In: 2012 IEEE International Symposium on Mixed and Augmented Reality (ISMAR), pp. 59–64. IEEE (2012)
55. Lebeck, K., et al.: Towards security and privacy for multi-user augmented reality: foundations with end users. In: 2018 IEEE Symposium on Security and Privacy (SP). IEEE (2018)
56. Sun, M., Zheng, M., Lui, J.C.S., Jiang, X.: Design and implementation of an android host-based intrusion prevention system. In: Annual Computer Security Applications Conference (2014)
57. Wu, W.-C., Hung, S.-H.: Droiddolphin: a dynamic android malware detection framework using big data and machine learning. In: Research in Adaptive and Convergent Systems (2014)

Impact of Smartphones on Self-Rated Health of Rural Older Adults Using the PSM Method

Yue Li, Chengmeng Zhang, Chengye Huang, Haoyu Suo, Na Liu, Xinyue Hu, Yang Li, and Gong Chen[✉]

Institute of Population Research, Peking University, Beijing 100871, China
chengong@pku.edu.cn

Abstract. Smartphones equipped with the internet have made Smart Senior Care to be true, providing older adults with more efficient and targeted high-quality services. However, ageing rural populations tend to have low smartphone usage, limiting their ability to enjoy the convenience brought by technological advancements in the internet era. This study investigates the impact of smartphones on self-rated health (SRH) of older adults in rural areas. Using the rural elderly service survey data from Yueyang County, Hunan Province in 2022, the Logit model was used to explore the influencing factors of smartphone use among older adults in rural areas. Propensity Score Matching (PSM) was used to examine the potential effects of smartphones on older adults' SRH, while also dealing with confounding factors. The results show that age, education, marital status, annual income, and number of children significantly influenced smartphone use among older adults in rural areas. Furthermore, controlling for the same confounding variables, smartphone use had a positive effect on SRH of older adults in rural areas, improving SRH of 6.4% of older adults. Theoretically, this study may enrich and expand the data acquisition methods and theoretical framework of health influence effect studies. Practically, this study may provide a basis for improving SRH of older adults in rural China and promoting the implementation of Smart Senior Care.

Keywords: Self-rated health · Propensity score matching · Influencing factors · Smartphones

1 Introduction

China has entered an aging society, and the degree of aging continues to deepen. According to the data from the seventh national census, the population aged 60 and above in China reached 264 million, accounting for a high proportion of 18.70% of the total population. According to estimates, around 2035, the population of older adults aged 60 and above will exceed 400 million, accounting for more than 30% of the population entering a severe aging stage [1]. The aging of the population brings many problems, among which the health problems of older adults are the most prominent [2]. Older adults are a particular group in the population structure. The health status of older adults not only has a significant impact on family economics and care resources, but also has a significant impact on social development, social security, and medical resources [3].

H. Jin et al. (Eds.): IAIC 2023, CCIS 2060, pp. 306–319, 2024.
https://doi.org/10.1007/978-981-97-1332-5_24

With the development of Smart Senior Care and the continuous upgrading of smartphone functions, the use of smartphones has become a social trend. It has been widely popularized and applied in urban and young populations [4, 5]. However, in rural areas, the coverage and promotion efficiency of smartphones are relatively low, gradually becoming "internet refugees" in the era of mobile internet. At the same time, presenting various usage characteristics and influencing factors [6]. On the other hand, extending the life expectancy of older adults, pursuing quality of life in later years, and efforts to maintain good health have become the most urgent needs of older adults. Therefore, the demand for smartphones used by older adults not only reflects the profound impact of Smart Senior Care, but also has a particular impact on the health of older adults [7]. Therefore, this study explores the characteristics of smartphone use among rural older adults and their impact on the health in order to explore methods and paths to improve older adults' health. This is an important topic that needs to be discussed in the process of achieving healthy aging in China.

Self-rated health (SRH) of older adults refers to their perception and understanding of their physiological and psychological conditions [8]. It reflects subjective and objective health aspects and is a comprehensive indicator of older adults' health [9]. SRH is frequently used in surveys and is easy to measure, has become a widely used method for measuring health status internationally. Many scholars have studied the impact of smartphones on SRH of older adults. Regarding research methods, most studies have mainly used descriptive statistics, difference analysis, multiple regression, and variance analysis to analyze the relationship between smartphone use and older adults' SRH. However, these studies did not consider the endogeneity and selection bias issues caused by the heterogeneity among samples, which may lead to significant bias in the estimated propensity value. At the same time, older adults are a special group in the population structure, and their SRH should not be overlooked.

To alleviate the endogeneity and self-selection in the model, Rosenbaum and Rubin [10] first proposed the propensity score matching (PSM), which they defined as the effect of an intervention on an individual's outcome, given their predetermined attribute characteristics. Since then, PSM has been widely used in many research fields, including demographic studies, socioeconomics, and business management [11]. For example, Shahidi et al. [12] used the Canadian community health survey data from 2009 to 2014 to evaluate the effect of unemployment benefits on SRH among the unemployed using PSM. Okamoto et al. [13] extracted a sample of 1288 men aged 60 years and over from the Japanese national survey of older adults and used PSM to evaluate the effect of paid work beyond retirement age on health. Using the PSM method can reduce endogeneity and self-selection problems, and explore whether the results vary by age, education, marital status, and annual income, thereby verifying the robustness of the research conclusions and obtaining treatment effects with higher practical value.

In summary, research on the use of smartphones among older adults has mainly focused on the whole population. While urban and rural populations are considered older adults, the differences in their living environments, cultural backgrounds, and economic status make the smartphone use of rural older adults more unique. Although there is much theoretical research on the relationship between smartphones and SRH of older adults, there is a lack of research focused on smartphones among rural older adults and its impact on SRH, especially in terms of empirical research. Moreover, due to the differences in smartphone use among different groups, it is also worthwhile to explore whether there are differences in the health benefits. Therefore, it is crucial to understand the relationship between smartphones and SRH of rural older adults. In this study, we investigated the use of smartphones by older adults in rural areas and their SRH through questionnaires and interviews. Based on the data analysis of the samples, the study focused on the impact of smartphones on SRH of older adults, aiming to explore the characteristics and influencing factors of smartphone use among older adults in rural areas.

2 Methods

2.1 Data Sources and Analysis

Yueyang County, Hunan Province, is an urbanization level of 31.9%. According to the data from the seventh national census, Yueyang County has a resident population of 561888, with 21.68% of older adults aged 60 or above. Yueyang County was selected as the survey site for this paper. On the one hand, because one of the author grew up in Yueyang County, he is familiar with the living conditions of rural older adults and it is relatively easy to carry out the survey work. On the other hand, because the income level of rural residents in Yueyang County is roughly comparable to the per capita income level of rural residents in Hunan Province, it can reflect the situation of older adults in rural areas.

This study aims to explore the impact of smartphones on SRH of rural older adults, and relevant modules in the questionnaire were selected, including: personal information (A), living condition (B), illness (C), and basic activity ability (D). After data processing, the final valid sample aged 60 and above was 376. The variables and descriptive statistics are shown in Table 1.

Dependent variable. According to the question "Would you say your health is very good, good, fair, poor, or very poor?", the responses were measured by assigning a value from 1 to 5 respectively. Since the number of samples with "poor", "very poor", "good", and "very good" were small, "poor" and "very poor" were combined into "poor", and "good" and "very good" were combined into "good". Meanwhile, to make the results more understandable, these codes were reversed. The higher the response assignment, meaning that the better SRH of rural older adults.

Independent variable. The independent variable is whether older adults use smartphones, i.e., older adults have owned and continuously used smartphones in the past year. If using a smartphone, it is recorded as 1; otherwise, it is recorded as 0.

Table 1. Descriptive statistics of variables.

Variable	Criteria	Variable	Criteria
A1. Gender	1: Male	B1. Smoking	1: Never
	2: Female		2: Used to smoke, now quit
A2. Age	1: 60–74		3: Frequent
	2: 75–89		4: Occasional
	3: ≥ 90	B2. Drinking	1: Never
A3. Education	1: Illiterate		2: 1–2 times/week
	2: Primary school		3: ≥ 3 times/week
	3: Junior high school		4: Frequently intoxicated
	4: High school and above	B3. Exercise	1: Never
A4. Marital status	1: Married		2: Once a week
	2: Unmarried/widowed		3: 1–2 times/week
A5. Annual income	1: < 3000 yuan		4: 3–5 times/week
	2: 3000–4999 yuan		5: ≥ 6 times/week
	3: 5000–9999 yuan	B4. Quality of sleep	1: Very good
	4: 10000–19999 yuan		2: Good
	5: ≥ 20000 yuan		3: Fair
A6. Number of children	1: 0		4: Poor
	2: 1–3		5: Very poor
	3: >3	C1. Chronic Diseases	1: Yes
			2: No
D1. Toileting	1: No help	D4. Continence	1: No help
	2: Need some help		2: Need some help
	3: Need help		3: Need help
D2. Feeding	1: No help	D5. Walking	1: No help
	2: Need some help		2: Need some help
	3: Need help		3: Need help
D3. Dressing	1: No help	D6. Personal hygiene	1: No help
	2: Need some help		2: Need some help
	3: Need help		3: Need help

From the seventh census, the rural male population accounted for 51.90% and the female population accounted for 48.10%, with a gender ratio of 107.90 (female = 100). The age structure of rural older adults in the 60–74, 75–89, and 90 years old and above accounted for 73.65%, 24.61%, and 1.74%, respectively. From the sample data, the male population accounted for 51.86% and the female population accounted for 48.14%, with a gender ratio of 107.73 (female = 100). The age structure in the 60–74, 75–89, and 90 years old and above accounted for 58.24%, 38.56%, and 3.19%, respectively. The survey sample is similar to the gender and age distributions of rural older adults in China, which can reflect the situation of rural older adults to a certain extent. For education, most older adults have a low level of education, generally in primary and junior high school. For marital status, most older adults are married. For annual income, with annual income below 3000 yuan and 20000 yuan and above. For the number of children, rural older adults generally have 1–3 children.

The analysis of variance uses hypothesis testing to determine whether the influencing factors can explain the dependent variable. Since both smartphone use and older adults' SRH are categorical variables, chi-square test was used to analyze the differences of smartphones on SRH of rural older adults. The results are shown in Table 2.

Table 2. Chi-square test results of smartphones on SRH of older adults.

	SRH: 1-Poor	SRH: 2-Fair	SRH: 3-Good	Total	χ^2	P
Smartphone use: 1-Yes	50 (27.62%)	78 (43.09%)	53 (29.28%)	181	17.870	<0.001
Smartphone use: 0-No	85 (43.59%)	46 (23.59%)	64 (32.82%)	195		
Total	135 (35.90%)	124 (32.98%)	117 (31.12%)	376		

From Table 2, SRH of older adults in rural areas was poor, and the majority of older adults did not use smartphones. The proportions of rural older adults who used smartphones were 27.62%, 43.09%, and 29.28% for poor, fair, and good health. The proportions of rural older adults who did not use smartphones were 43.59%, 23.59%, and 32.82% for poor, fair, and good health. There was a statistically significant effect of smartphones on SRH of rural older adults ($\chi^2 = 17.870$, P = 0.000 < 0.001).

2.2 Modeling of the Impact of Smartphones on SRH Based on PSM

Since the impact of smartphones on SRH of rural older adults is not randomly determined, we need to consider other influencing factors. The regression equation on SRH is as follows:

$$H_i = \alpha + \beta use_i + \gamma X_i + \varepsilon_i \tag{1}$$

where the dependent variable H_i represents the ith older adult's SRH; i is the individual; the independent variable use_i represents whether the ith older adult uses a smartphone; X_i represents the ith control variable; α is a constant; β indicates the degree of influence of smartphones on SRH; γ indicates the degree of influence of control variable on SRH; ε_i is the error term.

The impact of smartphones on SRH of rural older adults is actually to compare the difference between SRH of rural older adults who use smartphones and who do not use smartphones. However, because of their different living environments and interpersonal relationships, different older adults will make different choices. The results analyzed by traditional linear regression are subject to sample selectivity bias [14]. To alleviate endogenous problems, PSM was used to evaluate the impact of smartphones on SRH of rural older adults. PSM does not need to assume the function form, parameter constraints and error distribution in advance, nor does it need to explain the exogenous variables, so it has obvious advantages in solving the endogenous problems of variables [15, 16]. The main steps are as follows:

Calculate Propensity Score. The Logit model was used to predict the probabilities of rural older adults using smartphones, so as to obtain the propensity score $P(X_i)$ of each individual, as shown in Eq. (2).

$$P(use_i|X_i) = \frac{\exp(\gamma X_i)}{1 + \exp(\gamma X_i)} \tag{2}$$

where $use_i = 1$ represents the ith individual uses smartphone; $use_i = 0$ represents the ith individual does not use smartphone.

Conduct PSM. The treatment group (older adults who use smartphones) and the control group (older adults who do not use smartphones) were tested for matching quality based on propensity values. If there is no significant deviation between the two groups, it is considered to meet the requirements of ignorability and overlap assumption.

Calculate Average Treatment Effect. In the matched samples, the differences between older adults who use and do not use smartphones were compared. The average treatment effect includes three categories: average treatment effects on treated (ATT), average treatment effects on untreated (ATU), and average treatment effect (ATE). In this paper, we studied the impact of smartphones on SRH of older adults and therefore ATT was used for analysis, expressed as in Eq. (3).

$$ATT = E(H_1|T = 1) - E(H_0|T = 1) = E(H_1 - H_0|T = 1)$$
$$= E\{E[H_{1i}|use_i = 1, \ P(X_i)] - E[H_{0i}|use_i = 0, \ P(X_i)]|use_i = 1\} \tag{3}$$

where H_1 represents SRH of older adults who use smartphones; H_0 represents SRH of older adults who do not use smartphones; $T = 1$ indicates the treatment group.

3   Results

3.1   Analysis of Influencing Factors of SRH

To estimate the average treatment effect of smartphones on SRH of rural older adults, the Logit model was used to estimate the propensity score, so as to find the suitable matching objects for rural older adults using smartphones in the control group. In this paper, the control variables were included in the regression model to analyze the influencing factors of older adults using smartphones. The estimated results are shown in Table 3.

Table 3. Regression estimation results based on the Logit model.

Variable	Coefficient	Standard error	Z value	P value	Variable	Coefficient	Standard error	Z value	P value
A1	0.465	0.340	1.37	0.171	B4	−0.087	0.103	−0.84	0.400
A2	−1.134	0.256	−4.42	0***	C1	0.223	0.297	0.75	0.452
A3	0.635	0.154	4.13	0***	D1	−16.295	3187.846	−0.01	0.996
A4	−0.689	0.310	−2.22	0.026**	D2	0.000	(omitted)		
A5	0.173	0.076	2.29	0.022**	D3	−35.466	2246.264	−0.02	0.987
A6	−0.502	0.276	−1.82	0.069*	D4	31.641	3483.946	0.01	0.993
B1	0.061	0.180	0.34	0.734	D5	48.765	2759.639	0.02	0.986
B2	−0.042	0.142	−0.30	0.768	D6	−34.790	2318.739	−0.02	0.988
B3	0.084	0.066	1.27	0.203					

Number of obs = 366, Log likelihood = −200.983, Pseudo R^2 = 0.208, LR chi^2 = 105.370

Note: * * *, * *, and * indicate that the estimation results are significant at the 1%, 5%, and 10% levels, respectively

From Table 3, age, education, marital status, annual income, and number of children are statistically significant. Pseudo R^2 = 0.208, it shows that all control variables have a strong explanatory on the use of smartphones by rural older adults. For the personal information, age has a significant negative impact on the statistical level of 1%, indicating that younger older adults are more likely to use smartphones. Education has a significant positive impact on the statistical level of 1%, indicating that the higher the education, the greater the probability of using smartphones. Marriage status has a significant negative impact on the statistical level of 5%, indicating that married older adults have a high probability of using smartphones. The annual income has a significant positive impact on the statistical level of 5%. The higher the income of older adults, the greater the probability of using smartphones. The number of children has a significant negative impact on the statistical level of 10%, indicating that older adults with fewer children have a greater probability of using smartphones. For the living condition, illness, and basic activity ability, the relationships between these variables and dependent variables did not pass the significance at all levels.

3.2 Impact of Smartphones on SRH of Rural Older Adults

Analysis of Common Support Domain and Matching Results. To ensure the matching quality of sample data, we analyzed the density function graph after obtaining the propensity score to test the matched common support domain. There are many PSM methods, and there is no absolutely good method that works in all cases. In this paper, nearest neighbor, caliper, and kernel matching methods were used to verify the robustness of the matching results [17]. The density function plots based on different matching methods are shown in Fig. 1, Fig. 2, Fig. 3.

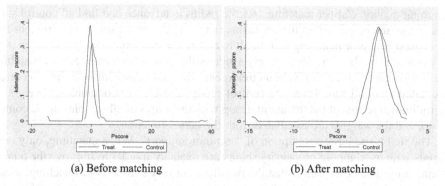

(a) Before matching   (b) After matching

Fig. 1. Density function plot based on the nearest neighbor matching method.

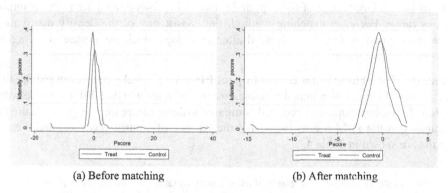

(a) Before matching   (b) After matching

Fig. 2. Density function plot based on the caliper matching method.

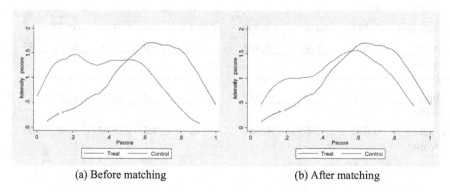

(a) Before matching   (b) After matching

Fig. 3. Density function plot based on the kernel matching method.

Nearest neighbor matching: based on the propensity score, we find the sample with the closest score to the treatment group in the control group and form a pair. All individuals can make full use of the treatment group information, but if there is a significant difference in the distribution of PS values between the two groups, it will affect the matching quality. Caliper matching: set the radius in advance and find all control samples in the unit circle within the set radius range. This method solves the problem that the nearest neighbor matching is difficult to ensure the matching quality, but the matching requirements become increasingly strict as the radius decreases. Kernel matching is a nonparametric method that pairs treatment group samples with an estimation effect calculated from all samples in the control group. The effect is obtained by weighting the individual scores of the treatment group with the scores of all samples in the control group, and the weights are calculated from the kernel function.

The first is the basic assumption of "common support". When matching, only individuals with overlapping propensity scores are retained in order to improve the quality of matching. The best result is that the two lines are very similar after matching, which means that the "common support" has a relatively large range. As can be seen from the results in Fig. 1, Fig. 2, Fig. 3, Fig. 4, the propensity scores of the treatment and control groups have a large range of overlap, and most of the observations are in the common values range. For each matching method, the consistency of the density distribution between the two groups was enhanced after matching, which can ensure that the data has good characteristics.

Analysis of Matching Balance and Impact Effects. The balance test can require that the mean differences of matched variables are insignificant and that differences between matched sample groups have reduced compared to those before matching. To ensure the reliability of the PSM results, we tested the balance of control variables, and the test results are shown in Table 4.

Table 4. Balance test results of explanatory variables before and after PSM.

Matching method	Pseudo R^2	LR chi^2	P > chi^2	Mean Bias	Med Bias	B	R	%Var
Unmatched	0.197	98.51	0.000	22.1	11.9	77.9*	0.33*	53
Nearest neighbor matching	0.015	6.89	0.808	4.9	3.0	28.6*	0.93	35
Caliper matching	0.011	5.28	0.917	4.1	2.3	25.0	1.51	35
Kernel matching	0.012	5.46	0.907	4.8	4.2	25.4*	1.03	35

From Table 4, after nearest neighbor, caliper, and kernel matching methods, there are no significant systematic differences in control variables between the treatment and control groups (P > 0.05). After sample matching, the standardized bias of explanatory variables was reduced from 53% to 35%, and the total bias was significantly reduced. Pseudo R^2 decreased from 0.197 to 0.011–0.015 after matching. The chi-square statistic (LR chi^2), the mean standard bias (Mean Bias), and the median standard bias (Med Bias) all decreased significantly. According to the analysis of the above results, it can be seen that PSM can reduce the differences in the distribution of variables between the treatment and control groups, as well as eliminate the estimation bias caused by sample self-selection.

Figure 4 shows the standardized bias of each control variable based on different matching methods. It can be found that each variable changes from a more dispersed state to a relatively concentrated state after matching. The standardized biases of most variables are significantly reduced (less than 10%) and the distribution is around 0, which further verifies that the matching results are stable.

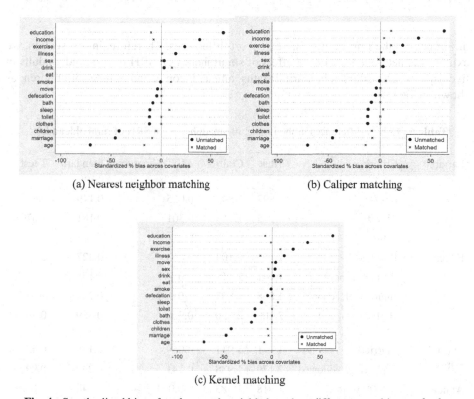

(a) Nearest neighbor matching   (b) Caliper matching

(c) Kernel matching

Fig. 4. Standardized bias of each control variable based on different matching methods.

We measured the average treatment effect of smartphones on SRH of rural older adults, and the estimated results are shown in Table 5. The results obtained by using three different matching methods are basically consistent, which indicates that the sample data has good robustness. The ATT_difference represents the difference in the dependent variable due to different values of the binary variable when the other matching variables are consistent. The average treatment effect was positive, indicating that the treatment group (using smartphones) improved SRH of rural elderly by 6.4% compared with the control group (not using smartphones).

Analysis of Cohort Differences. The previous analysis in this study can reflect the impact of smartphones on SRH of rural older adults, but it cannot reflect the structural

Table 5. Impact of smartphones on SRH of rural older adults.

Matching method	Treat	Control	ATT_difference	Standard error	T-test
Nearest neighbor matching	2.029	1.956	0.073	0.131	0.56
Caliper matching	2.029	1.955	0.074	0.105	0.71
Kernel matching	2.029	1.985	0.045	0.126	0.35
Average	2.029	1.965	0.064	0.121	0.540

differences in the impact effects, i.e., cohort differences. Therefore, in this section, we explore the heterogeneity of the effect of smartphones on SRH of rural older adults in five aspects: age, education, marital status, annual income, and number of children, as shown in Table 6.

Table 6. Cohort differences in the impact of smartphones on SRH of rural older adults.

Variable	Criteria	Treat	Control	ATT_difference	Standard error	T-test
Age	60–74	1.897	1.843	0.055	0.149	0.370
	75–89	2.265	1.963	0.301	0.181	1.670*
	≥ 90	\	\	\	\	\
Education	Illiterate	2.167	2.104	0.063	0.327	0.19
	Primary school	1.941	2.118	−0.177	0.158	−1.12
	Junior high school	1.868	1.887	−0.019	0.257	−0.07
	High school and above	2.400	2.050	0.350	0.389	0.90
Marital status	Married	1.971	1.911	0.059	0.138	0.43
	Unmarried/widowed	2.179	1.938	0.241	0.242	0.99
Annual income	<3000 yuan	1.918	1.983	−0.064	0.248	−0.26
	3000–4999 yuan	2.091	1.909	0.182	0.366	0.50
	5000–9999 yuan	2.000	2.000	0.000	0.457	0.00
	10000–19999 yuan	2.040	1.780	0.260	0.274	0.95
	≥ 20000 yuan	2.138	2.017	0.121	0.276	0.44
Number of children	0	\	\	\	\	\
	1–3	1.953	1.922	0.031	0.142	0.22
	>3	1.889	2.208	−0.319	0.253	−1.26

Ageing not only has a significant impact on smartphone use but also indirectly affects SRH of rural older adults after they begin using smartphones. The use of smartphones by older adults aged between 75 and 89 years significantly improved their SRH by 30.1%. Therefore, the impact of smartphone use on this age group is even more pronounced after older adults have overcome the technological challenges and have realized the convenience of smartphones.

Smartphone use is also affected by the level of education, which in turn affects SRH of older adults. Older adults with high school or higher education have a greater impact on their SRH after using smartphones, with their health improving by 35%. Older adults with low education can obtain some basic functions from smartphones, while those with higher education can obtain more scientific knowledge and gain a deeper understanding of spirituality.

Regarding marital status, unmarried or widowed older adults are more likely to be affected by smartphones compared to married older adults, with their SRH improving by 24.1% after using smartphones. This is likely because unmarried or widowed older adults tend to experience loneliness more often, and the companionship and social functions of smartphones can compensate for the loneliness caused by the absence of a partner. This can ultimately improve their SRH.

About the annual income, older adults with higher incomes are more likely to benefit from smartphone use. Smartphones, as an important channel for accessing information, can improve older adults' awareness of health and greatly broaden the channels for higher-income older adults to learn about and purchase health-related items. Many low-income older adults face physical and social barriers in accessing healthy diets. In contrast, older adults with higher annual incomes have better socioeconomic status and relatively better health care coverage, and smartphones bring more prominent health benefits to this group of older adults. Therefore, the higher the annual income, the greater the positive impact of smartphone use on their SRH.

In terms of the number of children, the lower the number of children, the more significant the impact of smartphones on SRH of rural older adults. This is because the number of children may affect the mental health of older adults, and a decrease in the number of children implies a weakening of the family retirement function. In contrast, smartphones can effectively compensate for loneliness, help older adults gain perceived social support through entertainment and social functions, and enable older adults with fewer children to obtain timely health help, thereby improving their SRH.

4 Conclusions

Using the rural elderly service survey data from Yueyang County, Hunan Province in 2022, we explores the impact of smartphones on SRH of rural older adults in the context of Smart Senior Care by using Logit regression and PSM method. Through the regression analysis, it is found that the control variables including age, education, marital status, annual income, and the number of children affect older adults' willingness to use smartphones.

Based on the different matching methods, including the nearest neighbor, caliper, and kernel matching methods, the impact of smartphones on SRH of rural older adults was analyzed. The results show that smartphone use has a positive effect on SRH of rural elderly. Using the PSM method can reduce the distribution difference between the treatment and control groups, verify the robustness of the matching results and the validity of the sample data.

By analyzing the cohort differences in SRH of rural older adults, we found that the promotion effect of smartphones on SRH of rural older adults was mainly among those aged 75–89 years, high school and above, unmarried/widowed, with higher annual income and fewer children. For example, SRH of older adults with higher education is more positively affected by smartphone use. Compared with the married elderly, the unmarried/widowed older adults have significant improvement in SRH after using smartphones. With the steady growth of annual income, the use of smartphones will promote SRH of rural older adults. This study not only enriches the research scope in this field, but also provides the positive results and reliable basis for the development of Smart Senior Care.

SRH can be used as an important indicator to evaluate the health status of older adults. In the future work, the author will consider adding the next round of survey data to further explore the effects of different factors on SRH of older adults, so as to obtain more detailed and rigorous empirical conclusions.

Acknowledgments. This research was funded by the Strategic Research and Consulting Project of the Chinese Academy of Engineering, grant number 2022-XBZD-30.

References

1. Bai, C., Lei, X.: New trends in population aging and challenges for China's sustainable development. China Econ. J. **13**(1), 3–23 (2020)
2. Alanazi, H., Daim, T.: Health technology diffusion: case of remote patient monitoring (RPM) for the care of senior population. Technol. Soc. **66**, 1–11 (2021)
3. Ybarra, M., Suman, M.: Reasons, assessments and actions taken: sex and age differences in uses of Internet health information. Health Educ. Res. **23**(3), 512–521 (2008)
4. Farkiya, R., Tiwari, D., Modi, N.C.: A study on impact of internet usage on quality of life of senior citizens. Jaipuria Int. J. Manage. Res. **4**(2), 52–58 (2018)
5. Zhang, Y., Wu, B., Chen, P., Guo, Y.: The self-rated health status and key influencing factors in middle-aged and elderly: evidence from the CHARLS. Medicine **100**(46), e27772 (2021)
6. Zhang, Y.: Measuring and applying digital literacy: implications for access for the elderly in rural China. Educ. Inf. Technol., 1–20 (2022)
7. Ma, X., Zhang, X., Guo, X., Lai, K., Vogel, D.: Examining the role of ICT usage in loneliness perception and mental health of the elderly in China. Technol. Soc. **67**, 1–9 (2021)
8. Zadworna, M.: Pathways to healthy aging - exploring the determinants of self-rated health in older adults. Acta Physiol. (Oxf) **228**, 1–9 (2022)
9. Dermatis, Z., Lazakidou, A., Anastasiou, A., Liargovas, P.: Analyzing socio-economic and geographical factors that affect the health of the elderly. J. Knowl. Econ. **12**(4), 1925–1948 (2021)
10. Rosenbaum, P.R., Rubin, D.B.: The central role of the propensity score in observational studies for causal effects. Biometrika **70**(1), 41–55 (1983)
11. de Zwart, P.L., Bakx, P., van Doorslaer, E.K.A.: Will you still need me, will you still feed me when I'm 64? The health impact of caregiving to one's spouse. Health Econ. **26**, 127–138 (2017)

12. Shahidi, F.V., Muntaner, C., Shankardass, K., Quiñonez, C., Siddiqi, A.: The effect of unemployment benefits on health: a propensity score analysis. Soc. Sci. Med. **226**, 198–206 (2019)
13. Okamoto, S., Okamura, T., Komamura, K.: Employment and health after retirement in Japanese men. Bull. World Health Organ. **96**(12), 826–833 (2018)
14. Lam, L., Lam, M.: The use of information technology and mental health among older caregivers in Australia. Aging Ment. Health **13**(4), 557–562 (2009)
15. Kim, K., Choi, S., Lee, S.: The effect of a financial support on firm innovation collaboration and output: does policy work on the diverse nature of firm innovation? J. Knowl. Econ. **12**, 645–675 (2021)
16. Alemu, A., Ganewo, Z.: Impact analysis of formal microcredit on income of borrowers in rural areas of Sidama region, Ethiopia: a propensity score matching approach. J. Knowl. Econ., 1–21 (2022)
17. Stuart, E.A.: Matching methods for causal inference: a review and a look forward. Stat. Sci. **25**(1), 1–21 (2010)

DBSCAN Clustering and Kleinberg Algorithm-Based Research on Autism Hotspots and Frontier Trends in China

Zixu Hao[1], Siyu Sun[1], Rongwei Leng[2], and Xu Xu[1(✉)]

[1] School of Health Management, Changchun University of Chinese Medicine, Changchun 130117, China
yaxiai@163.com
[2] Modern Education Center, Changchun University of Chinese Medicine, Changchun 130117, China

Abstract. Objective The research used DBSCAN and Kleinberg algorithm to described the research hotspots of autism in China and explore the frontier trend of autism research. **Methods** Studies on ASD published from 2001 to 2023 from the CNKI. DBSCAN clustering algorithm and Kleinberg burst word detection algorithm were used to detect keyword clustering and mutant words, respectively, and the research hotspots, development context and frontier trends in this field were summarized. **Results** The number of publications in this field showed an overall upward trend. Low density of scholar and inter-institutional collaboration; The keyword clustering results could be summarized into three categories: the etiology and influencing factors of autism, autism diagnosis technology, autism intervention and rehabilitation training. **Conclusion** The research focuses on intervention in children with autism. The mechanism of the influence of social support on family function in autism is a frontier field in autism research.

Keywords: Autism · DBSCAN Clustering · Kleinberg Algorithm · Research hotspots · Frontier

1 Introduction

Autism spectrum disorder (ASD) refer to a group of early-onset, lifelong, heterogeneous neurodevelopmental conditions with complex mechanisms of emergence [1]. In the 40s of the 20th century, the American physician Kanner first described and reported the condition; In 1966, the British scholar LOTTER [2] first Population-based epidemiological investigation of ASD; Shi Huifeng et al. have done a meta-analysis on the prevalence of ASD in children, and the results show that the prevalence of ASD in children aged 0–6 in China increased significantly [3]. The study of Zhao Yanan et al. found that the rehabilitation intervention expenditure directly related to rehabilitation treatment in autistic families accounted for 85.56% of the total annual household income [4]. Therefore, various intervention methods for autism have emerged one after another, which have attracted widespread attention in the government, society and families. The research of

H. Jin et al. (Eds.): IAIC 2023, CCIS 2060, pp. 320–328, 2024.
https://doi.org/10.1007/978-981-97-1332-5_25

Liu Zhizhi and others found that the development of special groups including autism and academic research on special groups in China has entered a more in-depth stage [5].

This research to sort out the autism research literature included in the core database of Chinese through bibliometric methods and used DBSCAN algorithm and Kleinberg algorithm to detect keyword clustering and mutant words, analyze the research hotspots in this field, and explored the cutting-edge trends in this field.

2 Materials and Methods

2.1 Data Collection and Search Strategy

Taking the Chinese literature on the theme of "autism" or "autism" retrieved by CNKI as the object, the source categories of the literature were locked as "Peking University Core" and "CSSCI", the search period was set to 2000–2023, duplicate documents, information reports and other documents were eliminated, and finally 1804 valid documents were obtained and the authors, institutions and keywords were econometrically analyzed through COOC software, and visual maps were drawn to analyze the hot spots and frontier trends of autism research in China.

2.2 Salton Exponential Method

The Salton index method can encapsulate the keyword co-occurrence matrix to obtain a correlation matrix that represents the similarity of the two keywords, indicating the closeness of the relationship between the two keywords.

The calculation formula is:

$$S = N_{ij}/(N_i * N_J) \tag{1}$$

N_i and N_j indicate the frequency of the keywords i and j, respectively, and N_{ij} represents the frequency of occurrence of the keywords i and j.

Due to the excessive value of 0 in the correlation matrix, it is easy to produce large statistical errors, in order to avoid this situation, this research uses "1-S" to obtain a dissimilarity matrix representing the degree of difference between the two keywords.

2.3 DBSCAN Algorithm

DBSCAN (Density-Based Spatial Clustering) is "density-based clustering", and this research applies this algorithm for keyword clustering analysis. The DBSCAN algorithm is based on the assumption that the cluster structure can be determined by the compactness of the sample distribution. It can not only find clusters with any shape, but can also be used to detect outliers [6]. The DBSCAN algorithm defines a density function through distance, calculates the density near each sample, and finds those areas where the samples are relatively concentrated according to the density value near each sample, which are the clusters we are looking for. The basic idea is that data points that are tightly distributed should be classified as a class, while data points with no or very few points around them may be outliers.

The formula is:

$$D(x) \sum_{i=1} w_i K \left[\frac{\|X - X_i\|}{\overline{d}h} \right] [7] \tag{2}$$

$\overline{d} = \frac{2}{n(n-1)} \sum_{i<j} \|x_i - x_j\|$, $\overline{d}$ is the average distance between two points.

2.4 Kleinberg Burst Word Detection Algorithm

This research is based on the burst word detection algorithm proposed by Kleinberg [8]. Kleinberg's algorithm definition [9]: there are "n" groups of data, and the data of group "t" has a total of "dt" articles, of which "rt" articles contain sudden words; Let R = $\sum_{t=1}^{n} r_t$, R is the number of all papers related to a topic in "n" batches of data; D = $\sum_{t=1}^{n} dt$, D is the number of all papers in the field in the "n" batch data. Define the burst weight index to show how prominent keywords are [10].

The formula is:

$$weigh = \sum_{i=t_1}^{t_2} (\sigma(0, r_t, d_t) - \sigma(1, r_t, d_t)) \tag{3}$$

3   Outcome

3.1   Trends in Literature Output

Changes in the number of publications can reflect the development of a research field. Since the 21st century, although the output of autism research literature in China has fluctuated, the overall trend is still on the rise (see Fig. 1), and the output of literature has increased significantly in the past 10 years, transitioning from the embryonic stage to the growth period.

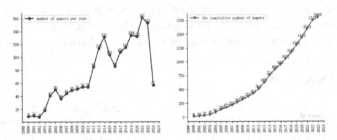

Fig. 1. Publication trend of domestic autism intervention research since the 21st century

3.2 Collaboration Between Authors and Institutions

Collaboration is an important means of academic research. The research was statistically aggregated 3976 authors and 1961 publishing institutions. Liu Jing of Peking University Sixth Hospital, published the most articles, with 40 articles. The School of Public Health of Har-bin Medical University published the most articles, with 33 articles. According to Price's law, there are 185 core authors in this field, with a total of 1581 articles, accounting for 87.6% of the total literature, more than 50% of all literature, so the field has formed a core author group. Authors (see Fig. 2) and institutions (see Fig. 3) with a frequency greater than 10 were extracted to draw co-occurrence maps, although authors and publishing institutions in this field had some cooperation, the scale and closeness of cooperation were still at a low level.

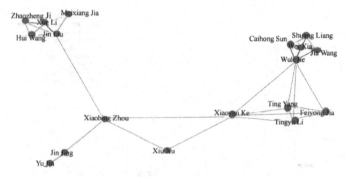

Fig. 2. Author co-occurrence diagram

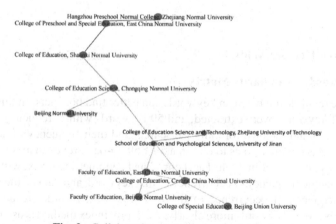

Fig. 3. Collaboration between issuing agencies

3.3   Research and Development Context

Draw the keyword attention evolution map (see Fig. 4), retaining the top ten high-frequency keywords in word, "autism" as the theme words of this search, too high word frequency affected the effect of the map so it was eliminated. According to the analysis of the effect of the figure, it is found that in the early stage of the study (2002–2009), the attention of "psychological theory" was obviously high, and the research focus at this stage was more concentrated, and the research on the psychological theory ability of children with autism was emphasized. The number of keywords in the late stage of the study (after 2010) increased, and a large number of scholars began to conduct relevant research from different angles, and the attention at this stage was relatively average, and no high attention was seen for a keyword.

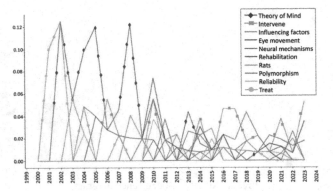

Fig. 4. Keyword attention evolution map year by year

3.4   Research Hotspot Analysis

3.4.1   Keyword Dissimilarity Matrix

Studying the relationship between keywords can better find hot spots. In this research, a total of 3311 keywords were extracted, and 59 keywords with a frequency greater than 10 were merged and retained for pairwise search, and their frequency of occurrence in the literature was counted, so as to obtain a 59 * 59 co-occurrence matrix (see Table 1). The co-occurrence matrix in Table 1 is a correlated, symmetric matrix, with data on the diagonal being the frequency of occurrence of that keyword, and data on the nondiagonal being the frequency of occurrence of the two keywords. In order to understand the relationship between keywords more clearly, the Salton index method is used to construct a keyword dissimilarity matrix (see Table 2). The larger the data values in Table 2, the farther apart the two keywords, the more distant the connection, and the worse the similarity.

3.4.2 Keyword Clustering Map

The DBSCAN algorithm was applied for clustering, and the clustering results were shown as 10 classes (see Table 3). Keywords can be regarded as a concentrated embodiment of the research focus of a certain field, and keywords with high frequency represent research hotspots in the knowledge network system. Using VOS viewer software to draw a keyword clustering map (see Fig. 5), combined with the keyword dissimilarity matrix, the clustering results are sorted out, and the research hotspots are summarized as follows:

First, the intervention of autistic patients: At present, the cause of autism spectrum disorder is unknown, and the intervention of autism spectrum disorder is a world problem. The core keywords of clusters C1, C3 and C9 were "intervention", which intervened in the stereotyped behavior, sleep and social disorders of autistic patients through common attention, music therapy, acupuncture, integrated education and other means.

Second, autism diagnosis and treatment technology: early screening of autism has become a consensus, and early detection, early diagnosis and early treatment are essential for the alleviation of autism disorder symptoms. The main keywords contained in clusters C4, C7 and C8 are diagnosis, treatment, efficacy, prevalence and other frequent and closely related, and the analysis shows that the early diagnosis and treatment of children with autism is one of the hot spots in this field.

Third, the etiology and influencing factors of autism: autism is a complex heterogeneous disease, which is the result of the combination of genetic factors and environmental factors, but the pathogenesis is still unclear. The main keywords contained in clusters C2, C5, C6 and C10 are risk factors, influencing factors, genes, polymorphisms, etc., reflecting that scholars have carried out research on the etiology and influencing factors of autism from the perspectives of genetics and neurobiology.

Table 1. High-frequency keyword co-occurrence matrix (partial)

High-frequency	Autism	Children autism	Children	Autism disorder	Intervention	Diagnosis
Autism	1168					
Children autism	9	292				
Children	171	0	268			
Autism disorder	1	1	83	108		
Intervention	72	11	7	2	97	
Diagnosis	22	4	7	5	3	32

Table 2. High-frequency keyword dissimilarity matrix (partial)

High-frequency	Autism	Children autism	Children	Autism disorder	Intervention	Diagnosis
Autism	0					
children Autism	1.0000	0				
Children	0.9997	1.0000	0			
Autism disorder	1.0000	1.0000	0.9986	0		
Intervention	0.9997	0.9998	0.9999	0.9999	0	
Diagnosis	0.9997	0.9998	0.9996	0.9993	0.9995	0

3.5 Frontier Trend Analysis

The Kleinberg burst word detection algorithm was used for mutation word detection to retain the top 20 mutant words in burst weight (see Fig. 6). Before 2013, the main keywords that emerged are: theory of mind, executive function, polymorphism, and rehabilitation. From 2013 to 2019, the main keywords that emerged were: eye movement, intervention, case-control studies, and influencing factors. After 2020, there are 3 keywords that emerged: music therapy, social support, and MRI. Music therapy and social sup-port are both autism intervention methods, so it is speculated that autism intervention will still be a research hotspot in this field in the future, and the influence mechanism of social support on autistic family function will become the frontier of autism research.

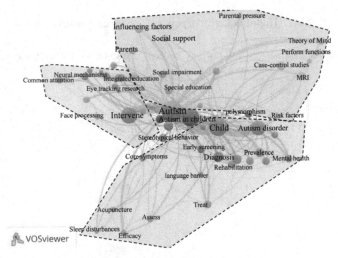

Fig. 5. Keyword clustering map

Table 3. Keyword clustering results

Cluster labels	Core keywords
C1	Intervene; Common attention; eye tracking; Neural mechanisms; etc
C2	Autism in children; Polymorphism; Gene; Language barrier; etc
C3	stereotypical behavior; symptoms; music therapy; Sleep disturbances; etc
C4	Child; Regression analysis; prevalence; Behavior; Mental health
C5	parenting pressure; influencing factors; Parents; Social support; etc
C6	executive functions; case-control studies; High-functioning autism; etc
C7	Treat; Efficacy; assess
C8	Infants; early screening; diagnosis
C9	Rehabilitation; social impairment; Review
C10	Meta-analysis; Risk factors

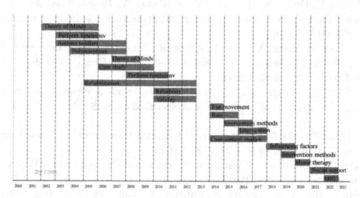

Fig. 6. Keyword mutation map

4 Conclusion

This research analyzes the research hotspots and frontier trends in this field by extracting the authors, institutions and keywords of domestic autism research literature since the 21st century, draws the following conclusions:

First, the research on autism in China is on the rise, and a core author group has been formed in the field, but the scale and closeness of cooperation between authors and publishing institutions are still at a low level. The research hotspots can be summarized into three categories: intervention of autistic patients, etiology and influencing factors of autism, and diagnosis and treatment techniques of autism.

Second, research on early intervention in children with autism is the focus of research in this area. Studies have shown that high-intensity behavioral intervention training in the early science of ASD can significantly improve symptoms. The high-attention keywords have undergone a shift from "treatment" to "intervention", indicating that scholars pay more attention to alleviating the symptoms of autism in children through various interventions, rather than simple medical treatment.

Third, the influence mechanism of social support on the quality of life and family function of children with autism will become the frontier of autism research. The social support and family function of autistic families are lower than those of healthy families, and favorable subjective social support can promote the positive play of family functions, which also suggests that scholars should consider parents and family conditions in follow-up research in addition to paying attention to autistic patients themselves.

References

1. Jiang, M., et al.: Autism spectrum disorder research: knowledge mapping of progress and focus between 2011 and 2022. Front. Psychiatry **25**(14), 1096769 (2023)
2. Lotter, V.: Epidemiology of autistic conditions in young children. Soc. Psychiatry **1**(3), 124–135 (1966)
3. Zou, Z., Liu, Y., Huang, H., Liu, C., Cao, X., Zhang, Y.: Research and strategies of family intervention on the prevalence of autism spectrum disorder in children. Chin. J. Gen. Pract. **23**(08), 900–907 (2020)
4. Zhao, Y., Luo, Y., Wang, X., Fan, H., Zhang, R., Zheng, X.: Study on the direct economic burden of families of autistic children aged 2–6 years in China. Chin. J. Disease Control Control **25**(09), 1085–1090 (2021)
5. Liu, Z., Jiang, K., Wang, Z.: Research hotspots and trends of mental health in special populations in the past decade——Visual analysis based on WoS. Med. Philos. **43**(07), 46–51 (2022)
6. Lu, C., Deng, S., Ma, W., et al.: Cluster federated learning method based on DBSCAN clustering. Comput. Sci. **49**(S1), 232–237 (2022)
7. Fu, J., Ding, J.: Comparison of visualization principle of Citespace and VOSviewer software. Libr. Inf. Technol. Agric. **31**(10), 31–37 (2019)
8. Kleinberg, J.: Bursty and hierarchical structure in streams. In: Proceedings of the 8th ACM SIGKDD International Conference on Knowledge Discovery and Data Mining, pp. 91–101. ACM Press, New York (2002)
9. Zhou, S.: Research on Chuci literature burst information monitoring based on Kleinberg algorithm. Comput. Knowl. Technol. **11**(04), 86–89 (2015)
10. Xu, S., Xu, D., Han, S., Yang, Y.: The application of SemRep and burst monitoring algorithm in bibliometric analysis – taking the development trend of drug therapy for diseases as an example. J. Inf. Sci. **40**(07), 745–755 (2021)

FSANet: A Lightweight Network for Tobacco Grouping Using Multi-scale Convolution and Attention Mechanism

Yongzhou Su[1], Kaihu Hou[1(✉)], Jie Long[2], Xiaolei Gai[3], Yiwu Zhang[2], and Xiaowei Zhang[2]

[1] College of Mechanical and Electrical Engineering, Kunming University of Science and Technology, Kunming 650504, China
576521346@qq.com, 2317509093@qq.com

[2] Yunnan Tobacco Quality Inspection and Supervision Station, Kunming 650106, China

[3] Yunnan Tobacco Leaf Co., Ltd., Kunming 650217, China

Abstract. To address the issues of slow classification speed, group confusion, and difficult feature extraction in the flue-tobacco grouping task using single-size convolutional kernels in traditional convolutional neural networks, we propose a lightweight network model called FSANet. This model combines a multi-scale convolutional structure and attention mechanism. Firstly, we design a multi-scale feature fusion module (FSAConv) to extract and fuse multi-scale features from flue-treated tobacco images. The module utilizes feature partitioning strategies to assign feature subsets for parallel convolution and pooling operations simultaneously, enhancing the network's sensitivity to different semantic targets. Additionally, the subset features are adaptively calibrated through the SE (squeeze and excitation) channel attention mechanism. Furthermore, FSAConv introduces the Ghost convolution module and HardSwish activation function to maintain model expressiveness while reducing parameter count. Finally, based on the inverted residual of MobileNet, we construct a lightweight network called FSANet by replacing 3 × 3 convolutions with FSAConv modules. Using our self-constructed flue-cured tobacco dataset, experimental results demonstrate that compared with GHostNet and MobileNetV3 models, FSANet exhibits an increase of 13.8% and 20% in parameter count respectively while achieving accuracy improvements of 5% and 10%. The constructed FSANet model enhances classification accuracy for flue-cured tobacco with superior overall performance suitable for actual production scenarios with limited storage resources and low hardware capabilities.

Keywords: Flue-tobacco grouping · Lightweight network · FSANet · Multi-scale features · Attention Mechanism

1 Introduction

Flue-cured tobacco is the most important raw material for cigarette production, and quality grouping of flue-cured tobacco is crucial for quality control in the cigarette production process [1]. The Chinese flue-cured tobacco standard divides the quality

H. Jin et al. (Eds.): IAIC 2023, CCIS 2060, pp. 329–343, 2024.
https://doi.org/10.1007/978-981-97-1332-5_26

of flue-cured tobacco leaves into positive and negative groups from the perspective of color. The color of the regular group of flue-cured tobacco is uniform, including orange, lemon yellow, reddish brown groups, and mature leaves. The characteristic of sub group flue-cured tobacco is uneven color, including variegated, smooth leaves, slight green and green yellow, etc. The variegated group is further subdivided based on the proportion of variegated areas. By analyzing and judging the color of different groups of flue-cured tobacco, the maturity and quality level of tobacco leaves can be accurately evaluated [2], thereby implementing standardized quality control in the cigarette production process.

Currently, the grouping of tobacco leaves mainly relies on manual operations [3]. However, due to factors such as fatigue and subjective preferences, manual grouping often suffers from errors and low efficiency. In recent years, computer vision-based automatic classification methods for tobacco leaves have been widely researched and applied. Computer vision algorithms can automatically analyze visual features such as color and shape in tobacco leaf images, construct feature vectors as inputs to classification models, and achieve automatic recognition and classification of tobacco leaves [4]. Deep learning algorithms have been extensively used in various tobacco classification tasks including disease identification [5], leaf drying [6], leaf grading [7], etc., during the tobacco production process. However, in most studies on quality classification of tobacco leaves, auxiliary group leaves are manually removed first before performing automatic classification on main group leaves. This approach ignores the existence of auxiliary group leaves and fails to achieve automated classification for all samples of tobacco leaves. To achieve comprehensive automated grading of tobacco leaves, it is necessary to develop a novel algorithm that can simultaneously identify and classify both main group and auxiliary group tobaccos; this is also one of the current research challenges in this field.

In recent years, scholars have conducted research on the classification of tobacco leaves and made some progress in the field of deep learning [8]. The key to achieving good results in deep learning lies in effectively representing features [9], which requires models to understand and represent the diversity and complexity of tobacco leaf characteristics. To this end, researchers have explored various approaches for extracting detailed features of tobacco leaves. For example, Lu [10] et al. recognized the importance of local features of tobacco leaves and proposed a network called A-ResNet-65 for extracting global features of tobacco leaves, while using the ResNet34 network to extract local features. In addition, Chen [11] et al. combined MobileNetV2 with Swin Transformer to extract tobacco leaf features from different levels. In terms of exploring fine-grained information about tobacco leaves, some researchers have explored improving image resolution. For instance, Xin [12] et al. constructed a high-resolution dataset of tobacco leaves and proposed an efficient grading method based on deep dense convolutional networks. Furthermore, Lu [13] et al. modified the B-CNN framework by designing a dual-input image model and dual CNN feature extractor for tobacco leaf image classification; this model takes undistorted global images as well as high-definition local images into account to fully utilize both coarse-grained and fine-grained information carried by original leaf images. These methods employ dual-input network structures to extract features at different scales, enabling more comprehensive learning and representation of fine-grained characteristics of tobacco leaves and enhancing automatic classification accuracy for cured tobaccos. However, these methods often increase computational complexity and are difficult to meet real-time grouping requirements in practical production processes [14]."

To address the issues of large model parameters and long training times in deep learning models, researchers have begun exploring the application of lightweight models in tobacco grouping tasks. For example, Wang [15] et al. introduced a channel attention mechanism SE module based on ShuffleNetV2 to improve the representation ability of network models, and fully integrated the exposed features and global information of tobacco by embedding pyramid pooling modules PPM. However, this model only performed binary classification and did not meet the detailed requirements for tobacco leaf grading. In order to achieve more refined tobacco leaf grading, He [16] et al. conducted a more detailed classification of tobacco categories and improved the model's expression ability for high-scale key features through channel weighting and dynamic loss adjustment methods. In addition, Xu et al. [17] used separable extended convolution to increase the receptive field of convolutional kernels, while maintaining lightweight and improving tobacco classification performance. The goal of these studies is to reduce model parameter volume, shorten training time, and improve classification performance in tobacco grouping tasks by adopting lightweight models. These methods have partially addressed the issue of large computational complexity in deep learning models, making them more suitable for deployment in resource-constrained environments.

In these studies, how to further improve the accuracy and robustness of tobacco leaf grading remains a challenge. To address this issue, this paper proposes a lightweight network model called FSANet, which combines multi-scale convolutional structures and attention mechanisms to enhance the performance of tobacco leaf grading. The main contributions of this paper are as follows: Firstly, to tackle the problems of irregular distribution and inconsistent feature sizes on tobacco leaves' surface, we design a multi-scale feature fusion module (FSAConv) for extracting and fusing multi-scale features from tobacco leaf images. This module adopts a feature partitioning strategy that assigns subsets of features to multiple parallel convolutional and pooling branches in order to process and capture multi-scale features simultaneously, enhancing the sensitivity of the network towards different semantic targets. Secondly, due to the regional nature of tobacco leaf features and their high similarity with neighboring groups of leaves, we introduce an SE (squeeze-and-excitation) channel attention mechanism module [18] that adaptively recalibrates target features while suppressing irrelevant information interference. Additionally, the FSAConv module also incorporates Ghost convolution modules [19] and HardSwish [20] activation functions to maintain model expressiveness while reducing parameter count. Finally, based on MobileNet inverted residual blocks, we replace 3×3 convolutions with FSAConv modules and reduce the depth of the network structure to construct FSANet—a lightweight network that supports fast fine grouping of tobacco leaves.

2 Experimental Method

2.1 Ghostconv

The core idea of the Ghost module is to generate more feature maps [21]. Each Ghost feature map is obtained by applying a certain non-linear transformation (such as convolution, ReLU, etc.) to one or more real feature maps. In this way, the Ghost module can generate more feature maps using fewer computational resources. The Ghost convolution can be divided into three steps: in the first step, a small number of ordinary convolution kernels are used to extract feature information from the input feature map and generate partial feature maps Y'; in the second step, a cheaper linear operation $\phi_i()$ is used to perform linear operations on the feature maps generated in the first step to produce ghost feature maps; in the third step, the final feature map is generated by concatenating the feature map Y' with the ghost feature map. Figure 1 illustrates the implementation process of Ghost convolution.

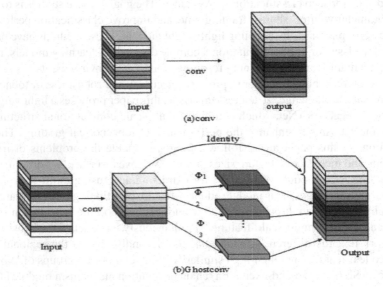

Fig. 1. Ghost convolution

Given an input image $X \in \mathbb{R}^{c \times h \times w}$ (with c channels, h height, and w width), and a total of n convolutional kernels. First, we perform convolution operation on the image using a subset of the kernels, resulting in partial feature maps.

$$y' \in \mathbb{R}^{h' \times w' \times n} = X * f + a \qquad (1)$$

In the equation, f represents the convolution operation, and a represents the bias term. By performing a linear transformation on $y' \in \mathbb{R}^{h' \times w' \times n}$, we obtain its remaining feature maps.

$$y_{ij} = \phi_{i,j}(y'_i), \quad \forall i = 1, \ldots, m, j = 1, \ldots, s \qquad (2)$$

Among them, m represents the number of channels in Y', y' is the i-th channel, and j represents the j-th linear transformation performed by y'.

In the case of having the same number of channels, Ghost convolution significantly reduces the computational cost compared to standard convolution. Taking $c \times h \times w$ as the input feature map, $k \times k$ as the convolution kernel, and $d \times d$ as the linear operation kernel, let $h' \times w' \times n$ be the output feature map. The computational cost of standard convolution is denoted as $C = n \times h' \times w' \times c \times k \times k$. When using Ghost convolution with an output channel number $n = m \times s$, y' undergoes $s - 1$ linear transformations. Therefore, its computational cost can be expressed as

$$G = \frac{n}{s} \times h' \times w' \times c \times k \times k + (s - 1) \times \frac{n}{s} \times h' \times w' \times d \times d \tag{3}$$

The compression ratio of computational complexity is:

$$R = \frac{C}{G} = \frac{sc}{s + c - 1} \approx s \tag{4}$$

From Eqs. (3) and (4), it can be observed that compared to conventional convolutions, Ghost convolution offers an s-fold improvement in both computational speed and parameter reduction.

2.2 FSAConv

The features of tobacco leaf images exhibit a multi-scale distribution in space, which requires the feature extraction model to adapt and effectively extract these features. Fixed receptive field sizes of convolutional kernels are not flexible enough to handle situations with targets of different scales, as small targets may be filtered out or cannot be recognized effectively [22]. MobileNetv2 is a lightweight CNN model, but it struggles to capture multi-scale features in tobacco leaf images when using single-size 3×3 convolutional kernels for grouping tasks [23]. To address this issue, this paper proposes a multi-scale convolution module called FSAConv. This module replaces the standard 3×3 convolutional kernel in MobileNetV2 and integrates SE attention mechanism, multi-scale convolutions, feature segmentation, and different pooling methods. The structure of FSAConv is shown in Fig. 2.

In the FSAConv module, a feature map is first divided into multiple subsets using subspace partitioning. Each subset can extract features within different receptive fields, thereby capturing multi-scale information in tobacco leaf images [24]. The input feature tensor is split into three sub-feature tensors, each with one-third of the original channels. They interact and extract features among channels within the feature subspace.

In the FSAConv module, different scales of convolution kernels (including 3×3, 5×5, and 7×7) are introduced and various pooling operations are applied to different scale feature maps. For each subspace, different convolution branches are iteratively traversed. This allows each subset to extract features within different receptive fields, capturing multi-scale information in tobacco leaf images. Not only can features be extracted within different receptive fields, but also the spatial size of feature maps can be reduced while preserving important information. Then, the feature maps processed through convolution

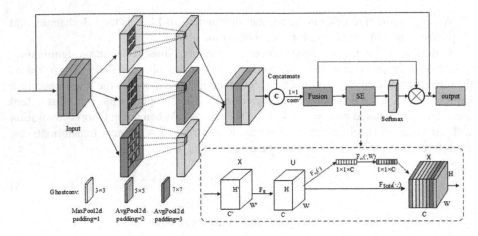

Fig. 2. FSAConv structure

and pooling operations at different scales are concatenated along the channel dimension to obtain a comprehensive feature map containing multi-scale information. However, as the kernel size increases, there is a significant increase in parameter count. To handle input tensors with different kernel sizes without significantly increasing computational complexity, Ghost convolutions are introduced. Finally, the obtained features Fi from different branches are concatenated, and the concatenated feature tensor represents the comprehensive information of all subspace pooled features.

$$F = Cat([F1, F2, F3]) \tag{5}$$

Secondly, the comprehensive feature maps are processed using the SE attention mechanism. Firstly, the features F are passed into the SE module and undergo a compression of channel dimension (sq) through global pooling, followed by an excitation operation (ex) consisting of two fully connected layers. The weighted (Scale) operation is then applied to multiply the weights with the input feature map F, resulting in an output feature map Z that contains information with varying degrees of importance. The SE attention mechanism can be represented as follows:

$$F_{sq}(u_c) = \frac{1}{H \times W} \sum_{i=1}^{H} \sum_{j=1}^{W} u_c(i,j) \tag{6}$$

$$F_{ex}(s, W_0) = \theta(g(s, W_0)) = \theta(W_2 \psi(W_1 s)) \tag{7}$$

$$F_{scale} = (u_c, e_c) = e_c u_c \tag{8}$$

$$Zi = SEweight(Fi), i = 1, 2, 3 \tag{9}$$

Next, by using Softmax to recalibrate the channel attention vectors, we obtain recalibrated multi-scale channel weights. Finally, these recalibrated weights are applied element-wise multiplication with their corresponding feature maps to generate a refined feature map with richer multi-scale feature information as the output.

$$\text{att}i = Soft\max(Zi) = \frac{\exp(Zi)}{\sum_{i=0}^{S-1} \exp(Zi)}, S = 3 \tag{10}$$

$$\text{att} = \text{att0} \oplus \text{att1} \oplus \text{att2} \tag{11}$$

Among them, att represents the multi-scale channel weights after attention interaction. Then, the recalibrated multi-scale channel attention weights att_i are multiplied with their corresponding feature maps Fi, and summed up to obtain the multi-scale output Y.

$$Y = \sum_{i=0}^{S-1} Fi \cdot \text{atti} \tag{12}$$

Finally, residual connection is introduced to add the input feature tensor X to the concatenated feature tensor Y. If the shape of the input feature tensor does not match the shape of the concatenated feature tensor, interpolation is performed on X to adjust it to have the same spatial size as Y. By residual linking, the information of the original features is preserved and the feature representation is enriched. This helps improve the representation and learning abilities of the model. With this design, FSAConv module achieves efficient and effective multi-scale feature extraction, resulting in better performance and robustness when processing tobacco leaf images.

2.3 FSANet

As shown in Fig. 3, a new module called FSABlock is obtained by replacing the 3×3 convolution with an FSAConv module at the corresponding position in the bottleneck block of MobileNet. The HardSwish activation function is adopted instead of the traditional RELU activation function. Compared to RELU, HardSwish provides stronger non-linear modeling capability while requiring relatively less computation, which enables the model to maintain high performance and higher computational efficiency. By using the HardSwish activation function, FSABlock can better capture the non-linear features of input data and enhance the model's expressive power. Therefore, FSANet can adaptively readjust weights across channel dimensions, playing a crucial role in improving model representation ability.

Through the stacked FSABlock module, a novel and efficient backbone network called FSANet is proposed. Table 1 presents the network design of FSANet. In this table, it can be observed how FSANet constructs the network by stacking different FSABlock modules. These modules have different parameter settings, such as channel expansion factor t and stride, to adapt to feature extraction tasks of varying complexities.

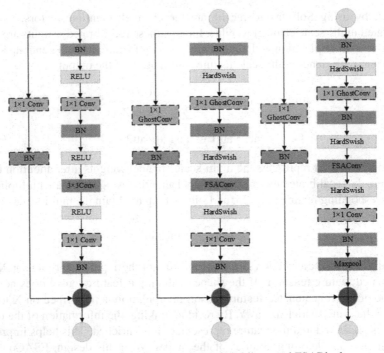

Fig. 3. Illustration and comparison of Mobilent, and FSABlocks

Table 1. Network design of FSANet

Layer	Input_size	Input channels	Output channels	t	stride
Conv2d	224 × 224	3	36	–	2
FSABlock	112 × 112	36	24	1	1
FSABlock	112 × 112	24	48	2	2
FSABlock	56 × 56	48	64	2	2
FSABlock	28 × 28	64	96	3	2
FSABlock	14 × 14	96	128	32	2
Conv2d	7 × 7	128	480	–	–

3 Analysis of Experimental Grouping for Tobacco Leaves

3.1 Instruments and Equipment

The tobacco leaf image collection work is carried out at the Yunnan Tobacco Quality Supervision and Testing Station, with tobacco leaves sourced from multiple tobacco purchasing stations in Yunnan Province. In order to obtain high-quality tobacco leaf images, this experiment utilizes a self-developed tobacco leaf grading platform for image acquisition. This platform was built in collaboration between our research team and the

Yunnan Tobacco Quality Inspection Station. The image collection takes place inside a camera dark box to eliminate interference from ambient light sources. The entire system includes an industrial CMOS camera, fixed-focus lens, LED light source, conveyor belt device, PLC controller, etc. Figure 4 shows the setup of the acquisition equipment.

This article uses the MV-CA050-11UC 5-megapixel CMOS industrial camera to capture images. The CCD chip size is 6.6 mm × 8.8 mm, with a pixel size of 3.45 μm and a diagonal length of 2/3 in. The obtained tobacco leaf image has a maximum resolution of 2448 × 2048. The imaging system uses the M0824-MPW2 fixed-focus lens with a focal length f of 8 mm and an aperture of 2.6. Figure 5 illustrates the imaging principle of this image acquisition system.

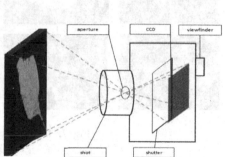

Fig. 4. Construction diagram of acquisition equipment

Fig. 5. Schematic diagram of imaging principle of tobacco leaves

3.2 Tobacco Leaf Dataset

According to the Chinese tobacco grading standard GB2635-1992, tobacco leaves can be classified into eight categories: F (orange-yellow), L (lemon), R (red-brown), H (fully ripe), K (mixed color), V (slightly greenish), GY (green-yellow) and S (smooth). Among them, fully ripe(H) and smooth(S) tobacco leaves are relatively rare under natural conditions. In addition, the Chinese flue-cured tobacco standard divides mixed color(K) tobacco leaves into three grades based on the percentage of mixed color area: 20%–30%, 30%–40%, and above 40%.

Therefore, this dataset mainly includes eight types of tobacco leaves: F, L, R, K1, K2, K3, V, and GY. Using the image acquisition system introduced in Sect. 3.1, a total of 9571 images of these eight types of tobacco leaves were collected from different varieties across Yunnan province. Figure 6 shows examples of the constructed datasets for each category.

The dataset was randomly divided into training set, validation set and test set in a ratio of 6:2:2. The tobacco leaf dataset is shown in Fig. 7.

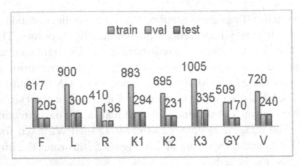

Fig. 6. Dataset statistical chart

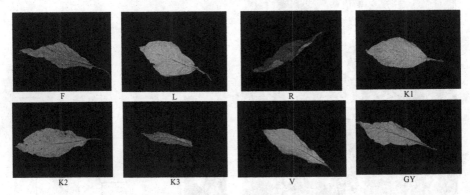

Fig. 7. Different types of tobacco leaves

3.3 Analysis and Application of Experimental Results

3.3.1 Experimental Environment Setup

The experimental environment for this study is Ubuntu 20.04 operating system, Python 3.8, with 30 GB of memory and an Intel(R) Xeon(R) Gold 6330 CPU. The GPU used is NVIDIA GeForce RTX 3090 with a capacity of 24 GB. The deep learning framework employed is PyTorch 2.0.1, utilizing CUDA Toolkit version 11.7 and CUDNN V8.5.0 as the accelerated training package for models.

This study is based on the proposed FSANet model, with the original resolution of tobacco leaf images collected at 2384 × 1528. The images were resized to 224 × 224 and normalized. The network model training was performed using cross-entropy loss function and Adam optimizer, with an initial learning rate of 1e−3 for a total of 60 iterations. To improve convergence, avoid overfitting, enhance training stability and performance, image augmentation techniques such as rotation, flipping, and color enhancement were applied to the training set while the test set and validation set remained unaltered. A learning rate decay strategy was employed where every 10 epochs resulted in a decay factor of 0.1 being applied to the learning rate.

3.3.2   Comparative Experiment

To evaluate the performance of FSANet algorithm, experimental comparisons are being conducted with current mainstream algorithms, including Mobilenetv3_ Small, Mobilenetv3_ Large, ghost, resnet18, shufflenetv2. And the accuracy of the validation set, total parameters, and parameter size are used as evaluation indicators for the model.

Table 2. Comparison of experimental results

Models	Accuracy	Total params	Params size (MB)
Mobilenetv3_small	74.71%	1,526,056	5.82
Mobilenetv3_large	77.53%	4,212,280	16.07
Ghostnet	82.65%	3,911,756	14.92
resnet18	83.43%	11,180,616	42.65
shufflenet_v2	82.28%	2,278,604	8.69
FSANet (ours)	87.63%	3,367,744	12.85

From Table 2, it can be seen that compared to the other four lightweight models, FSANet exhibits significantly higher classification accuracy than MobileNetV3, Ghost-Net, shufflenet_v2, and ResNet18, with improvements of 10%, 5%, 4%, and 5% respectively. At the same time, FSANet also has significantly lower total parameter count and parameter size compared to the lightweight models MobileNetV3_large and GhostNet, with parameter sizes increasing by 13.8% and 20% respectively. Although FSANet has a slight increase in parameter count compared to Mobilenetv3_small and shufflenet_v2, it achieves improved accuracy. FSANet possesses significant advantages in terms of detection speed and model size, making it more suitable for deployment on embedded devices for tobacco intelligent grading models.

The confusion matrix is a common evaluation metric in multi-classification tasks. The performance of the test set was tested by using the trained model, resulting in the confusion matrix shown in Fig. 8.

From the confusion matrix, it can be observed that shufflenet_v2 and ghost exhibit relatively accurate classification of primary group tobacco leaves, but their performance in classifying adjacent secondary group tobacco leaves is poor. FSANet demonstrates good classification results for primary group tobacco leaves. However, there still exists a small amount of confusion among neighboring categories such as green-yellow tobacco leaves (GY), slightly green tobacco leaves (V), and various degrees of mixed color tobacco leaves. This could be attributed to two factors: firstly, these categories of tobacco leaves have minimal differences in features, making it challenging to accurately extract distinguishing characteristics; secondly, the dataset was collected from multiple regional sites where factors like variety and curliness of the tobacco leaves may influence their features, further complicating feature extraction. Future research should consider enhancing the model's capability for feature representation and adaptability.

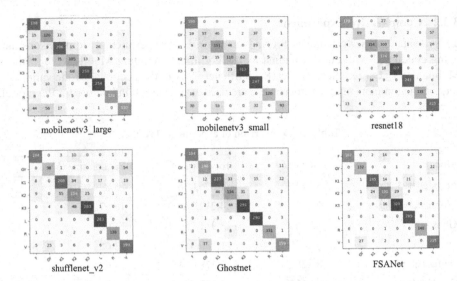

Fig. 8. Confusion matrix

3.3.3 Visualization Analysis

In order to verify the ability of FSANet model to extract subtle features of flue-cured tobacco, experiments were conducted on flue-cured tobacco samples with varying degrees of feature differences for thermal map visualization, including variegated samples with different feature differences and blue-green flue-cured tobacco samples with smaller feature differences. The thermal diagram is based on the Grad-CAM method and visualized for analysis as shown in Fig. 9.

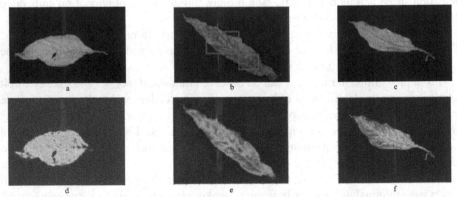

Note: The red box in the figure is for better display of the heat map effect, and the actual training process is not marked

Fig. 9. Thermal diagram

It can be clearly observed from the thermal diagram that the FSANet model can accurately capture the features of different color blocks, while also responding well to feature blocks of different sizes. For example, for the heat map of variegated flue-cured tobacco, the model successfully focused on different color areas such as light brown and dark brown. For green flue-cured tobacco, the model also clearly identified small local green patches. This indicates that the FSANet model has strong robustness in feature extraction and can capture subtle differences in tobacco images.The FSANet model has the ability to extract multi-scale features, allowing the model to accurately express the detailed features of flue-cured tobacco, which is beneficial for improving the performance of tobacco classification.

4 Conclusion

This article proposes and implements a new lightweight convolutional module, FSANet, to improve model performance in tobacco grouping tasks. FSANet captures and extracts richer feature information through multi-scale convolution combined with feature segmentation and attention mechanisms, while using the Ghost convolution module to reduce the model's parameter count. By designing a natural state tobacco leaf grouping experiment, a tobacco leaf grouping dataset consisting of 8 categories was constructed. The experimental results show that the FSANet proposed in this paper improves parameter efficiency and significantly improves classification performance compared to other lightweight networks. This validates the effectiveness of this module in expressing complex tobacco features and is more suitable for deploying intelligent tobacco grading models on embedded devices.

However, there is still a certain degree of confusion when FSANet deals with some similar tobacco categories. This may be due to the small differences in tobacco leaf features among these categories, which makes it difficult to obtain accurate features, as well as the presence of factors such as variety and curl degree in tobacco leaves, which increases the difficulty of feature extraction. These issues will become the focus of future research, including considering how to further enhance the model's feature expression ability and adaptability to improve the performance of tobacco classification tasks.

Acknowledgement. This work was supported by China Tobacco Corporation Yunnan tobacco company science and technology plan key projects [grant number: 2020530000241003 and grant number: 2021530000241012] and Kunming University of Science and Technology 2022 Student Extracurricular Academic and Technological Innovation Fund Project [grant number: 2022KJ117].

References

1. Luo, H., Zhang, C.: Features representation for flue-cured tobacco grading based on transfer learning to hard sample. In: 2018 14th IEEE International Conference on Signal Processing (ICSP), August 2018. https://doi.org/10.1109/icsp.2018.8652385
2. Man, Z.: Effect of body and color on quality of tobacco leaves in technology grading. J. Anhui Agric. Sci. (2013)
3. Yang, S., Dong, C., Wang, F., Zhou, M., Yuan, M., Huang, J.: Fisher's tobacco leaf grading method based on image multi-features. In: 2022 International Conference on Artificial Intelligence and Computer Information Technology (AICIT), September 2022. https://doi.org/10.1109/aicit55386.2022.9930167

4. Harjoko, A., Prahara, A., Supardi, T.W., Candradewi, I., Pulungan, R., Hartati, S.: Image processing approach for grading tobacco leaf based on color and quality. Int. J. Smart Sens. Intell. Syst. **12**(1), 1 (2019). https://doi.org/10.21307/ijssis-2019-010

5. Lin, J., et al.: CAMFFNet: a novel convolutional neural network model for tobacco disease image recognition. Comput. Electron. Agric. 107390 (2022). https://doi.org/10.1016/j.com pag.2022.107390

6. Wu, J., Yang, S.X.: Modeling of the bulk tobacco flue-curing process using a deep learning-based method. IEEE Access **9**, 140424–140436 (2021)

7. Lu, M., Jiang, S., Wang, C., Chen, D., Chen, T.: Tobacco leaf grading based on deep convolutional neural networks and machine vision. J. ASABE **65**(1), 11–22 (2021). https://doi.org/10.13031/ja.14537

8. Lin, H., Bi, Y., Zhang, X., et al.: Deep Tobacco Leaf Grading Using Self-Adaptive Attention. Available at SSRN 4414972

9. Wei, X.S., Wu, J., Cui, Q.: Deep learning for fine-grained image analysis: a survey, arXiv preprint arXiv:1907.03069 (2019)

10. Lu, M., Jiang, S., Wang, C., et al.: Tobacco leaf grading based on deep convolutional neural networks and machine vision. J. ASABE **65**(1), 11–22 (2022)

11. Chen, D., Zhang, Y., He, Z., Deng, Y., Zhang, P., Hai, W.: Feature-reinforced dual-encoder aggregation network for flue-cured tobacco grading, SSRN, February 2023. https://doi.org/10.2139/ssrn.4355545

12. Xiaowei, X., Huili, G., Ruotong, H., et al.: Intelligent large-scale flue-cured tobacco grading based on deep densely convolutional network. Sci. Rep. **13**(1), 11119 (2023)

13. Lu, M., Wang, C., Wu, W., et al.: Intelligent grading of tobacco leaves using an improved bilinear convolutional neural network. IEEE Access **11**, 68153–68170 (2023)

14. Barman, U., Choudhury, R.D., Sahu, D., Barman, G.G.: Comparison of convolution neural networks for smartphone image based real time classification of citrus leaf disease. Comput. Electron. Agric. (2020). https://doi.org/10.1016/j.compag.2020.105661

15. Wang, H., Gu, W., Liu, X., et al.: Classification algorithm for natural state tobacco primary and secondary groups based on lightweight SE-PPM. J. Northwest A&F Univ. (Nat. Sci. Edn.) **01**, 1–11 (2024). https://doi.org/10.13207/j.cnki.jnwafu.2024.01.006. Accessed 24 Oct 2023

16. He, Z., He, P., Zhang, Y., et al.: Real-time grouping of tobacco through channel weighting and dynamic loss regulation. Ind. Crops Prod. **195**, 116427 (2023)

17. Ming, X., Jinfeng, G., Zhong, Z., et al.: Multi-channel and multi-scale separable dilated convolutional neural network with attention mechanism for flue-cured tobacco classification. Neural Comput. Appl. **35**(21), 15511–15529 (2023)

18. Hu, J., Shen, L., Albanie, S., Sun, G., Wu, E.: Squeeze-and-excitation networks. IEEE Trans. Pattern Anal. Mach. Intell. **42**(8), 2011–2023 (2020). https://doi.org/10.1109/tpami.2019.2913372

19. Han, K., Wang, Y., Tian, Q., Guo, J., Xu, C., Xu, C.: GhostNet: more features from cheap operations. In: Proceedings of the 2020 IEEE/CVF Conference on Computer Vision and Pattern Recognition (CVPR), June 2020. https://doi.org/10.1109/cvpr42600.2020.00165

20. Howard, A., et al.: Searching for MobileNetV3. In: Proceedings of the 2019 IEEE/CVF International Conference on Computer Vision (ICCV), October 2019. https://doi.org/10.1109/iccv.2019.00140

21. Longlong, L., Zhifeng, W., Tingting, Z.: GBH-YOLOv5: ghost convolution with BottleneckCSP and tiny target prediction head incorporating YOLOv5 for PV panel defect detection. Electronics **12**(3), 561 (2023)

22. Li, M., Lu, Y., Cao, S., et al.: A hyperspectral image classification method based on the nonlocal attention mechanism of a multiscale convolutional neural network. Sensors **23**(6), 3190 (2023)

23. Hoang, V.T., Jo, K.H.: PydmobileNet: improved version of mobilenets with pyramid depthwise separable convolution, arXiv preprint arXiv:1811.07083 (2018)
24. Zhang, H., Zu, K., Lu, J., Zou, Y., Meng, D.: EPSANet: an efficient pyramid split attention block on convolutional neural network, arXiv preprint, May 2021

Author Index

H. Jin et al. (Eds.): IAIC 2023, CCIS 2060, pp. 345–348, 2024.
https://doi.org/10.1007/978-981-97-1332-5

Printed in the United States
by Baker & Taylor Publisher Services